寰宇文献 Universal Library | SINOLOGY 系列

SELECTED WORKS OF BERTHOLD LAUFER

劳费尔著作集

第五卷

[美] 劳费尔 著

黄曙辉 编

中西書局
ZHONGXI BOOK COMPANY

图书在版编目(CIP)数据

劳费尔著作集 / (美) 劳费尔著；黄曙辉编. —上
海：中西书局，2022
(寰宇文献)
ISBN 978-7-5475-2015-4

Ⅰ.①劳… Ⅱ.①劳… ②黄… Ⅲ.①劳费尔 – 人类
学 – 文集 Ⅳ.①Q98–53

中国版本图书馆CIP数据核字（2022）第207067号

第 5 卷

083

玉——中国考古学与宗教的研究

附：书评一则

PUBLICATION

OF

FIELD MUSEUM OF NATURAL HISTORY

ANTHROPOLOGICAL SERIES
Vol. X.

Chicago, U. S. A.

1912

FIELD MUSEUM OF NATURAL HISTORY.

PUBLICATION 154.

ANTHROPOLOGICAL SERIES.   VOL. X.

JADE

A STUDY IN CHINESE ARCHÆOLOGY AND RELIGION

BY

BERTHOLD LAUFER

68 Plates, 6 of which are colored, and 204 Text-figures

The Mrs. T. B. Blackstone Expedition

CHICAGO, U. S. A.
February, 1912.

PREFACE

AT the close of the year 1907 the Field Museum honored me with the task of carrying on research and making collections in Tibet and China for a period of three years, extending from 1908 to 1910, under an endowment liberally provided by Mrs. T. B. Blackstone of this city. The results of the expedition, accordingly, cover two distinct fields — an ethnological collection bearing on Tibet and neighboring regions, inclusive of an abundance of material relative to Lamaism (paintings, images, masks, objects of the cult), and an extensive collection illustrating the archæology of China. The latter has been planned in such a way as to afford an insight into the development of all phases of life in China's past. In conformity with the tendency of this Institution, this group of collections is not by any means intended to illustrate the development of art but of culture in China. The results of the work in Tibet are designed to be brought out in six volumes. It has been proposed to work up the Chinese material in a series of monographs, the first instalment of which is the present publication. The mortuary clay figures, the bronze and iron age of China, Buddhist stone sculpture, are the subjects contemplated for the next issues.

This volume does not pretend to be a contribution to sinology. Its general scope is explained in the Introduction. Written, in the first place, to furnish the necessary information on the jade collection in the Field Museum, it applies to students of archæology and religion in that it furnishes a great deal of new material and research on the early development of religious and artistic thought in ancient China. All specimens, with the exception of the jade and stone implements on Plates IX–XIII, and of the jade book on Plate XIV, are in the Field Museum and form part of the Mrs. T. B. Blackstone collection.

The lack of Chinese types is met by two expediencies. In all illustrations derived from Chinese books, the Chinese titles accompanying them have been reproduced in facsimile, so that the Chinese designations for the various types of jade will be easily recognized. Wherever necessary, in using Chinese words romanized, references are made in parenthesis to the number of the particular character in Giles's Chinese-English Dictionary, which is on the desk of every student in this field. The second revised and enlarged edition, as far as it has appeared (four fascicules), is quoted throughout; for the remainder the first edition is referred to. Professor Giles deserves hearty congratulation

i

on the completion of this monumental work, which is marked by progress and new results on every page, and for which every student of Chinese is largely indebted to him. As our archæological research advances, the necessity of a special archæological dictionary will be more and more felt.

The generosity of Mrs. T. B. Blackstone in contributing a fund toward the printing of this volume is gratefully acknowledged.

I have to thank Dr. O. C. Farrington, Curator of Geology, for the mineralogical definitions of many specimens, and Mr. H. W. Nichols, Assistant Curator of Geology, for manifold advice on technical questions.

The manuscript was completed in October, 1911.

BERTHOLD LAUFER,
Associate Curator of Asiatic Ethnology.

CHICAGO, February, 1912.

CONTENTS

iii

LIST OF ILLUSTRATIONS

TEXT–FIGURES [1]

[1] Chinese drawings with no indication of source are derived from the book of *Wu Ta-ch'êng* (see Introduction, p. 12).

v

PLATES

INTRODUCTION

> "The Illustrated Mirror of Jades says:
> In the second month, the plants in the
> mountains receive a bright luster. When
> their leaves fall, they change into jade.
> The spirit of jade is like a beautiful
> woman."

The two minerals nephrite and jadeite, popularly comprised under the name jade, belong to the hardest and most cherished materials of which primitive man availed himself in shaping his chisels, hatchets, ornaments, amulets and many other implements. Such objects, partially of considerable antiquity, have been found in many parts of the world—in Asia, New Zealand, in prehistoric Europe and America. The geographical areas occupied by nephrite are so extended that it may almost be classified with flint as one of those mineral substances utilized to a large extent by a great number of peoples. Since the localities where nephrite occurs in nature seem to be difficult to discover, and since nephrite was known for a long time as coming only from Asia and New Zealand, and jadeite merely from certain districts of Asia, the scientific world was being held in long suspense by what is known as the nephrite question. Though now a matter of historical interest, it may not be amiss to review it briefly for the instructive value which it bears on the development of science.

At the time when Heinrich Fischer[1] in Freiburg carried on his epoch-making investigations on the mineralogical and archæological sides of crude and wrought jades, no places of the occurrence *in situ* of nephrite were known in either Europe or America. The problem, therefore, pivoted around the question as to how the peoples of these two parts of the world had obtained the material for their nephrite and jadeite objects. It then was a matter of natural consequence that Fischer should elaborate the theory that the nephrite objects in pre-Columbian America had been transported there, owing to an influx and settlement of Asiatic tribes, and that those brought to light in Europe were accounted for by the assumption of a prehistoric commercial intercourse with Asia, or by the migrations of peoples.[2] This argument once

[1] See Bibliography at end.

[2] It should be remarked, by the way, that Fischer was a true scholar and most conscientious worker, and certainly greater in his lifelong error than many a minor demi-scholar who *post festum* celebrated a cheap triumph over the end of the nephrite question.

formed an important factor in the heated debate over the Indogermanic migrations and the original habitat of the group, and was exploited in favor of a supposition of an eastern origin of the neolithic culture of Europe.[1] In 1900, OSCAR MONTELIUS (Die Chronologie der ältesten Bronzezeit, p. 204) wrote: "There is much divergence of opinion on the often discussed question whether the numerous works of jadeite and nephrite excavated in southern and central Europe have to be explained as an importation from the Orient.   I am of opinion that at all events some of these works are to be attributed to such an importation." Fischer's theory was first attacked by A. B. MEYER in several papers, particularly "Die Nephritfrage kein Ethnologisches Problem" (Berlin, 1883; translated in *American Anthropologist*, Vol. I, 1888, pp. 231–242). First, MEYER thought it improbable that ready-made implements or unwrought stones should have been transported over oceans and wide tracts of land; an argument of no great validity, as there are, in the history of trade, numerous examples proving the contrary.   Secondly, he referred to localities, increasing from year to year, where jade occurring *in situ* had become known.   Captain I. H. Jacobsen brought nephrite from Alaska, where it is found as a mineral and worked by the natives into numerous objects.[2]   In Europe, strata of nephrite were discovered in the eastern Alps in the Sann valley, near St. Peter, and in the Murr valley, near Graz.   In Switzerland, bowlders of jadeite were sighted on Lake Neuenburg; even a sort of nephrite workshop was discovered in the vicinity of Maurach, where hatchets chiseled from the mineral and one hundred and fifty-four pieces of cuttings were found.   At the foot of Mount Viso, in Italy, jadeite was met with *in situ*. Single erratic bowlders of nephrite in diluvial deposits had been signaled in Germany at an earlier date, near Schwemmsal, Potsdam, and Leipzig, the latter weighing seventy-nine pounds (FISCHER, pp. 3–5 *et passim*).   Credner supposed that these three bowlders were transported through the ice from Scandinavia to their present localities.   Another geologist, Traube, succeeded in discovering nephrite *in situ* in small bands and greater layers near Jordansmühl in the serpentine strata of the Zobten mountains of Silesia.   A large serpentine hatchet instratified with bands of nephrite is known from Gnichwitz only two miles distant from Jordansmühl, and made from a material which agrees with the material of the latter locality.   Thus, Fischer's ingenious hypothesis of importation or migration could no longer be upheld in regard to Germany.

[1] MAX MÜLLER, Biographies of Words, Appendix II: The Original Home of Jade.   London, 1888.

[2] For further notes regarding the occurrence of jade in America, see article "Nephrite," by Mr. HODGE in Handbook of American Indians, Vol. II, pp. 55–56 (Washington, 1910).

The most important investigation in the direction of proving the indigenous origin of the jade objects of Switzerland was undertaken by A. BODMER-BEDER in Zürich.[1] He made a great number of careful examinations of stone implements from the Swiss lake-dwellings and by comparing these with the analyses of the raw material found in Switzerland, he arrived at the conclusion that the species of nephrite met with in the nephrite objects of Lake Zug occurs *in situ* in the territory of the Gotthard, whence it arrived in the district of Zug through the movement of glaciers and rivers, and that also the human products discovered in the other lakes are of autochthonous material. Scientists are generally convinced now that the end of the nephrite question has come, that such a question no longer exists. It seems perfectly safe to assume that the ancient jade objects of Europe and America were not imported from Asia, but, for the greater part, manufactured on the spot. Certainly, it does not now follow that each and every jade object wherever found *must* be a local production, and that an historical inquiry is gagged forever. The channels of historical development are manifold and complex, and the working after an easy schematic routine is fatal and infertile. No lesser archæologist than SOPHUS MÜLLER (Urgeschichte Europas, p. 21, Strassburg, 1905) sounded a timely warning to this over-enthusiasm by remarking: "It is true, one has succeeded in recent years after long search in pointing out nephrite and jadeite in the Alps, but merely under conditions which render it highly improbable that the extensive territories where such hatchets are found should have derived them from there. We may rather presume with certainty that at least a great part of the hatchets mentioned have been imported by commerce from the Orient,[2] whence the knowledge of polishing hatchets is derived. At a somewhat later time, an extensive trade in stone artefacts was carried on within the boundaries of Europe."

The naturalists have had their say in the matter, but the historian is entitled to push his reflections farther on. Further investigations will be required for a satisfactory solution of all questions bearing on the dissemination of jade implements. The origin, *e. g.*, of the jade hatchets unearthed by Schliemann in the oldest walled city at Hissarlik has, to my knowledge, not yet been determined. I certainly admit that the question in its widest range as raised by Fischer has been settled, but there are still other sides to it calling for attention.

[1] Petrographische Untersuchungen von Steinwerkzeugen und ihrer Rohmaterialien aus Schweizerischen Pfahlbaustätten (*Neues Jahrbuch für Mineralogie*, Vol. XVI, 1903, p. 166).

[2] Compare article by C. MEHLIS, Exotische Steinbeile (*Archiv für Anthropologie*, Vol. XXVII, 1902, No. 4, pp. 599–611).

If we consider how many years, and what strenuous efforts it required for European scientists to discover the actual sites of jade in central Europe, which is geographically so well explored, we may realize that it could not have been quite such an easy task for primitive man to hunt up these hidden places, even granted that geological conditions may then have been different or more favorable. Or was that primitive man so much keener and more resourceful than our present scientists? Or if not, we must grant him the same difficulties in the search for jade as to them. And if he overcame these difficulties and after all found jade, it seems to me that he must have been prompted by a motive pre-existing and acting in his mind; the impetus of searching for jade, he must have received somehow and from somewhere, in the same manner as was the case with our modern scientists who, without the nephrite question in their heads, would not have searched for it, and who finally found it, only because they sought it. This is the psychological side of the historical aspect of the problem.

Why did the Romans discover the Terra Sigillata on the Rhine and in other parts of Germany unknown to the indigenous population? Because they were familiar with this peculiar clay from their Mediterranean homes, because they prized this pottery highly and desired it in their new home. Let us suppose that we should not possess any records relating to the history of porcelain. The chief substance of which it is made, kaolin, is now found in this country, in Germany, Holland, France and England, all of which produce objects of porcelain; consequently, porcelain is indigenous to Europe and America, because the material is found there. By a lucky chance of history we know that it was made in neither country before the beginning of the eighteenth century, and that the incentive received from China was the stimulus to Boettger's rediscovery in Dresden. Of course, arguing *a priori*, the peoples of Europe and America could have made porcelain ages ago; the material was at their elbows, but the brutal fact remains that they did not, that they missed the opportunity, and that only the importation and investigation of Chinese porcelain were instrumental in hunting for and finding kaolinic clay. And, while there have been porcelains produced by local industry in Europe and America for the last two centuries, we have, side by side with them, numerous direct imports from China which continue despite the output of the home market.

Similar conditions may have prevailed also in an early stage of the history of Europe. Even if jade occurs there in a natural state in several localities, even if there is conclusive proof that it was dug and worked in various areas, we are entitled to question, — did the idea of searching for jade, working it and valuing it so highly, originate independently in all these quarters? Is it not possible, at least theoretically, that such an idea once arising was diffused from tribe to tribe or group to group, as the agency in the search for the prized material? Even in the present state of the question, it cannot be denied that the possibilities of a trade in jade pieces existed, as suggested by O. Montelius and Sophus Müller, whom HERMAN HIRT (Die Indogermanen, Vol. I, p. 317, Strassburg, 1905) joins in their view, as the sources from which they are derived are restricted to a few localities. Such a commerce in Europe was an easy transaction, if compared with the striking parallel in Asia moving on a much larger scale. For the last two millenniums, Turkistan has furnished to China the greater supply of her jade, wrought and unwrought, and the most colossal bowlders of the mineral were constantly transported from Khotan to Si-ngan fu and Peking over a trade-route unparalleled in extent and arduousness in Europe and requiring a four to six months' journey. There is, further, the example of the lively trade in jadeite from Burma overland into Yünnan Province, and the transportation of jeweled nephrite objects from India into China in the eighteenth century. These are all achievements of commerce and transportation compared to which the difficulties in the limited area of Europe dwindle into a nothingness. If bronze was bartered from the Orient into the northernmost part of Europe, if Prussian amber found its way to Italy, Greece and anterior Asia, and if obsidian was everywhere propagated by trade (SOPHUS MÜLLER, l. c., p. 48; R. DUSSAUD, Les civilisations préhelléniques, p. 77), it is reasonable and logical to conclude that the same opportunities were open to jade.

Nothing could induce me to the belief that primitive man of central Europe incidentally and spontaneously embarked on the laborious task of quarrying and working jade. The psychological motive for this act must be supplied, and it can be deduced only from the source of historical facts. From the standpoint of the general development of culture in the Old World, there is absolutely no vestige of originality in the prehistoric cultures of Europe which appear as an appendix to Asia. Originality is certainly the rarest thing in this world, and in the history of mankind the original thoughts are appallingly sparse. There is, in the light of historical facts and experiences, no reason to credit the prehistoric and early historic populations of Europe with any spontaneous ideas relative to jade; they received these, as everything else, from an outside source; they gradually learned to appreciate the value of this tough and compact substance, and then set to hunting for natural supplies.

The most extensive collection of jades in existence is the Heber R. Bishop collection, the greater part of which is now on view in the Metropolitan Museum of Art, New York. The object of this collection is universal and includes specimens of jades from all countries and ages. It is exceedingly rich in magnificent objects of Chinese art, of intrinsic value from an artistic point of view. The collection of Chinese jades which I had the opportunity to make for the Field Museum in Si-ngan fu, Shensi Province, on two occasions in 1909 and 1910, is chiefly gathered from an archæological standpoint to illustrate an important phase in the ancient culture-life of China, and represents by no means a duplicate parallel to the Bishop collection, as it includes a great deal of material which does not exist in the latter. Also in the department of eighteenth century jades, there is a marked difference between the Mrs. Blackstone and the Bishop collections, as pointed out here in the concluding chapter. Mr. Bishop's Chinese specimens of jade, I understand, were mostly procured in Peking or Shanghai, where the gorgeous modern art-work prevails. The greater opportunity for objects of archæological interest is afforded in Si-ngan fu, the centre of the old civilization where numerous primitive specimens are exhumed from the graves of the Chou and Han periods, and where the best private jade collections of Chinese connoisseurs exist. I had the good fortune to receive valuable instruction there from an expert scholar who had spent almost a life-time on the study of mortuary jades and other antiquities, and whose extensive collections, partly the fruit of his own excavations, were gradually acquired by me. The interesting jade amulets of the dead, chiefly derived from his collection, are here published for the first time with his explanations, nor have they ever been described before in any Chinese book. None of these types are in the Bishop collection. For this and other reasons, the two are not rival collections, but supplement each other in many respects.

The description of the Bishop collection appeared in two volumes under the title "Investigations and Studies in Jade," New York, privately printed, 1906. The preface of Heber R. Bishop is dated June, 1902.[1] The collection is stated there to number nine hundred specimens, which are catalogued under the three headings mineralogical, archæological, and art objects. Nearly two hundred pieces were presented to the Metropolitan Museum of Art, New York. Vol. I (277 pages) consists of five parts. Part I contains a general introduction with a survey of the nephrite question. Part II, entitled

[1] He died in December of the same year. A sketch of his life and aspirations written by G. F. Kunz will be found in *American Anthropologist*, 1903, pp. 111–117.

"Jade in China," is elaborated by S. W. Bushell. Mr. Bishop requested "a condensed article on jade by a native Chinese scholar,
treating upon its uses in China from the earliest period down to the
present day." Accordingly, Dr. Bushell commissioned a scholar,
T'ang Jung-tso by name, to write such an essay, the text of which is
reproduced in facsimile and provided with a translation by Bushell.
This document is composed of a conglomeration of literary quotations
from ancient texts which are rather inexactly and sometimes incom-
pletely cited. Dr. Bushell did not verify them from the originals,
and it is hard to see why a Chinese scholar of his standing agreed to
such a procedure, as he certainly knew that the Chinese cyclopædias,
above all the *T'u shu tsi ch'êng*, contain the completest possible collections of notes on jade. I do not see much sensible exposition of the
subject of jade from the Chinese standpoint in T'ang Jung-tso's essay,
which, though undeniably comprising a number of useful extracts,
lacks intelligent understanding. Another Chinese article written by
Li Shih-ch'üan and illustrated with sketches depicting the various
stages in the modern manufacture of jade is very instructive. Part
III, "Jade as a Mineral," edited by G. F. Kunz, presents the
most complete and thorough investigation of this subject carried
on by a whole staff of specialists. The questions treated in detail
in this section are the colors of jade, whether natural or the result
of weathering or staining; the translucency, lustre, opalescence,
sheen, and other qualities; the tenacity of jade; its resonant quality
which makes it valued by the Chinese as a material for musical
instruments; the relative hardness and specific gravity of nephrite
and jadeite, their different chemical constitution, and the localities
of their occurrence. Part IV is devoted to a sketch of the methods
of working jade. Part V, "Worked Jade," is intended to give a
general summary of this subject under two headings, "Prehistoric"
and "Historic." Vol. II (293 pages) contains the descriptive catalogue of the collection. Dr. Bushell has bestowed great pains in
fully describing the Chinese portion of it and explaining with careful
exactness the ornamentation and inscriptions displayed on the fine
Chinese specimens in which the collection abounds. Mr. Bishop
expressed in the preface the hope that his book might be found to be of
some value as a book of reference. His work is doubtless one of the
most beautiful and sumptuous books ever published in this country,
but it is a matter for profound regret that its valuable contents is
practically lost to science, owing to its unwieldy size and weight (one
hundred and twenty-five pounds) and its distribution in only ninety-
eight copies, none of which have been sold, but which have all been

presented to libraries, museums, and it is said, "to the crowned heads and other great rulers of the world." It would be desirable that the main bulk of the work might be republished in a convenient edition for wider circulation.

As the collection of Chinese jades in the Field Museum was made by me from another point of view than is the Bishop collection in New York, so the contents of this study is plainly distinct from the monumental work of Mr. Bishop. The subjects treated there have not been repeated here. The methods of working jade and the trade in jade from Turkistan and Burma to China are not discussed, as Dr. Bushell has thoroughly canvassed this ground. He has also worked up a map showing the jade-producing districts of Turkistan. New as the subject matter of this publication is, aside from the concluding chapter, which was necessary in order to trace the development of jade works down to the present time, so also is the presentation of the subject itself which is based on archæological methods. I have endeavored to furnish a piece of research-work in which jade is to yield the material to delineate cultural and chiefly religious developments in ancient China. I do not mean to deal with jade for its own sake, but as a means to a certain end; it merely forms the background, the leading motive, for the exposition of some fundamental ideas of Chinese religious concept which find their most characteristic expression and illustration in objects of jade. To trace their relation to thought was therefore my chief aim, and hence the result has rather become a contribution to the psychology of the Chinese.

A consideration of the Chinese sources utilized will give also occasion to speak on the methods pursued by me and some of the general results of the work.

Antiquities of jade have been treated by the Chinese in the following special works:

1. *K'ao ku t'u,* "Investigations of Antiquities with Illustrations," by LÜ TA-LIN, in ten chapters, first published during the Sung dynasty in 1092, and re-edited by HUANG SHÊNG in 1753 as an appendix to Wang Fu's *Po ku t'u.* In Chapter 8, a small collection of jades in the possession of Li Po-shih from Lu-kiang is figured, but without investigation.

2. *Ku yü t'u,* "Ancient Jades Illustrated," by CHU TÊH-JUN, in two chapters, published 1341, and republished in 1753 with the edition of the *Po ku t'u* mentioned. .The explanations appended to the illustrations are meagre; several of them have been reproduced in this paper.

3. *Ku yü t'u p'u,* "Illustrated Description of Ancient Jades" in 100 chapters, being the collection of jade belonging to the first emperor

of the Southern Sung dynasty, Kao-tsung (1127–1162 A. D.) and consisting of over seven hundred pieces, prepared by a commission of nineteen, including one writer and four artists,[1] headed by LUNG TA-YÜAN, president of the Board of Rites (*Li pu*) who also prefaced the work in 1176. Dr. BUSHELL, who has devoted a careful study to this book, gives also a translation of this preface (BISHOP, Vol. I, p. 32). The second preface (*Ibid.*, p. 33) by KIANG CH'UN, dated 1779, relates how "a manuscript copy of the book had been purchased in 1773, when the Emperor had issued a decree to search throughout the empire for lost books, and a copy sent to be examined by the library commission then sitting. This year I again read through the original manuscript and found the description clearly written and the illustrations cleverly executed, so that it was worthy of being compared with the *Süan ho Po ku t'u*. This book describes the ancient bronzes referred to in the *Chou li*, while our work describes the jade, so that we could not spare either. The *Po ku t'u* was reprinted several times and gained a wide circulation, while this book remained in manuscript[2] and attracted no notice, not being included in the Catalogue of Literature of the Sung History, nor quoted by older writers. Lung Ta-yüan, whose name is included in the chapter on Imperial Sycophants of the Sung History, died before the date of publication, but he is left at the head of the commission, in memory of the work done by him. His actions were not worthy, but that is no reason for suppressing his book. I venture to bring this book before the eye of the Emperor, that it may again be referred to the library committee for revision and be corrected by them, and have the honor of being reprinted under special imperial authority."

The verdict of the Library Committee seems to have been unsatisfactory, continues Dr. Bushell, for they criticize the book most severely in the Imperial Catalogue (*Se ku ts'üan shu tsung mu*, Ch. 116, pp. 7–9) on account of there being no references to it in later books, and of certain anachronisms in the list of members of the commission, and declare it finally to be a fraud, and not even a clever one; without any examination, however, of the contents—as Bushell adds, which could have hardly been the case.

[1] Liu Sung-nien, Li T'ang, Ma Yüan, and Hia Kuei. It was their task to reproduce the jade pieces in colors for the one original copy to be dedicated to the emperor.

[2] It has been said that the *Ku yü t'u p'u* was published in 1176 and republished in 1779 (HIRTH, *T'oung Pao*, Vol. VII, p. 500, and The Ancient History of China, p. 89). It is of importance to note that 1779 is the date of the first publication, and that the work was allowed to remain in manuscript up to that time, *i. e.* over 600 years. The original manuscript was then lost, and we have no means of judging in how far, or how correctly the illustrations of the original have been preserved in the printed edition.

If it is too much to say that the whole work is a fraud, I quite agree
with this criticism in that it contains a great deal of purely fictitious
matter. Fictitious are, in my opinion, all the ancient inscriptions
alleged to be inscribed on the jades which have never existed in ancient
times and are simply the invention of the T'ang or Sung periods.[1]
Even Dr. Bushell, who evinces confidence in this work, admits that
there are many tablets figured in it which have little pretension to the
great antiquity assigned to them, and that some of the inscriptions are
evidently copied from pieces of ancient bronze figured in archæological
books, and that, in fact, many of the specimens in the later parts of
the collection seem to be derived from a similar source — the fountain-
head of almost all Chinese decorative art.

It is also suspicious that many pieces, e. g., all the tablets of rank in
the first chapter ascribed to the Hia dynasty, have been inscribed on
the back as having belonged to the T'ang and Southern T'ang dynasties.

The work opens with two oblong jade tablets ascribed to the mythical emperor Yü from the supposed resemblance of the two undeciphered
characters on the upper side with the so-called tadpole characters on
the alleged inscription of Yü (*Kou lou pei*). On the back, we find an
inscription reading "Dark-colored tablet (*kuei*) of Yü the sovereign
who regulated the waters. Collection of the Imperial Treasury of
the period *K'ai yüan* (713–741 A. D.) of the Great T'ang dynasty."
The Sung authors tell us that these two pieces came to light in the
period *Chih ho* (1054–55), in the river Ts'i when its waters were dried
up, and that these were both found inside of large bronze kettles
(*ting*) each weighing over a hundred catties; the walls of these
urns were covered with inscriptions identical in character with
those on the tablets. The Sung authors suppose that they had been
thrown into the river during the T'ang dynasty as an offering to the
river-god, to restore the river to normal conditions. But would genuine
relics of the Hia dynasty have been used for this purpose? It seems
rather plausible that these two alleged tablets of Yü were fabricated
at the time of the T'ang dynasty, possibly with the idea of serving as
offerings to a river-god.[2] The name of Yü as the ruler of water was

[1] Inscriptions, particularly those containing dates, on jade pieces are suspicious
in any case. In the best archæological collection of jades, that of Wu Ta-ch'êng
(see below), there is among two hundred and fifteen ancient pieces, not one inscribed,
nor is there one in my collection. A few pieces in the Bishop collection with alleged
Han inscriptions are, for this and also for other reasons, highly suspicious. The
dating of jade objects became a fashion only in the K'ien-lung period.

[2] Also WU TA-CH'ÊNG (see below under 5) states in the preface of his work that
under the T'ang and Sung many imitations of jade objects were made which can
hardly be distinguished from the ancient genuine ones. The same author's judgment on the *Ku yü t'u p'u* is: "Its drawback consists in the indiscriminate choice
of a confused mass of objects, nor does it betray intelligence."

appropriately chosen for such a purpose, and the two characters were modeled after the curious style of that doubtless ancient inscription which a later age has associated without foundation with the name of Yü. It goes without saying that at the time of Yü, if such a personage ever existed, jade tablets of this type had not yet made their début, for these were purely a creation of the official hierarchy of the Chou dynasty. Thus, it is likewise a legendary anachronism, if the *Shu king* (Ch. *Shun tien*, 7) and *Se-ma Ts'ien* (CHAVANNES, Vol. I, p. 61) ascribe the five insignia of rank (*wu jui*) to the mythical emperor *Shun*, as these are connected with the five feudal princes and the whole system of feudalism and investiture of the Chou period; and CHAVANNES is certainly right in saying that this consideration demonstrates the legendary character of the accounts relative to Shun. There is, further, no ancient text describing a jade tablet of the type here referred to, and if it were by any means an object really going back to times of great antiquity, it would be incredible that the T'ang people should have been so idiotic as to fling such a precious relic down to the bottom of a river. The entire story of the Sung authors, gifted with a lively imagination, is open to grave doubt and suspicion, and may, after all, be a concoction made up by them *ad hoc*. It does not betray much critical acumen on their part to make these two pieces contemporaneous with Yü.[1]

Our confidence in this production is not increased by considering the two following jade tablets also very generously attributed to the Emperor Yü. The former of these is adorned with ten unexplained and unexplainable characters shaped into strange figures of insects, fish, and birds; on the back, there is an inscription (in *li shu*) calling this specimen "a tablet with seal-characters of Yü" (*Yü chuan kuei*) and giving the period *Shêng yüan* (937–942 A. D.) of the Nan T'ang (reign of Li King). Nevertheless, we are assured that the writing of Yü cannot be doubted. The latter Yü tablet is provided with ten seal characters, and the same inscription on the reverse as the preceding one. Then we advance to a tablet with twenty-one characters of a different style, said to resemble those on the bells of the Shang dynasty, while the reverse is adorned with the sentence: "Jade tablet of prosperity of the rulers of the Shang." Three more Shang tablets follow, marked on the reverse as "preserved in the treasury of the period *T'ai ts'ing* of the Liang dynasty" (Wu ti, 547–550 A. D.). These alleged Shang tablets are just as fictitious as those of Yü; tablets of this kind did not exist under the Shang, and if they had existed, would not have been

---

[1] One of these is figured in CONRADY's China (*Pflugk-Harttung's Weltgeschichte*, p. 528).

engraved with inscriptions in general nor with these particular inscriptions, as also the *kuei* of the Chou dynasty were never provided
with them. Curiously enough, the number of Yü tablets is four, and
the same number applies to the Shang tablets; there is also numerical
systematization in forgery.

The Chinese epigraphists have justly passed these inscriptions over
with silence, and I am not aware that any one of the numerous
Chinese works on inscriptions and ancient characters has ever availed
itself of the services of the *Ku yü t'u p'u*. It is unnecessary to continue this criticism, as we shall have ample occasion to come back to
this work in dealing with the single types of ancient jades. While it
is entirely untrustworthy for archæological studies, it has a certain
value in presenting a grammar of ornaments and giving the names for
these, as they were current in the Sung period. We shall see in the
course of this investigation that many of these designs are strongly
influenced or even directly created by the pictorial style of the Sung
artists, and that they represent a more interesting contribution to
the art of the Sung than to any former period.

This case will also sufficiently show how much criticism is required
for judging a Chinese illustration of an ancient art-work, which should
not be utilized before its sources and merits are critically examined and
ascertained. Also he only can use it who has seen and handled actual
specimens of an identical or similar type; the imperfection and inac-
curacy of Chinese drawings will always lead astray one who has missed
those opportunities. The favorite method of culling engravings of
bronzes from Chinese books and building far-reaching conclusions on
this material as to the development of ornamentation cannot be
accepted and will always lead to grave disappointments in the end.

4. *Tsi ku yü t'u*, "Collection of Ancient Jades with Illustrations,"
a small work published in 1341 during the Yüan period. In all probability, this book is now lost; it is quoted occasionally in the *San ts'ai
t'u hui*, published in 1607.

5. The most recent and valuable Chinese contribution to the
study of antique jades is entitled *Ku yü t'u k'ao*, "Investigations into
Ancient Jades with Illustrations," in two quarto-volumes published
in 1889 by the well known scholar and statesman, WU TA-CH'ÊNG,[1]
who was born in Su-chou in 1833. He graduated as *tsin shih* in 1868
and became a member of the Han-lin College. In 1884 he went to
Corea as Commissioner, then served as Governor of Kuang-tung
Province and, appointed subsequently Governor of Hunan, made a
vain attempt to introduce the telegraph there. In 1894 he was ordered

---

[1] His other works are enumerated by PAUL PELLIOT, *T'oung Pao*, 1911, p. 448.

to Tientsin to assist Li Hung-chang against the Japanese; his efforts, however, were not rewarded with success, and he has since been living in retirement; he is said to be an enlightened man and well-disposed towards Europeans.[1] In his work on jades, two hundred and fifteen pieces are illustrated in outline, as a rule reduced to seven-tenths of their original size, described as to their coloring, identified with their ancient names and explained with quotations from ancient literature, among which *Chou li*, *Li ki*, *Shi king*, *Tso chuan* and the dictionary *Shuo wên* are conspicuous. The text is a facsimile reprint of Wu's own expressive and energetic handwriting. The engravings in his work are far above the average of similar accomplishments of the Chinese and executed with care and in good proportionate measurements. As most of these jade objects and their designs are flat, the Chinese draughtsman had a much easier task with them than he encounters, *e. g.*, with bronze vessels; Chinese art is one of linear designs in which it excels, while objects of bodily dimensions are always apt to be misdrawn to a certain extent. I was forced to reproduce the material of Wu almost in its entirety, owing to its great archæological importance.[2]

It will be noticed at a glance how widely different this material is from that published in the former Chinese works. All his ancient specimens have a spontaneously archaic character. It is a truly archæological collection, explained with great erudition and acumen, and reflects the highest credit on the modern school of Chinese archæologists. Wu Ta-ch'êng is not bound by the fetters of the past and not hampered by the accepted school-traditions. With fair and open mind, he criticizes the errors of the commentators to the *Chou li*, the *Ku yü t'u p'u* and many others, and his common sense leads him to new and remarkable results not anticipated by any of his predecessors. Because my own collection is a counterpart of his, being made from an archæological, not an artistic point of view, I could choose no better guide for the interpretation of this collection than him; I have followed him with keen admiration and stand to him in the relation of a disciple to his master. If I have been able to write the chapters on the jade symbols of sovereign power and the jade images of the cosmic deities, my lasting thanks and acknowledgments are due to this great scholar whose ingenious investigations have furnished the basis for this research; but for his efforts it would have been impossible to attack these complicated problems with any chance for success.

[1] After GILES, Chinese Biographical Dictionary, p. 889.
[2] All text-figures where no special source is indicated are derived from the book of Wu.

Only a few among us at present have an idea of the extent and depth of fruitful archæological work now carried on by Chinese scholars. The opinion still largely prevails in our circles that the whole archæological Chinese wisdom is bound up with the Sung catalogues of the *Po ku t'u* and *Ku yü t'u p'u*, to which the superficial *Si ts'ing ku kien* and the brilliant *Kin-shih so* are possibly added.   But there are many dozens of modern well illustrated catalogues of bronzes and other antiquities accompanied by keen and clever disquisitions which do not shun discrediting or even refuting the worn-out statements of the *Po ku t'u*.   It has been almost entirely overlooked in Europe that the latter work, however valuable it may be in many respects, presents nothing but the traditions of the Sung period relative to objects of the Chou and Han periods; it must be thoroughly examined in each and every case in how far those claims are founded, in how far they agree or disagree with the traditions handed down in the contemporaneous texts of antiquity, and in how far they may be biased by the peculiar conditions of art and artists obtaining under the Sung dynasty.   The so-called monster *t'ao-t'ieh* certainly existed as a decorative design on bronze vases of the Chou; but whether in all cases, when the *Po ku t'u* points this design out on Chou bronzes, it is really intended in the minds of the Chou artists is another question which requires special critical examination; for there are many designs of other conventional monsters on those ancient bronzes.

To cite only one example as to how far modern Chinese archæologists go in contradicting the old, beloved school opinions, there is now the unanimous opinion in China and Japan that the so-called metal mirrors with designs of grapes, birds, lions and horses cannot come down from the time of the Han dynasty, as asserted by the *Po ku t'u* without the shadow of an evidence; they originated shortly before the T'ang dynasty, probably in the fifth and sixth centuries, and I may add, under Persian influence, as indeed the composition of this pattern first appears in the Sassanidian art of Persia, but never in Greek art as hitherto believed in Europe.   For all serious future investigations into Chinese antiquities, it will be incumbent on us to pay due attention to the works, opinions and results also of modern Chinese (as well as Japanese) archæologists.   The time has gone when only the *Po ku t'u* and the very weak *Si ts'ing ku kien*, which is of small value, may be ransacked at random and haphazardly by the foreign inquirer. Studies exclusively based on such books, without regard to the world of reality, deserve, in my opinion, no acknowledgment and are practically worthless.

Aside from these monographs, there are numerous other books

devoting a chapter or two to art-works of jade and cyclopædias giving extracts and quotations on the subject. The *Ko ku yao lun* by Ts'AO CHAO, published in 1387, and the *Po wu yao lan* by KU YING-T'AI, published between 1621 and 1627, are especially noteworthy. The great cyclopædia *T'u shu tsi ch'êng* of 1726, a copy of which is in the John Crerar Library (C 750), contains eight chapters on jade in its section on National Economy, Ch. 325–332. There is also a great amount of useful information in the *Yen kien lei han* (Original Palace edition of 1710 in 140 Vols. in the Newberry Library, N 36), *Ko chih king yüan* published in 1735 by CH'ÊN YÜAN-LUNG (1652–1736), *Pên ts'ao kang mu* by LI SHIH-CHÊN (completed in 1578), and certainly in the *P'ei wên yün fu* (Ch. 100 A). To enumerate all Chinese sources is unnecessary, since the sinological reader knows where to turn, while no advantage would accrue from such a task to the general reader.

As our collection relates to the cultural conditions of antiquity, we are certainly obliged to consult the ancient texts in which its ideas are reflected. The classical Book of Songs (*Shi king*) and the Book of History (*Shu king*) are prominent among these.

The principal sources bearing on the ancient religious cult and containing ample material on the ceremonial usage of jade are the three great Rituals, the *Chou li*, the *Li ki*, and the *I li*.[1] Of the former, we possess the excellent translation by EDOUARD BIOT[2] which is a monument of stupendous and sagacious erudition and remains the only work of Chinese literature heretofore translated into any foreign language with a complete rendering of all commentaries. In a great number of passages, I was prevented from following any of the accepted translations, especially in those cases where archæological objects and questions are involved. If it is true that Chinese archæology must be based on the knowledge of Chinese texts with the same method as classical archæology, it is no less true that the interpretation of the ancient texts will have a great deal to learn from the facts of archæological research and its living objects of stone, clay or metal which are harder than any paper-transmitted evidence. In the light of revived antiquity, we shall learn better to understand and appreciate the ancient Rituals in particular. If I am obliged, most reluctantly, to deviate from such authorities as Biot, Legge and Couvreur, I beg my critics not to interpret this necessity as arrogance or a mania for

---

[1] I availed myself of the Palace Edition published 1748 by order of the Emperor K'ien-lung in 182 Vols. (John Crerar Library, Nos. 213–215). As to the illustrations, I did not always quote them from this edition, which but few readers may have at their disposal, but rather from current European books easily accessible to every one.

[2] See Bibliography at end.

knowing better on my part, but as a suggestion intimated by a consideration of the new material here offered. This advance in our
knowledge is not my merit, but merely the consequence of favorable
opportunities granted me by a fortunate chance. I cannot dwell here
on a literary discussion of the three Rituals. It is well known that
the *Chou li* was not put together until under the Han dynasty; nevertheless, it reflects the peculiar culture of the Chou period in such a
complete and systematic manner as could have only been written
at that time. It is a state handbook expounding in minutest detail
the complex organism of the governmental institutes of the Chou
emperors. There is no doubt that the book has been touched and
worked over, perhaps also interpolated as the *Li ki*, under the Han
editorship; but substantially and virtually, it is the property of the
Chou time.[1] The Han commentators were no more able to explain
intelligently many passages in it, as the culture of the Chou had perished
before the hatred and persecution of the Ts'in, and, as we now see to
our great surprise, interpreted quite wrongly most of the ceremonial
utensils of the Chou, which were no longer within the reach of their
vision.

Here we must briefly touch one of the curious results of the following investigation which will interest sinologues and archæologists
alike. It seems that the Chinese commentators attempted to render
an account of the appearance of ceremonial and other antiquarian
objects either on the ground of oral traditions, or from hearsay, or,
in the majority of cases, on reconstructions evolved from their own
minds; but their comments are not based on a real viewing of the
objects concerned. This state of affairs is easily evidenced in general
by a glance at the so-called Illustrations to the Rituals, as the *San li t'u*
of Nieh Tsung-i of the Sung period (962 A. D.), or the illustrated volumes of the K'ien-lung edition, which pretend to picture all objects of
importance mentioned in the ancient texts. It was always a source of
wonder to me how the Chinese got hold of these weak drawings which
bear the indelible stamp of unreality and depict many objects as,
e. g., weapons, carriages and houses, in a way which we must decry
as utterly impossible from a purely technical viewpoint; and there
is likewise reason to wonder that such figures could find their way
into foreign books (Biot, Pauthier, Zottoli, Legge, Couvreur) to illustrate ancient Chinese culture, and be passed as the real thing without
a word of comment or criticism. A comparison of these reconstructive
or purely imaginary pictures with the actual specimens of the Chou

---

[1] EDKINS's criticism of the *Chou li* in his paper Ancient Navigation in the Indian
Ocean (*Journal R. Asiatic Society*, Vol. XVIII, p. 19) deserves special attention.

period now at our disposal will show that, in the plurality of cases, there is hardly a shadow of resemblance between the two. Wu Ta-ch'êng has taken the lead in this new field of research, and we are indebted to him for the restoration of the truth in the place of romanticism with regard to archæological objects of primary importance. The image of the Deity Earth has been mistaken for the part of a chariot wheel-nave which never existed in this form, while the former was construed in the shape of an eight-pointed star-figure, going back to a misunderstanding read by the commentators into their texts.

The most instructive examples of this kind are the *ku pi* and *p'u pi*, *i. e.* jade disks with "grain" and "rush" pattern.

The ancient Chinese texts are clad in a brief and laconic style, never wasting a word on the description of objects then known to everybody. They simply give the names of numerous vases, weapons, insignia, etc., without further details, so that there is plenty of room for the commentators to expand. These were scholars alien to the world, of versatile intuition perhaps, but lacking in the knowledge and observation of life and reality. It was not found unreasonable to answer that the jade disks *ku* had a bunch of cereals, and those called *p'u* a design of rushes engraved on them; the Sung artists accepted this comment, and quite characteristic of the pictorial tendency of their time, reconstructed those disks by drawings with realistic representations of the respective plants. For two thousand years, the Chinese have groped absolutely in the dark as regards the true nature of these disks. Now we know that such designs never existed in the Chou period, that the disks *ku* were covered with concentric rows of raised dots, an ornament called "grain," and that the disks *p'u* were decorated with a mat impression consisting of hexagons, the pattern receiving its name from a rush-mat. This and many other examples revealed on the pages to follow will furnish much food for reflection.

First, in regard to the methods of the archæologist. A net distinction should always be made between the wording of the ancient texts and the additional utterances of commentatorial wisdom. The commentators, very often, may certainly be right and reasonable, but should be held up as suspects under all circumstances and acquitted only on close trial. Their thoughts are usually afterthoughts, reflections, adjustments, compromises, evasions. It all depends upon the length of time by which the editor is separated from the time of the original. The singular world of the Chou was shattered in the period of the Han, and the Han scholars knew little in fact about that bygone age. Chinese later illustrations to the classics can be consulted and utilized only after a most painful scrutiny of the subject in question, and in almost all cases, they will then be found worthless because fanciful. The sinologue is confronted with this problem,— is the Chinese language really that clear and logical structure such as has been given out by a certain school of philologists, if the Chinese themselves, and even their best scholars of the remote Han period, were liable to misunderstand their ancient classical texts step for step? And what is the cause for these misunderstandings? To one initiated into the ethnological mode of thinking, it is not far to seek. Indeed, it would be unjust to brand the Chinese with special reproach in this matter, and to expose their working-methods to unfair reflec- tions.

What developed in China along this line, is a subconscious factor which has dominated the cultural life of all peoples of the globe from the dim beginnings of mankind until the present day. It is the pre- vailing tendency of the human mind to account for the reason of existing customs and traditions, and to seek, with the advance of individual conscious reasoning, for rational explanations of phenomena purely emotional and ethnical at the outset. This method results in a new association of ideas which has nothing in common with the origin of the notion in question, and may be the outcome of pure speculation. In China, where the bent to systematizing speculation was always strong in the minds of individual thinkers, the effects of this mental process come more intensely to the surface than in smaller communities more strongly tied by a uniformity of tribal thoughts, and for this reason, the Chinese offer the best imaginable material for a study of the psychological foundations of ethnical phenomena.

The errors in the interpretation of ancient customs and notions committed by Chinese commentators and editors, their failures in their attempts at a reconstruction of the past, and their positive pro- ductions of newly formed ancient artistic designs, never existing in times of antiquity, are not logical blunders to be imputed to their intellectual frame, but emanations of their psychical constitution evolved from a new process of association. The problem moves on purely psychological, not on mental lines. To revert to our above example,— the trend of thought in the Chou time was symbolic, swayed by impressions and sentiments received from celestial and cosmical aspects of the universe, and strove for expression in geo- metrical representations, so much so that the singular art of the Chou cannot be better characterized than by the two words symbolic and geometric, or, as geometric symbolism. Round raised dots or knobs were suggestive, on mere emotional grounds, of a heap of grain-seeds.

This mode of observation became foreign to subsequent generations who, reflecting upon the peculiar traits of the Chou culture, could but realize that a real representation of grain in the manner of a living plant was intended. The Sung artists with their inspirations for naturalistic designs took possession of this notion and instilled it with life by sketching it on paper. Thus, they transformed a rational reflection by mere intuition into a permanent motive of art promulgated as the production of the Chou period.

This case is by no means unique, but it is due to such misinterpretations, reflections and afterthoughts that many hundreds of artistic motives (and certainly not only these, but also customs, habits, traditions, social conventions, moral principles, etc.) have arisen everywhere among mankind. But, whereas in other cultures it is not always easy to unravel the mystery of this development, the long history of China and the vast stores of her literary and artistic wealth will often allow us to peep behind the stage, and to grasp the human and psychic force of such transformations of thought. It must therefore remain one of our principal endeavors, in the treatment of Chinese archæological subjects, to penetrate into the psychical basis of motives,— not only because this procedure will simultaneously furnish most valuable contributions to the psychology of the Chinese, which is much needed, but also in order to attain to a correct understanding of the history of the motive itself, since otherwise our knowledge would be an utter fallacy and self-deception.

We must grasp the nature-loving spirit of the impressionistic Sung artists to appreciate their very neat naturalistic designs of cereals and rushes on the Chou disks of jade. And we must understand, on the other hand, the complex organism of the world and life conception of the Chou period, which is quite a distinct and peculiar China in itself, to be prompted to the conclusion that the Chou design proposed by the intuition of the Sung artists cannot possibly have been an inheritance of the Chou. Then we must realize how the gigantic power of the superman Ts'in Shih Huang-ti had broken the ritualistic culture of the Chou, how a few remains and ruins of it only were exhumed and aired again by the revival activity of the Han, a movement of great earnestness and deep honesty of intention. Thus, we gain a basis for a judgment of the thinking and doings of the Han and later commentators, on whose shoulders the art-historians, art-critics and compilers of art-catalogues of the Sung period stand.

By thus joining link for link in this long chain and carefully listening to each tradition, we may finally hope to learn something of the development of Chinese ideas and art. Certainly, if handled by this

analytic method, Chinese archæology and art-history is not so easy
as it may appear to the outsider from some popular books and light
essays to which the public has been treated in recent years by authors
who, light-minded, take everything in a storm, because they are unable
to recognize the complexity and weight of a problem. "To sit on
the bottom of a well, and to say that Heaven is small," curtly remarks
a Chinese proverb. I utterly fail to see of what avail it is to us to
build the roof before the ground-pillars, of what advantage all these
discussions on subjective evolutions of motives, on analysis of style
and esthetics of Chinese art will be in the long run, as long as we do
not know the solid basis, the meaning and history of these motives,
and as long as such phantoms will be easily destroyed by every serious
investigation. For certainly not by intuition or opinions derived
from a general or vague knowledge of art can we hope to reach the goal,
but only by the most absorbed method of research consulting the views,
traditions and sentiments of those people who created the monuments
which we desire to understand.

First of all, we must understand the works of Chinese art, before
we can judge them, and that is the most difficult side of the question.
Certainly, I am not an advocate of seclusion or monopolization. A
sound open-door policy in this field carried out in a fair-minded spirit
of sympathetic coöperation would be a desirable and refreshing pro-
gram which may lead to fruitful results. The field is new and wide,
and there is room for many platforms. A sane and whole-hearted
exposition of any Chinese subject through an experienced art-student
will always be welcome; a fair and impartial criticism or suggestion
of an outsider or newcomer to this branch of science may prove as a
stimulus to greater efforts. We are all inquirers and seekers for the
truth, and everybody has to learn, and everybody is liable to commit
errors where the field is virgin. But dictatorial positiveness of judgment based on insufficient material and facts is surely detrimental
to the good cause.

In this paper, antique objects of jade are dealt with in so far as
they are living realities, being represented by palpable specimens in
Chinese collections or in our own. Numerous jade objects are mentioned in the ancient texts, none of which, however, have survived.
These should be taken up, whenever necessity arises and such objects
will actually be discovered. Thus, we hear _e. g._ of jade and other
stone mirrors[1] and screens, and even of jade shoes discovered in ancient

[1] Stone mirrors were known in ancient Peru, but no specimen exists in any
museum. In YUAN Y ULLOA's Voyages (Vol. I, p. 482) the following is on record:
"Stone mirrors are of two sorts. One of the 'Ynca stone,' the other of the gallinazo
stone. The former is not transparent, of a lead color, but soft. They are generally

graves; but ·no such specimens have been preserved in any Chinese collections.

SIAO TSE-HIEN, the author of the *Nan Ts'i shu* ("Annals of the Southern Ts'i dynasty") narrates that in Siang-yang (Hupeh Province) brigands opened an ancient tumulus which according to tradition was that of Chao, king of Ch'u (B. C. 515–489); in the tomb were buried footgear of jade (*yü li*) and a screen of jade (*yü p'ing-fêng*), the latter being very curious. In another report it is remarked that these jade shoes were the jade clogs of eunuchs (*kung jên yü ki*), and another tradition in the *Ts'i ch'un ts'iu* ("Spring and Autumn Annals of Ts'i") has it that people of Siang-yang opened an ancient tumulus, in which there was a jade mirror and ancient records written on bamboo tablets, the characters of which could not be deciphered; only the Buddhist monk K'ien Shan could read them, and it is supposed that this grave is identical with the one mentioned in the "Annals of the Ts'i dynasty."[1] The rifling of this tomb may be referred to the year 479 A. D.[2]

There are also reports on the discovery of cuspidors of jade, one of which was found with two bronze swords and sundry articles of gold and jade in the grave of Siang, king of Wei (B. C. 334–286).[3] Of the interesting jade casks buried with the corpse in the Han period[4] to guard the flesh against decay, none has as yet come to light, unfortunately, and there are numberless other types of burial jades of which

of a circular form and one of the surfaces flat with all the smoothness of a crystal looking glass. The other oval and something spherical, and the polish not so fine. They are of various sizes, but generally 3 or 4 inches in diameter; though I saw one a foot and a half. Its principal surface was concave and greatly enlarged objects, nor could its polish be exceeded by the best workman among us. The gallinazo stone [obsidian?] is very hard, brittle as flint and black color." — Copper mirrors were also used, as reported by GARCILASSO DE VEGA, Royal Commentaries of Peru (Book II, Ch. XVI, 1688, translated by P. RYCAUT). "The looking-glasses which the ladies of quality used were made of burnished copper; but the men never used any, for that being esteemed a part of effeminacy, was also a disgrace if not ignominy to them."

[1] *Ts'i kuo k'ao*, Ch. 8, pp. 7b, 8a, 10a. — The *Ts'i kuo k'ao* "Investigations into the Seven States" (which are Ts'in, Ts'i, Ch'u, Chao, Han, Wei and Yen) was compiled by TUNG SHUO of the Ming dynasty, in fourteen chapters. Reprinted in Vols. 40–41 of the Collection *Shou shan ko ts'ung shu*. It contains extracts and notes of culture-historical interest regarding these seven feudal principalities. — Compare on the above passage also HIRTH, Chinese Metallic Mirrors (*Boas Anniversary Volume*, p. 216), and, in regard to another case of the burial of jade mirrors, DE GROOT, The Religious System of China, Vol. II, p. 414.

[2] Compare A. TSCHEPE, Histoire du royaume de Tch'ou, p. 280 (Shanghai, 1903).

[3] DE GROOT, The Religious System of China, Vol. II, p. 397.

[4] They are frequently mentioned in the Annals of the Han Dynasty. LÜ PU-WEI, who died in B. C. 235, reports in his book *Lü-shih Ch'un Ts'iu:* "Pearls are placed in the mouth of the dead, and fish-scales are added; these are now utilized for interment with the dead." The Commentary to this passage remarks: "To place pearls in the mouth of the dead (*han chu*) means, to fill the mouth with them; the addition of fish-scales means, to enclose these in a jade casket which is placed on the body of the deceased, as if it should be covered with fish-scales."

we can form no clear idea. In Chinese antiquarian studies, we must always be mindful of the incompleteness and deficiency of material. Only stray fragments and heaps of ruins have been transmitted, while the best and most glorious monuments of the ancient civilization have become the prey of natural decay or wilful destruction. What is preserved is a trifle compared with what is gone. The losses are immense and irreparable, and no Varus will ever return us these legions. This consideration should never be lost sight of to guard us against premature conclusions and hasty combinations. Only the most extensive series of types covering wide local and temporal areas will allow us to reach a fairly satisfactory result, and even then reservations must be made and judgment restricted in view of the thousands of gaps sadly existing in our knowledge. At the best, we may hope in the end for a reconstruction of the ancient culture-life, as we may piece together and supplement a jar in shreds; sound skepticism will keep from joining the fragments wrongly.

The word *yü* "jade" most frequently occurs in the oldest texts and is said to have been known to the legendary Emperor Huang-ti (alleged B. C. 2704).[1] The Chinese word is just as general and comprehensive as our word "jade," which may therefore be freely used as its equivalent, and includes nephrite, jadeite, bowenite, sometimes also in ancient pieces special beautiful kinds of serpentine, agalmatolite and marble.[2] In the times of antiquity, the number of species and varieties called *yü* was doubtless much greater than at the present time, as we see from a series of manifold names occurring in the oldest texts (*Shi king* and *Shu king*), many of which remain unexplained. Li Shih-chên, the great Chinese naturalist of the sixteenth century, recognizes fourteen varieties of jade, most of them being distinguished from their colors and localities. At present, it is only nephrite and jadeite that is acknowledged as true jade by the Chinese, all other stones receiving special names.

[1] HIRTH, The Ancient History of China, pp. 13, 91.

[2] It is not correct, as WILLIAMS (The Middle Kingdom, Vol. I, p. 309) remarks, that white marble, ruby, and cornelian all come under it. Jade is, as the Chinese say, a species in itself; also agate is considered as *sui generis*. — It is well known that our word *jade* is derived from Spanish *piedra de hijada*, "stone of the loin," because the stone was supposed to cure pain in the loin. Another etymology is offered by F. GRENARD (Mission scientifique dans la Haute Asie, Vol. II, p. 188) who proposes to derive the word from Turkish *yada-tchi*, "a sorcerer who is able to produce rain and fine weather by means of a magical stone." *Yada* is the name of this stone (W. RADLOFF, Wörterbuch der Türk-Dialekte, Vol. III, Col. 207, 210). It is impossible for two reasons to accept Grenard's suggestion. First, there is no evidence for the word *yada* to denote *jade* or exclusively *jade*, the proper Turkish designation of which is *kash* (RADLOFF, Vol. II, Col. 389). Secondly, our word *jade* is but a recent introduction first brought to England by Sir Walter Raleigh (1552—1618) who always uses the Spanish name for the stone in his books; the word does not appear in our literature before this time, as we should expect, if Grenard's surmise were correct.

While from about the Christian era Turkistan became the chief source for the supply of jade to China, to which Yünnan and Burma were later added, neither Turkistan nor Yünnan come into question in very early times. The jades used in the period of the Chou, and most of those of the Han dynasty, were quarried on the very soil of China proper, as we know from the accounts of the Chinese, and as we can still ascertain from the worked jade pieces of those periods which in quality and color are widely different from any produced in Turkistan and Burma. In BISHOP's work (Vol. I, p. 9) it is said: "Jade has not yet actually been seen *in situ* by any competent observer in any of the eighteen provinces of China proper, and it is permissible, meanwhile, to doubt its occurrence and to await more certain evidence. The interior of China is almost unexplored from a geological or mineralogical point of view. . . . There may have been ancient quarries which have long since been exhausted; the material of some of the older carved pieces is certainly different in many respects from anything produced now, and seems to point to lost sources of supply." This supposition is quite correct and is confirmed by the results of many inquiries which I had occasion to make at several times in Si-ngan fu and other places of Shensi Province: all Chinese questioned by me, experts in antiquarian matters, agree in stating that the jades of the Chou and Han dynasties are made of indigenous material once dug on the very soil of Shensi Province, that these quarries have been long ago exhausted, no jade whatever being found there nowadays. My informants pointed to Lan-t'ien and Fêng-siang fu as the chief ancient mines.

As early as in the *Shu king* (Tribute of Yü, 19) and in the geography of the *Chou li*, a trade is mentioned consisting of jade and other minerals in the territory of Yung-chou comprising the northern part of the present province of Shensi between the river Wei in the south and the Ordos region in the north (HIRTH, *l. c.*, p. 122). As we now have an opportunity of studying a great number of ancient Chou and Han specimens of jade in the Bishop and Mrs. Blackstone collections, we may now establish the fact with a high degree of certainty that the Chou jades without exception, and the greater part of the Han jades, are made of indigenous material, quarried on the domain of the earliest settlements of the Chinese which they had naturally well explored. This conviction agrees very well with the traditions of the Chinese, as we shall see presently. It requires but little experience and routine work along these lines to distinguish these ancient jades with their salient characteristics of structure and color, and their additional historic qualities acquired in the graves, from the Turkistan and

Burmese nephrites and jadeites. Their appearance will be described in the following chapters. It was doubtless the Chinese themselves who, being acquainted with jade in their country, probably for millenniums, gave the impetus to the jade fishing and mining industries of Turkistan. Also this case may throw a side-light on the nephrite question of Europe; home-sources do not exclude imports, and scarcity or exhaustion of sources may favor them.

The mountains south of Si-ngan fu in Shensi Province produced jade, gold, silver, copper and iron in the first century B. C., as expressly stated in the "Annals of the Former Han Dynasty" (*Ts'ien Han shu*) under the reign of Wu-ti; it is mentioned also in the "Biography of Tung-fang So" (see *Ta Ts'ing i t'ung chi*, Ch. 181, p. 14 b). The jade of Lan-t'ien[1] enjoyed a special reputation. The distinguished physician T'AO HUNG-KING (452–536 A. D.), the author of a treatise on *Materia medica* (*Ming i pieh lu*), states that the best jade comes from that locality; he mentions also the occurrence of jade in Nan-yang, Honan Province, and in the Lu-jung River of Tonking,[2] also that brought from Khotan and Kashgar; if translucent and white as hog's lard, and resonant when struck, it is genuine. In the eleventh century, however, it is positively asserted by SU SUNG, an able student of natural science, that in his time no more jade was quarried in those home quarters nor in Tonking, and that it was only found in Khotan. His lifetime may therefore be regarded as the date when the native output of the mineral had come to an end. The high value of the ancient jades is, consequently, enhanced considerably by their material no longer existing or being found in its natural state.

There are indications that, aside from the Province of Shensi, other localities of jade may have existed or may still exist in China. Mr. BISHOP (Vol. I, p. 9) mentions four pebbles procured in Shanghai from a Mohammedan dealer in stones, who said that they were found in the bed of the Liu-yang River in Hunan Province, and which are of interest, as he says, as suggestive of another jade locality in China proper.

The Chinese "Gazetteer of Sze-ch'uan Province" (*Sze ch'uan*

---

[1] Lan-t'ien, "the Blue Field," received its name from the jade quarries, as expressly stated in the Chronicle of the place (*Lan-t'ien hien chi*, Ch. 6, p. 17, edition of 1875). According to T'ao Hung-king, it produced white and green jade. The "Jade Mountain" (*Yü shan*) was situated 43 *li* south-east of the town. As an analogy to the exhaustion of the jade mines, the Chronicle quotes the fact that in former times also silver ore (according to the *Wei shu*), as well as copper and iron were exploited there, all of which no longer occur. An exploration of this site may be recommended to our geographers.

[2] According to the Chinese description of Annam, jadeite (*fei ts'ui*, Devéria's translation *jade serpentine* is not to the point) is a production of that country (G. DEVÉRIA, Histoire des Relations de la Chine avec l'Annam, p. 88, Paris, 1880).

t'ung chi, Ch. 74, p. 43) mentions a "white jade-stone (*pai yü shih*) resembling jade" produced in the district of Wên-shan in Mou-chou; the natives avail themselves of it to make implements. A variety of green jade called *pi* (GILES No. 9009) from which arrow-heads can be made is ascribed to the district of Hui-wu or Hui-li in Ning-yüan fu (*Ibid.*, p. 24, and *Shu tien*, Ch. 8, p. 5). The latter work on Sze-ch'uan (written by CHANG CHU-PIEN in 1818, reprinted 1876) makes mention also of "black jade" (*hei yü*) after the cyclopædia *T'ai p'ing yü lan;* its color is black like lacquer, and it hence receives also the name "ink jade" (*mo yü*)[1] and ranks low in price. It occurs in western Sze-ch'uan and is identical with the jet or gagate now produced in the district of Kung (in Sü-chou fu), out of which implements are carved. It certainly remains doubtful whether these stones represent real jades, as long as there is no opportunity for identifying specimens of them. According to the "Description" or "Gazetteer of Kansu Province"[2] (Ch. 20, p. 7 b), jade is obtained from the river *Hung shui pa*[3] in Su chou. The geographical work *Huan yü ki*, published by Lo Shi between 976 and 983 A. D., mentions a kind of "brilliant jade" (*ming yü*) occurring in Kuang-chou, Ju-ning fu, Honan Province. In Nan-yang Prefecture of the same province, jade may have been dug from the fifth to the eleventh century A. D., as we noticed before.

Chinese sources refer to the production of jade in the prefecture of Kuei-lin, Kuang-si Province (G. DEVÉRIA, Histoire des Relations de la Chine avec l'Annam, p. 95, Paris, 1880). But this remains somewhat doubtful, as the designation in this case is *yü shih*, "jade-stone" (instead of *yü*) which may refer and usually refers to only jade-like stones.

Jade seems to have reached China also from the kingdom of the Caliphs. It is reported under the year 716 A. D. that the Emir Suleimân, who died in 717, sent an ambassador to China to present a robe woven of gold threads (brocade) and a flask of jade ornamented with jewels; the flask is called *sha-ch'ih* (or *sha-ti*), an inexplicable term (CHAVANNES in *T'oung Pao*, 1904, p. 32). A Chinese envoy who visited Bagdad in 1259 reports that the palace of the Caliph was built of fragrant and precious woods, and that its walls were constructed of black and white jade (BRETSCHNEIDER, *Chinese Recorder*, Vol. VI, 1875, p. 5).

---

[1] In Japanese *boku-giyoku* = jet (GEERTS, p. 234).

[2] *Kan-su t'ung chi*, last edition published in 1736. When passing through the capital of the province, Lan-chou, in January, 1909, I was informed that a new, revised and largely increased edition of this now scarce work was in course of preparation and was expected to be ready at the end of the summer of the same year.

[3] "The Embankment of the Red Water."

The Jesuit missionaries of the eighteenth century seem to have been under the impression that jade was produced in the provinces of Shensi and Shansi (see AMIOT in *Mémoires concernant les Chinois*, Vol. VI, p. 258). Father DU HALDE (A Description of the Empire of China, Vol. I, p. 16, London, 1738) sums up as follows: "The *Lapis Armenus* [his designation of jade] is not very dear in Yün-nan,[1] where it is found in several places, differing in nothing from what is imported into Europe. 'Tis produced also in the Province of Sze-ch'uan, and in the district of Ta-t'ung fu, belonging to Shansi, which furnishes perhaps the most beautiful *Yü-she* (jade) in all China; 'tis a kind of white *Jasper*, the white resembling that of *Agat;* 'tis transparent, and sometimes spotted when it is polished."

The city of Si-ngan fu is still the distributing centre for the unwrought pieces of jade arriving from Turkistan, and seems to have been so also in former times. Particularly fine bowlders are sometimes kept and guarded as treasures. The bowlder of whitish jade reproduced in Fig. 2 on Plate I was preserved as a precious relic in the Buddhist temple *Hing-lung se* of Si-ngan fu, where I acquired it for the Field Museum. I was informed there that it had come from Khotan, Turkistan, a long time ago. It measures 19.5 cm in length, 10.5 cm in width and 14.1 cm in height, and has a weight of eleven pounds.

The other water-rolled bowlder of natural polish in Fig. 1 of the same Plate was found in a dried-up river bed in the northern part of the province of Shensi and represents, also in the opinion of the Chinese, a kind of jade used during the Han period. The correctness of this statement is borne out by worked jade pieces of the Han period in our collection, exhibiting the same material. It is a bluish-green jade clouded with white and leaf-green speckles and sprinkled with large brown and black patches. The lower side is almost entirely occupied by an ivory-white, brown and russet coloring intersected by black strips and veins, almost producing the effect of an agate. This piece weighs somewhat over seven pounds, is 22.5 cm long, 15 cm wide and 5 cm thick. Such bowlders of so-called Han jade have occasionally turned up in the eighteenth century and were then worked into vases or bells or other objects. We shall come back to this point in Chapter VIII.

The color of jade was found to be permanent and unchangeable. The *Li ki* (*Yü tsao* III, 32) describing the qualities of a brave soldier

---

[1] According to Chinese statements, jade is found in Ch'êng-kiang fu of Yün-nan Province (G. DEVÉRIA, Histoire des Relations de la Chine avec l'Annam, p. 91, Paris, 1880).

EXPLANATION OF PL. I.

Fig. 1. Water-rolled Pebble of Jade of the Han Period.
Fig. 2. Bowlder from Khotan, Turkistan.

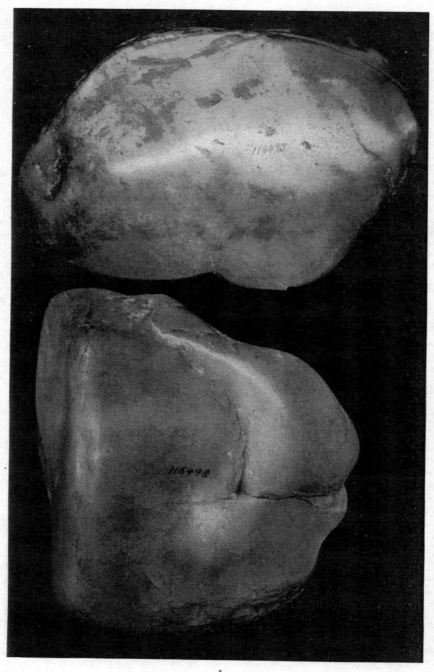

1
2
NEPHRITE BOWLDERS.

says of his countenance that it is always the same as jade.[1] But like every substance in nature, jade is subject to discoloration and decomposition. This effect is noticeable particularly in the ancient burial pieces of the Chou and Han dynasties which have sometimes undergone considerable changes during their subterranean history in the grave, and softened to such a degree that they may be mistaken for steatite.

"The Chinese are perfectly familiar with the disintegration of the surface and the gradual softening and decomposition of the material which occur in jade long buried in the ground. Jade in its crudest state always contains a greater or less proportion of iron, and this, gradually becoming oxidized by process of time, causes staining of the surface, the color of the stain often extending inward, especially where there happens to be any flaw or vein in the material. All kinds of variegated 'iron-rust' tints are produced in this way, passing from amber-yellow to the deepest brown, and sometimes becoming almost black" (BISHOP, Vol. I, p. 232).

A peculiar characteristic of the white jades of the Han period is that sometimes thick masses of chalk-white clayish matter permeate the jade substance. The Chinese call them "earth spots" (*t'u pan*) and attribute their presence to mercury absorbed by the stone while buried. It is impossible to accept this view, as the clay is embodied under the polished surface and must have been present there, before the piece in question was worked and polished. But apparently, these stones were intentionally cut in such a way that the clay became visible through the transparent surface. It seems to me that the Han people may have evinced a particular predilection for this natural phenomenon which usually occurs in ornamental pieces.

The character of the following research is such that one investigation is based on the other, and that the single chapters are mutually dependent. Chapter II cannot be understood without the knowledge of the first chapter, and the fundamental religious ideas expounded in Chapter V have their basis in the discussions of Chapter II. Again, Chapter VIII cannot be appreciated without an insight into the complex subjects of the preceding chapter, as, naturally, the ornaments of the dead are an echo of those of the living ones. In the culture of the Chou, to which a great and indeed the principal part of this material

---

[1] LEGGE (Li Ki, Vol. II, p. 26) translates: "His complexion showed (the beauty and strength of) a piece of jade," which is apparently not to the point; the *tertium comparationis* is the unchangeability of the color of jade which is likened to the firm and steady expression of the soldier's face (compare COUVREUR, Li Ki, Vol. I, p. 722).

refers, everything is system and consistent systematization, and each object and the idea illustrated by it must be perceived in close relation to, and in permanent context with, the entire system of the peculiar world-conception of that period. My prospective critics will doubtless appreciate this state of affairs and recognize that it would be vain to select at random the one or other object or point for ready attack, if not taken up and properly understood in connection with the whole subject; whoever is willing to further these studies, must consent, I regret to say, to digest first this material in its entirety.

The following pages contain only a small portion of my notes on jades. It would have been easy to increase them to double and more of their present extent, and to present them in a more readable form, if I had the privilege of the leisure of an author. The daily demands made by the immediate task of cataloguing and installing a large collection are not favorable to literary activity, and the necessity of working up in the near future an appalling quantity of other materials did not allow me to delve in this particular subject with that copiousness of detail which would have been desirable. Though dealing with polished jade, these notes will be found more crude than polished, and indeed pretend to be nothing more than chips and shavings from a workshop. May others take up and pursue the threads where they dropped from my hand.

I. JADE AND OTHER STONE IMPLEMENTS

The number of stone implements hitherto discovered on Chinese soil is exceedingly small, a fact to be accounted for in several ways. First it is due to the lack of systematic archæological searchings and excavations on sound methods handicapped by the prejudices of the people; secondly, the indifference of the Chinese towards these seemingly trifling objects which bear no inscriptions and therefore offer no antiquarian interest to them. While they have delved at all times in the graves of their ancestors to their hearts' delight to revel in antiquities of bronze, jade, or pottery, they left unnoticed or carelessly threw aside minor objects of stone and bone or small fragments which seem to us of primary scientific importance. A third reason, and probably the most weighty of all, will be found in the fact which we shall establish in the course of this investigation that, as far as the present state of our archæological knowledge and the literary records point out, the Chinese have never passed through an epoch which for other culture-regions has been designated as a stone age.

We can merely assert at the present time with some degree of certainty that at some remote indefinable period stone implements have been in use to a certain extent within the boundaries of what we now call the Chinese empire; this does not yet mean that they have been manufactured and employed by Chinese peoples themselves, as many other groups of tribes related and unrelated to the Chinese have been inhabiting the empire. It is therefore safe only to speak of stone implements of China, whereas it is not warranted to speak of Chinese stone implements. The evidence for such stone implements is furnished by three sources: (1) by a number of actual specimens which have come down to us, (2) by references made to such implements in Chinese records, and (3) by survivals of such plain implements in more elaborate ceremonial implements of later ages usually made of jade, or of other materials like copper and bronze. We shall take up these subjects *gradatim*. Before discussing a considerable amount of new material here published for the first time, it may be advisable to sum up briefly what has become known of such stone implements in our literature.

On April 30, 1884, the *Proceedings of the American Antiquarian Society* contained a paper by the mineralogist HEINRICH FISCHER in Freiburg "On Stone Implements in Asia," in which a survey of stone implements then known from India, China, Siberia and Japan is given.

29

"Of Chinese stone celts and implements," remarks the author, "very little is known. Despite a correspondence extending over almost the whole of Europe, I was able to discover but a single specimen which exists in the private museum of John Evans in London. It is described as being of nephrite; but upon an examination which I made of it, I think I am not mistaken in determining it to be of fibrolithe. What claimed my attention, was its edge, which did not bevel gradually, but was straight, and which circumstance is attributed to a continued sharpening of the edge."

The most interesting kind of objects mentioned by Fischer is a group of stone celts in the South Kensington Museum, London. "They are said to be made of jade; their mineralogical diagnosis, however, is as yet not definitely secured. They show colors varying from a dark coffee-brown to a yellowish green, and at least four out of every six are remarkable for being engraved with antique Chinese characters,— the names of their former possessors. Their Chinese name is 'yao-chan,' medicine spattles, and they were used for cutting drugs. They are of almost quadrangular shape, perforated near their bases either conically or perpendicularly; their edges run partly in a straight line and present sharp angles; where they are crescent-shaped the angles are rounded. There are specimens which exhibit different colors, as a clear whitish green at the bases, and in the middle a dark, dirty or a black green, and others present a grayish cloudy basis and are of a coffee-brown color in the middle. In this connection a question arises, which has been often asked but has not yet received due consideration, viz.: Has China ever been explored sufficiently to know whether she had a stone epoch? And if she had not, did the people who are now her inhabitants pass through such an epoch elsewhere, from which we should be able to consider the specimens just described as possible relics brought by them as souvenirs from their ancient homes?"

This latter question of Fischer seems superfluous; for the specimens mentioned by him have partly Chinese characters engraved upon them, which is sufficient proof for assuming that they originated on the very soil of China in an historical period.

JOHN ANDERSON [1] is supposed by many to have been the first to discover stone implements in China. But his material is open to criticism, and it is questionable whether it really belongs to the Chinese culture area. Noticing a stone implement exposed for sale on a stall in the bazar of Momien, Yün-nan Province, Anderson purchased it for the equivalent of a few pence. No sooner was his liking for such objects

---

[1] A Report on the Expedition to Western Yunan via Bhamô, Calcutta, 1871, pp. 410 et seq.: The Stone Implements of Yunan (i. e. Yün-nan), with five plates.

known than he was besieged by needy persons who willingly parted
with them for sums varying in value from four to eighteen pence each.
After his first investment, specimens to the number of about a hundred
and fifty were procured by different members of the expedition (Sladen,
Bowers, and himself); but all were purchased, none being discovered
by them.   Most were obtained at Momien, and a few in the Sanda
valley.   He was informed at Momien that stone implements were not
unfrequently turned up while ploughing the fields, and that they were
occasionally found lying exposed on the surface.   The belief prevails
that they and also bronze implements are thunderbolts, which after
they fall and penetrate the earth take nine years again to find or work
their way up to the surface.   "The high estimation in which they are
held, both in Yün-nan and Burma," continues ANDERSON himself,
"suggests the suspicion that the Chinese in former days did not neglect
to take advantage of the desire to possess those implements or charms
and made a profitable traffic in their manufacture. A consideration of the
character of some of the Yün-nan implements has led me to this conclusion.   A considerable percentage of them are small, beautifully cut
forms with few or none of the signs of use that distinguish the large
implements from the same localities, and, moreover, all of them are of
some variety of jade.   These facts taken in conjunction with their
elaborate finish, and the circumstance that jade was formerly largely
manufactured at Momien into a variety of personal ornaments, are the
reasons which have made me doubt the authenticity of many of the
small forms, and to regard them only as miniature models of the large
and authentic implements, manufactured in recent times as charms to
be worn without inconvenience."

It will be necessary to keep in mind and to share these doubts of
the discoverer, but even granted that the majority of pieces in his
collection is authentic, they are only of relative, perhaps merely typological, value for a study of Chinese archæology, as there is the greatest
likelihood of these specimens being productions of the Shan tribes of
Yün-nan, the earliest settlers in that region, and not of the much
later colonizing arrivals, the Chinese. I readily recognize in such
types of celts as figured by Anderson, e. g., on Plate II, 6 and 7, or
on Plate III, 12 and 13, striking analogies to Chinese celts found in
Shensi and Shantung; but such coincidences are not forcible as conclusive historical evidence, for numerous other analogies from the
most diverse parts of the world would admonish us to be cautious
in this respect.   Moreover, there is not a single specimen in the Anderson collection showing that characteristic perforation so prominent in
the jade implements of Shensi.   On the other hand, it may be argued

that also in the Shantung stone implements to be described by me this feature does not occur. But I am not going either to insist on the Chinese origin of the latter. We cannot separate archæological finds from their locality, and as the only conspicuous evidence available for the Anderson collection is that the objects forming it sprang up on the very territory occupied by the Shan, it will be safe to ascribe their origin to a non-Chinese culture-group, and to place certain restrictions on them in a consideration of Chinese archæology; they belong, not to the archæology of the Chinese, but of China in a geographical sense.

Recently, J. COGGIN BROWN,[1] following the path of Anderson, has examined and described twelve stone implements gathered in T'êng-yüeh or Momien, nine of which are said to be made from various jade-ites; he upholds the authenticity of the implements traded in that district in opposition to Anderson, but on grounds which are hardly convincing.

E. COLBORNE BABER[2] reports the discovery in a stone sarcophagus of a polished stone axehead of serpentine in Ch'ung-k'ing, Sze-ch'uan Province, and of a chisel of polished flint which he found in the posses-sion of an opium-smoker who was scraping the opium stains from his fingers with the edge of the implement; he said that he had found it, and another, in a stone coffin in a field near his house. "It is therefore undeniable," concludes BABER, "that these objects are found in con-nection with coffins, though what the connection may be is not clear. The natives call them *hsieh* 'wedges' and conceive that their use was to fasten down the lids of sarcophagi in some unexplained manner. A more plausible supposition is that they were buried with the dead in conformity with some traditional or superstitious rite; at any rate the theory is impossible that the people who hollowed out these ponderous monoliths worked with stone chisels, and left their tools inside." Unfor-tunately, the author does not give any description nor figures of his two specimens which he kept in his private collection, and I have no means of ascertaining what has become of them. Nevertheless, his account is valuable in that it shows the burial of stone implements with the dead in Sze-ch'uan, and we shall see that the same custom prevailed in Shensi.

A stone hatchet found by Williams in a mound forty feet high near Kalgan has been described by J. EDKINS.[3] The mound belongs to a large collection of graves, large and small, about seven miles east of the city of Yü chou, and 110 miles west of Peking. An ancient wall, nearly

[1] Stone Implements from the Têng-yüeh District, Yünnan Province, Western China (*Journal and Proceedings of the Asiatic Society of Bengal*, Vol. V, 1910, pp. 299–305, 2 plates).
[2] Travels and Researches in Western China, pp. 129–131 (in *Royal Geographical Society*, Supplementary Papers, Vol. I, London, 1886).
[3] Stone Hatchets in China (*Nature*, Vol. XXX, pp. 515–516, 1884). Compare the review by H. FISCHER in *Archiv für Anthropologie*, Vol. XVI, 1886, pp. 241–243.

round, twenty feet high and about eight miles in circumference, is still in existence there. The mound in which the hatchet was found is in the line of this wall — that is, the wall runs north-west and south-east from it. Hence the wall-builders did not regard the mound as sacred, for it would not in that case have been made to serve the purpose of a wall to their city on the south-west side. There is another large mound known as the grave of Tai Wang. It is a little to the east of the centre of the inclosed space once a city, and the principal road runs through the city by this mound from east to west. Rev. Mark Williams of Kalgan, who found the hatchet, and was the first foreigner to draw attention to the old city, was struck with the general resemblance of the mounds, the wall and the hatchet to what he is familiar with in Ohio. So close was the similarity that it seemed to him to require that the same class of persons who made the one should have made the other. Several pieces of broken pottery were found in the neighborhood of this mound, and their pattern is said to differ from modern crockery. The hatchet is about five inches long, and is made of a black stone not heavy. FISCHER concludes from this statement that it is likely to be a serpentine whose specific weight varies between 2.3 and 2.9. Nothing is said in regard to the shape and technique of the hatchet.

MARK WILLIAMS himself has given the following account of this find:[1]

"From Kalgan to Yü chou are ancient mounds in cluster on the plain or singly on eminences. These latter would indicate signal towers, while the former would suggest tombs. They are about thirty feet high, circular and oval in shape, and no arrangement can be observed in the clusters.

"At the base of a signal mound by the great wall of Kalgan I found a *stone axe*.

"The Chinese give no rational explanation of these mounds. I have as yet found no mention of them in ancient records. At Yü chou, one hundred miles south of Kalgan, is a cluster of forty mounds; four miles off are ruins of a city wall. Chinese cities have rectangular walls, with towers at short intervals. But this is a circular embankment with no remains of towers. The part of the remaining entrance is unlike the gate of a Chinese city. Records state that this was the seat of a Chinese prince who lived B. C. 200. In some places the wall is levelled, in other places it is perfect, making an acute angle at the summit. Cultivation has narrowed the bases of the mounds, but superstition prevents their destruction. To one familiar with the

[1] Ancient Earth-works in China. *Annual Report of the Smithsonian Institution*, 1885, Part I, p. 907, Washington, 1886.

works of the mound builders in the Mississippi Valley, the stone ax,
the mounds, circular wall, suggest a similar race."

Two flint arrow-heads, both without barbs, found by the well-known
naturalist ARMAND DAVID in Mongolia in 1866, have been published
by E. T. HAMY.[1] They are finely polished, recalling similar pieces
still in use at the time of the arrival of the first Russian explorers in
eastern Siberia. HAMY basing his evidence on a statement in David's
diary points out the comparatively recent origin of these finds which
have been made in a black diluvial soil together with small fragments
of pottery and metal instruments, and with the remains of recent
animals. It is therefore necessary, concludes HAMY, for the moment at
least, to abandon the theory of a Mongol quaternary man and the
ingenious considerations prematurely attached to it.[2] At all events,
these two arrow-heads rather seem to point in the direction of Siberian
than of Chinese antiquity. Dr. BUSHELL (in BISHOP, Vol. I, p. 29)
mentions one jade arrow-head in his private collection.

ENRICO H. GIGLIOLI[3] has described a stone implement found in
1896 by F. C. Coltelli in Yen-ngan fu, Shensi Province, and designated
as a yao ch'an "medicine spade." It is flat, of rectangular shape (22.8
cm long, 8.5 — 10.5 cm wide, 1.1 cm thick), with a perforation in the
upper end bored from one side only, with a diameter of 3.1 cm on the
one side and 2.3 cm on the other side, so that the perforation has the
shape of an obtuse cone. Altogether it resembles the types figured by
me on Plate V. Giglioli asserts that it did not serve as a battle-axe,
but as a mattock in husbandry. The material, he calls "fine jasper"(?)
and defines the colors of it as yellow, gray and white. For the rest,
he depends on the Anderson collection, eleven specimens of which are
reproduced and listed as Chinese.

In the Bishop collection, there is a small polished celt (No. 324)
made by BUSHELL (Vol. II, p. 106) previous to the Han dynasty and
described by him as "perforated for use as an amulet, with rounded
corners and bevelled rim, one face being perfectly flat, the other having
a bevelled cutting edge; in Burma as well as in southwestern China,
such amulets are supposed to make the wearer invulnerable." Another
celt in the same collection is decorated with the "thunder-pattern"
(meander) and the monster t'ao-t'ieh of which BUSHELL thinks it may

---

[1] Note sur les silex taillés d'Eul-Ché-San-hao (Bulletin du Muséum d'Histoire
naturelle, Vol. IV, pp. 46–48, Paris, 1898, 2 Figs.). See also the note by J. DENIKER,
The Races of Man, p. 362.

[2] On the glacial period in Mongolia see now G. MERZBACHER, Zur Eiszeitfrage
in der nordwestlichen Mongolei (Petermann's Mitteilungen, Vol. 57, 1911, p. 18).

[3] L'età della pietra nella Cina colla descrizione di alcuni esemplari nella mia
collezione, in Archivio per l'antropologia e la etnologia, Vol. XXVIII, p. 374, Firenze,
1898.

JADE CHISELS OF CHOU PERIOD, FROM SHENSI PROVINCE.

have been intended for a votive offering to a temple of the Thunder-god who was more worshipped in early days than now; but this supposi-tion is not supported by any Chinese text. Another celt in the Bishop collection bears the inscription *Hua-shih shêng ch'un*, "May the Hua family flourish like spring!" This inscription looks very suspicious and is certainly a recent additional improvement. Bushell calls these miniature celts by a Chinese name *yao ch'an* "medicine spades." [1]

The new material here submitted consists of two groups from two different localities; first, a collection of fifteen jade implements made by me in Si-ngan fu and illustrated on Plates II—VIII, and secondly, a collection of twelve stone implements originating from Ts'ing-chou fu in Shantung Province. The former lot has come to light from ancient graves in the province of Shensi, all situated west of the present city of Si-ngan along the road to the old town of Hien-yang. These graves are justly considered by the Chinese living in that locality as belonging to the period of the Chou dynasty (B. C. 1122–249), and the aspect of these implements found in them points to the same period, so that the internal evidence corroborates the historical tradition. They are all made of beautiful qualities of jade, highly polished and of most exquisite colors, such as is no longer mined, the supply having been exhausted long ago, but such as was found on the very soil of the province in that epoch to which these objects must be referred.

The stone hammer of dark-green jade, without perforation, repre-sented on Plate II, Fig. 1, is of particular interest, because only the blade is polished, while all other parts, also the lateral sides, are inten-tionally roughened to afford a firm grip to the hand clasping the ham-mer in using it. On the face shown in the illustration, the polished blade extends only 5 cm in length against 12 cm on the opposite face, while the total length of the implement amounts to 14.5 cm; its width over the back is 5 cm, over the blade 6.5 cm; its thickness is 2.5 cm near the butt and reaches 3.5 cm in the middle. Above the polished portion on the face visible in our plate a slight depression will be ob-served, apparently used for resting a finger in; there is another on the

---

[1] In the June number of *Man*, p. 81 (Vol. XI, 1911) there is a brief article by R. A. Smith on The Stone Age in Chinese Turkestan, illustrating on a plate twenty-four worked stones collected by M. A. Stein in the Lop-nor desert. Two jade celts and three arrow or lance-points are the only implements in this lot. The material is not such as to allow us to establish any historical connections, and is doubtless not associated with Chinese culture. — In the July number of *T'oung Pao* (1911, p. 437), Chavannes reviews a paper by Torii Ryuzo on his archæological explora-tion of southern Manchuria (in Japanese, Tokyo, 1910); he discovered prehistoric remains on the peninsula Liao-tung where he excavated stone axes and arrow-heads, fragments of pottery decorated with various geometric designs, stone weights and bone awls for the use of fishermen. I regret I have not yet had occasion to see this important paper.

upper left edge, and a still deeper round cavity on the opposite face
near the back.   These various features combine to show that this
hammer was not hafted, but freely worked with in the hand, and the
edge shows conspicuous traces of ancient use.   The edge is curved and
more rounded on one side than on the other.

The small axe shown in Fig. 3 of the same Plate (7.2 cm long, 3.9
cm wide over the back, 4.7 cm over the middle, and 4.4 cm over the
edge; greatest thickness 3.4 cm), of a finely polished light-green jade, is
of the same type, generally.   It is entirely polished except the butt.
There is an oblong piece with rough surface cut out of one of the lateral
sides for a finger-support, and there is a shallow round depression on the
lower face for the same purpose.   The cutting edge, very fine and sharp,
is almost straight, forming right angles with the lateral sides.

Fig. 2 on Plate II is a flat chisel, thick in the centre and gradually
sloping towards the edges.   It is perforated near the back.   All sides
are convex in shape, and the cutting edge runs in a big graceful curve.
The blade is not set off as in other pieces.   It measures in length
11.5 cm, 6–7.5 cm in width and 1.5 cm in thickness in the central portion.
The jade exhibits a leaf-green color containing various shades of green
intermingled with black streaks.

The large heavy hammer (weight 3½ *lbs.*) on Plate III is the
most remarkable specimen among these jade implements from the
graves of Shensi.   It is carved from a fine plant-green jade covered on
the lower face and the one lateral side visible in the plate with iron-
rust colored spots (black in the illustration).   In shape, it is unlike any
of the others, and though the blade is formed like that of a hammer
(compare Plate II, Fig. 1), it ends abruptly in a broad and blunt edge
(2.2 cm high) exhibiting a rough surface evidently much used for pounding.   The general shape of the implement is rectangular, the lower
face and the butt are almost plain, the lateral sides are straight.   The
upper surface is slanting in two planes towards the butt.   Two large
perforations are bored by means of a tubular drill[1] through the central
part of the body, side by side, separated only by a narrow strip 1 mm
in width and translucent when struck by the light.   The two holes
have been bored from the top where they form fairly regular circles
(the one 3.5 cm, the other 3.8 cm in diameter), while they are more
irregular on the lower face.   It is hard to see the purpose of these two
hollow cylinders, if it was not the object to diminish the weight of this
heavy piece.   The two tubes cut out were, of course, very useful to
yield the material for other carvings.   The perforation near the butt
was convenient when using the implement as a pounder; the palm

[1]Described in BISHOP, Vol. I, p. 203 and BUSHELL, Chinese Art, Vol. I, p. 144.

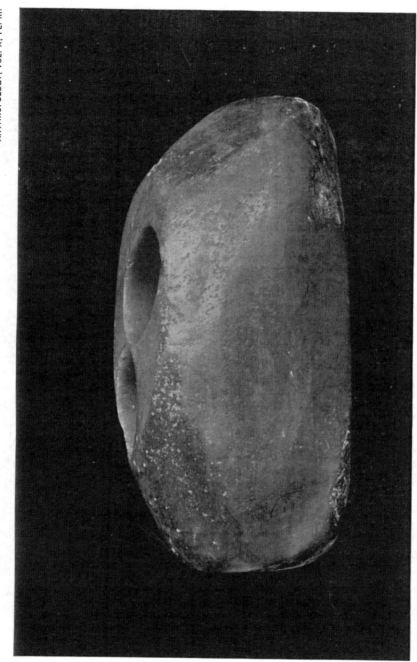

JADE HAMMER OF CHOU PERIOD, FROM SHENSI PROVINCE.

JADE CHISEL OF CHOU PERIOD, FROM SHENSI PROVINCE.

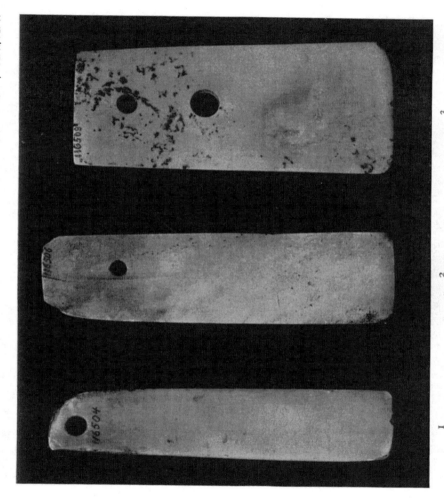

JADE CHISELS OF CHOU PERIOD, FROM SHENSI PROVINCE.

clutched around the butt, while thumb and middle finger caught into the hole from above and below.

The chisel on Plate IV is of extraordinary size and weight (2½ *lbs.*), 22 cm long, 11.7 cm wide above and 14.5 cm wide below, and 1 cm thick; it is somewhat thicker in the centre and gradually sloping from there towards the edges. The upper edge is somewhat slanting, the sides are slightly convex, and the blade makes an elegantly curved sweep. It is blunt, about 2 mm wide. The implement was accordingly ceremonial, as shown also by its size and weight. The large eye (2.5 cm in diameter) is just in the centre between the two lateral sides and has been drilled equally from the upper and lower face forming a projecting rim in the middle of the interior. The color of the stone is of an indistinct gray interspersed all over with deep-yellow spots (which show black in the plate).

On Plate V three chisels are shown. That in Fig. 1 is of oblong rectangular shape (15 cm long, 2.2–3.2 cm wide, and 1 cm thick) of milk-white opaque jade with yellow streaks and spots, perforated near the handle where a small piece is chipped off. The edge is but slightly rounded, almost straight and angular; it is blunt now, having apparently been long in use, and cracked, earthy matter having deeply penetrated into it. Fig. 2 represents the same type (15.5 cm long, 3.7–4.2 cm wide, and 9 mm thick in the middle) of a light sea-green jade. Above and below the perforation, a leaf-shaped cavity (7.5 cm long) in the surface will be noticed which offered a convenient resting-place for the second finger when the thumb and third finger clasped the narrow sides in using the instrument for cutting or scraping. The edge is still sharp, very little curved; the blade is gradually sloping towards the edge over 0.5 cm. The perforation has been effected from one face only, *i.e.* from the face opposite the one shown in the plate where the diameter is 1 cm while on the face shown it is only 0.5 cm.

The piece in Fig. 3 (14 cm long, 5.2–6.2 cm wide, 2–5 mm thick) is one of exceptional beauty because of the quality and color of its jade which has the appearance of ivory.[1] It forms a flat rectangle tapering, as also the others, towards the handle. The two eyes are drilled from the side shown in the plate (0.8 and 1.2 cm in diameter). The edge is slightly curved and bevelled off at the corners. This piece is so elegantly polished and of such elaborate material and workmanship that it cannot have been employed for ordinary use. The two perforations point also to its ceremonial character.

The chisel in Fig. 1, Plate VI, is made of a deep-black jade spotted in the lower portion with grayish-blue clouds as visible in the repro-

[1] Called by the Chinese "chickenbone-white" (*chi ku pai*).

duction.   It is 14.5 cm long, 3 cm wide over the back and 6.2 cm over the blade, 1.1 cm thick.   The eye is drilled from one side only, the one not shown in the plate where the diameter is 1.3 cm, while on the opposite face it is only 7 mm, so that the interior wall which is well polished assumes the shape of an obtuse cone.   It will be noticed that there is on the left lateral edge running along the lower face à projecting ridge which affords a support to the finger in handling the implement.   On the opposite lateral side, the lower edge is flattened out into a strip about 3 mm wide.   Both lateral sides are not vertical, but slightly slanting in such a way that a cut made latitudinally would show a trapezoid.   The blade with pointed curved edge is short, and as usual, bilateral, while the specimen in Fig. 2 has a unilateral blade, but otherwise resembling Fig. 1 in shape and having the same slanting lateral sides.   There is no perforation.   There are two curious incisions in the left lateral side, one in the upper left corner running from the middle of the back across to the lateral side, the other below it, a small segment of the size and shape of a thumb-nail being cut out of the lower face. It is easy to see that these two incisions afforded a hold to the finger-nails, and that this implement was handled in such a manner that the thumb lay flat on the right lateral side, the nail of the second finger fitting into the upper left incision and the nail of the middle finger into the lower incision on the left lateral side.   No doubt this specimen has been in actual use.   It measures 7 cm in length, 1.7 cm in width over the back, and 2.6 cm over the blade, and is 1.1 cm thick.   It is made of a jade of very peculiar coloration, a kind of soap-green, white light and dark blue-mottled, darker shades of green in more continuous masses and larger white spots being displayed over the lower face.

The knife-shaped object in Fig. 4, Plate VI, is carved out of a beautiful transparent leaf-green jade (only 3 mm thick) interspersed with masses of deep-black specks designated as "moss" (t'ai) by the Chinese. If held against the light, it presents a beautiful effect.   The very delicate character of this implement is sufficient proof for its having never been put to any profane use.   The blade (on the left side) is merely indicated and only 2 mm wide, while the cutting edge is blunt (½ mm thick) and of the same thickness as the two straight lateral edges.   In length it measures 12 cm, in width 3.6–3.9 cm.   The eye (diameter 5 mm) is bored from the two faces, and exactly in the middle of the interior wall a very regular ring is left, evidently with intention.   This implement, no doubt, was an emblem of power and rank and belongs to the category of objects discussed in the next chapter.

The jade chisel in Plate VII, Fig. 1, is of a similar type as that in Fig. 2, Plate VI, only somewhat longer and thicker (8.5 cm long, 2 cm

1   2
4   3

JADE CHISELS AND OTHER IMPLEMENTS OF CHOU PERIOD, FROM SHENSI PROVINCE.

Explanation of Pl. VII.

Fig. 1. Jade Chisel, *a* Front, *b* Side-View.
Fig. 2. Bronze Chisel for Comparison with Stone Type.

1, *a* and *b*   2
JADE AND BRONZE CHISELS OF CHOU PERIOD, FROM SHENSI PROVINCE.

1
2

JADE CHISEL AND KNIFE OF CHOU PERIOD, FROM SHENSI PROVINCE.

wide, 1.4 cm thick). It is shown in front (*a*) and side-view (*b*). It is grouped here with a bronze chisel (Fig. 2), trapezoidal in section (10.2 cm long), with long socket, to show the close agreement in form between the stone and metal chisels.

The stone (presumably jadeite) represented in Fig. 3 of Plate VI has been found in a grave of the Han period in the village *Wan-ts'un* west of *Si-ngan*. Nothing is known about the manner of its use in the grave. The Chinese call it "blood-stone" from the peculiar blood-red color covering the greater part of the upper surface, which besides shows layers of a deep black and along the left side a portion of a jade-white tinge. The lower face and the edge are black over which an indistinct stratum of red is strewn. The natural form of the stone is evidently preserved in it, and besides the high polishing, the effect of human work is visible in the deeply cut incision in the upper right corner where one small triangular piece has been sawn out. The traces of the saw are distinctly visible; the sawing was done along the slanting portion, and after sawing through, the piece was broken out, as can be recognized from the rough, irregular surface of the horizontal plane, while the slanting plane is smooth. Beside this, there is a shallow depression made rough which might have served for the insertion of the thumb, indicating that this stone was used as an implement for battering or pounding, the triangular point being held below; but the nature of the incision remains unexplained. I believe that the workman had some plan in mind of sawing and grinding this piece into shape; the beginning of his activity is here shown, and for some unknown reason, he was stopped or prevented from continuing his work. The implement is 7 cm long, 5.5. cm wide, and 1.6–1.9 cm thick. The blood-red color is explained by the Chinese as having originated from the blood of the corpse penetrating into the stone,[1] which is certainly fanciful.

The large jade knife in Fig. 1, Plate VIII, is a unique specimen of extraordinary dimensions, unfortunately broken in two pieces when found, two fragments being lost, without detracting from the possibility of realizing the original form. It is of rectangular trapezoidal shape, measuring in length over the central perforation 35 cm, over the back 34 cm, over the cutting edge 36.9 cm; the upper edge is 13.5 cm long, the lower 11 cm; the width varies between 11 and 13.2 cm, being 11.8 cm in the middle, on account of the concave cutting edge curved inwardly. The blade is 1.7 cm wide in the central portion and gradually diminishes in width towards both sides; it shows the same form and dimensions on both faces and the same angle of inclination. The thickness is only

[1] This view is expressed also by many authors, *e. g.* in the *Wu li siao shih* by FANG I-CHIH (edition of 1884), Ch. 7, p. 15 a.

4 mm along the back and increases from there in the direction towards the blade to 6 mm. The blade is transparent when viewed against the light. Many notches are visible in the cutting edge from which it might follow that actual cutting has been done with it. The back, and the upper and lower edges are carefully beveled. There are five perforations,[1] the three stretching in one vertical line parallel with the back having the same size (1.1 cm in diameter), while the central hole has a diameter of 1.5 cm and the one below of 0.5 cm; the boring has been executed from one face only, as can be seen also in the illustration; the projecting rings there visible are on the same level as the opposite face. The walls of these perforations are well polished as in all cases known to me. The fundamental color of the jade is light-green, full of black veins and spots and of white clouds, as may be recognized in the reproduction. Also this implement doubtless belongs to the emblems of power as described in the next chapter.

Figure 2, Plate VIII, represents also an extraordinary specimen 40.4 cm long, 5.2–6.2 cm wide, and 1.8 cm thick. It is a chisel cut out of a grayish silvery jade in which specks like silver clouds are strewn all over. The perforation has been drilled from both faces, the two borings not meeting exactly, and a projecting ring being left in the interior. The cutting edge is broken off, and apparently in times long ago. That no more than the edge is broken, can be seen from the lateral sides just tapering into a narrow strip above the breakage. The lateral edge, partially showing in the illustration, is hollowed out in a flat, long segment which is in a plane a bit lower than the remaining portion of this edge; this was perhaps done to afford a firmer grip to the second finger when handling the instrument.

A jade dagger, unique for its material, size and shape, is in the collection of H. E. Tuan Fang, Peking, and here reproduced in Plate IX from a photograph kindly presented by him to the author. It was dug up in 1903 not far from the old city of Fêng-siang fu in Shensi Province from a considerable depth and is, in all probability, older than the Chou period. Its substance is a peculiar light-reddish jade, such as I have seen in no other specimen, designated by the Chinese *hung pao yü* (GILES No. 5269). It is a two-edged dagger (92 cm long and 12 cm wide), both edges being equally sharp, running into a point bent over to one side, not central as in the later bronze daggers. Another peculiar feature is the flattening out of the two surfaces of the blade into four distinct zones running longitudinally. At the end of the blade a rectangular band, filled with cross-hatchings and surrounded on either side by four parallel incisions, is engraved. A rectangular, perforated hilt (16 cm long) is sharply set off from the blade, near which

[1] Compare a similar arrangement of four perforations in Fig. 40.

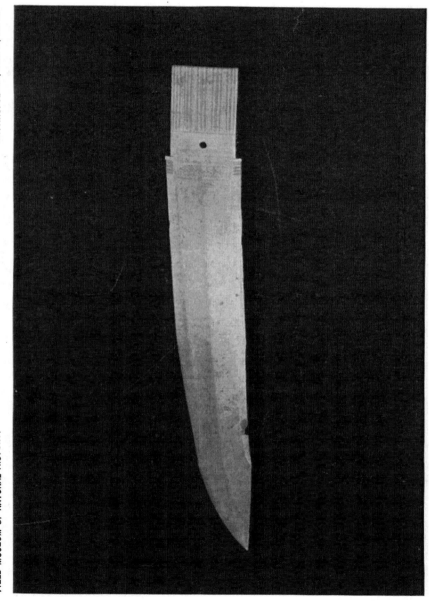

JADE DAGGER, PROBABLY SHANG PERIOD, FROM SHENSI PROVINCE, IN THE POSSESSION OF H. E. TUAN FANG, PEKING.

runs across a band filled with parallel oblique lines. Five bands, each consisting of four deeply grooved lines, are laid out in the opposite direction on the other side of the eye. The same ornaments are executed on both faces. It is evident that this elaborate and costly production was never destined for any practical purpose, but that it served either in some religious ceremony, or as an emblem of power, perhaps of sovereignty (compare Ch. II).

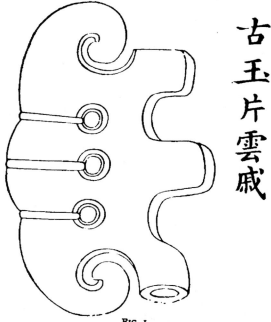

FIG. I.
Ancient Jade Axe (from *Ku yü t'u p'u*).

Ceremonial jade axes were still turned out at the time of the earlier Han dynasty, as we may safely infer from a report in the *Ku yü t'u p'u* (Ch. 28, pp. 6, 8), saying that in the period Shun-hua of the Sung dynasty (990–995 A. D.) a certain man opened the tomb of Huai-nan Wang [1] of the Han dynasty from which he obtained precious jades, and among these two jade axe-heads: "hence it is known," the author adds, "that they are objects from the beginning of the Han period." The two pieces are alike in shape and design, formed in what the Chinese call a "rolled or coiled-up cloud" (*k'üan yün*), *i. e.* the edge terminates on both ends in a convolute spiral; the blade has three round perforations arranged in a vertical row as in the corresponding bronze types after which these pieces were evidently modeled. There is a socket at the lower end for the reception of a handle (Fig. I).

Jade axes (*yü tsi*) [2] in connection with red-colored shields were used in the hands of dancers performing the dance *Ta wu* in the ancestral

---

[1] Title of the Taoist adept Liu Ngan, second century B. C.

[2] They are not "jade-adorned axes," as LEGGE (Li Ki, Vol. II, p. 33) translates, or "hache ornée de jade" (COUVREUR, Vol. I, p. 731) which, first of all, is not justified by the two simple Chinese words meaning only "jade axe," and secondly, what is and means an axe adorned with jade? No such thing exists or has ever existed, but there are jade axes made exclusively of jade (except the wooden handle) which it is doubtless easier to make than to adorn, *e. g.* a bronze axe with jade.

temple of Chou-kung in the kingdom of Lu (*Li ki, Ming t'ang wei,* 10),
but also by the Emperor in the temple of Heaven (*kiao miao*).   It is

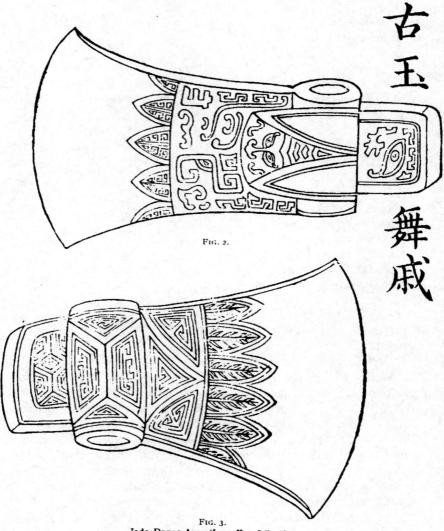

古玉
舞戚

FIG. 2.

FIG. 3.
Jade Dance-Axes (from *Ku yü t'u p'u*).

to this passage that the *Ku yü t'u p'u* (Ch. 27, p. 10) refers in describing
a ceremonial jade axe on which the head of a phenix is engraved.   "It
is an implement of the early Ts'in period and cannot come down from
the Han or Wei dynasties," is added in the descriptive text.

Figures 2 and 3 represent two other dance-axes illustrated in the *Ku yü t'u p'u,* both ascribed to the Han period. The one is made of jade pale-yellow and bright-white in color, without flaw and engraved with the monster *t'ao-t'ieh* (in the text called *huang mu* "the yellow-eyed") and a cicada pattern (*chan wên*) by which the leaf-shaped prongs are understood, so frequently displayed on the bronze beakers called *ts'un.* The other axe-head (Fig. 3) is of a bright-white jade with greenish speckles comparable to moss and decorated with "cloud and

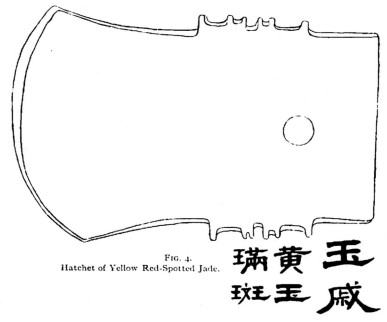

FIG. 4.
Hatchet of Yellow Red-Spotted Jade.

瑞黄玉
斑玉戚

thunder" (*yün lei*) patterns (meanders) and cicada designs with leaf veins.

It will be readily seen that these ceremonial jade hatchets of the Han are widely different in their artistic shapes and decorations from those of the Chou period, which are plain and unpretentious.

Another hatchet of jade is figured by WU TA-CH'ÊNG in his *Ku yü t'u k'ao* (reproduced in Fig. 4) and explained by him as an ancient dance-axe on the ground of the passages referred to. It will be recognized that this specimen is much simpler than any of the Han dynasty, and I am inclined to place it in the Chou period. Its rectangular shape, the form of its cutting edge, the perforation in the butt are all features occurring in the Chou celts, while the peculiar indentations in the lateral sides betray the ritualistic character.

In the *Kin-shih so*, two ancient bronze hatchets are well figured (Figs. 5 and 6), the one obtained from *Lo-yang* (Honan Province),

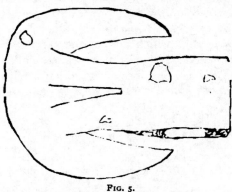

FIG. 5.
Ancient Bronze Hatchet (from *Kin-shih so*).

the other from *Lu-shih hien* (Honan Province). The latter (Fig. 6) is interesting with reference to the jade dance-axes in exhibiting a more primitive form of the triangular pattern, and it is very interesting to take note of the interpretation of the brothers *Fêng* that this ornament is a *yang wên* "a pattern of the male principle."[1] The piece in Fig. 5 is remarkable for its circular blade and the two lateral crescent-shaped barbs; the rectangular butt was stuck into the cléft of the wooden handle.

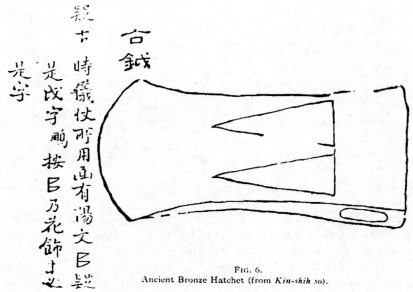

FIG. 6.
Ancient Bronze Hatchet (from *Kin-shih so*).

We are naturally led to the inquiry, what was the symbolical significance of jade chisels, knives and other implements buried in the grave with the dead? We noticed that the late Dr. BUSHELL, chiefly relying

[1] They expressly deny that it has the function of a written character. The Chinese wording certainly means in our language a phallic emblem.

on Anderson's statement, explained the miniature jade celts as protective amulets. This, however, is a very recent development. I am not aware of the fact that any such minute celts have ever been discovered in a grave; they result from surface finds, and many of them may be just a few centuries old.[1] The wearing of jade celts as personal adornments is not older than the Han period, as shown by two artistic specimens in our collection (Plate XXVIII, Figs. 3 and 4) found in Han graves. The first literary allusion to such charm celts occurs in the *Po wu chi* of the third century A. D. (see p. 64).

As early as the Shang dynasty (B. C. 1766–1122), the axe seems to have been the victorious emblem of the sovereign. In the sacrificial ode *Ch'ang fa* (LEGGE, *Shi king*, Vol. II, p. 642) in praise of the house of Shang, the founder of the dynasty T'ang is described as "the martial king displaying his banner, and with reverence grasping his axe, like a blazing fire which no one can repress."[2] The axe was accordingly a sovereign and martial emblem, and the emperors of the Chou dynasty had a pattern of axes embroidered on their robes (called *fu*, GILES No. 3630). This ornament was the eleventh among the twelve *chang* embroidered on the imperial state-robe (LEGGE, Chinese Classics, Vol. III, pp. 80–81).

Embroideries with representations of the axes *fu* were used on the altar of the God *T'ai-i* "the Supreme Unity," "the most venerable among the gods," as told in a hymn addressed to him in the "Annals of the Former Han Dynasty" (CHAVANNES, Se-ma Ts'ien, Vol. III, p. 618).

There was a constellation called "the Axes" which, being bright, foreshadowed the employment of axes, and when in motion, a levy of troops. The axes symbolize the events in the army and refer to the

---

[1] *E. g.*, in BISHOP (Vol. II, p. 208), a ceremonial axe (No. 637) for display on the altar of a Taoist temple is figured and described; it is attributed to the K'ien-lung period (1736–1795). Its back is straight, the sides concave, and the edge rounded and convex in outline. The figure of a lion stands on the top of the back, and two winged monsters covered with spiral designs are attached to the sides in *à jour* carving.

[2] A ditty in the *Shi king* (LEGGE, Vol. I, p. 240) reads thus: "In hewing the wood for an axe-handle, how do you proceed? Without another axe it cannot be done. In taking a wife, how do you proceed? Without a go-between it cannot be done." BIOT (in LEGGE's Prolegomena, p. 165) refers to the *Pi-pa ki*, a drama of the ninth century, in which the go-between presents herself with an axe as the emblem of her mission, and cites upon the subject this passage of the Book of Songs. The commentary does not say, remarks Biot, whether this custom of carrying an axe as an emblem be ancient; the go-between makes even a parade of her learning in explaining to the father of the young lady, whom she has come to ask for, why she carries an axe. In my opinion, this is merely a literary jest of the playwright. It does not follow either from the above passage that the negotiator of a marriage actually carried an axe as emblem; the making of an axe-handle by means of an axe is simply used jocosely by way of a metaphor, which occurs also in another song (*Ibid.*, p. 157).

execution made in times of war (SCHLEGEL, Uranographie chinoise, p. 298).

We shall see in the next chapter that the jade emblems of sovereign power were made in the shape of hammers, knives and other implements, and that these were connected with an ancient form of solar worship; this investigation will shed new light on the reasons for the burial of jade implements in the grave. Being emblems, and originally, in all probability, images of the solar deity, they shared in the quality of sun-light to dispel darkness and demons, and were efficient weapons in warding off from the dead all evil and demoniacal influences. [1]

Owing to the kindness of Mr. S. COULING, a medical missionary in the English Baptist Mission in Ts'ing-chou fu, Shantung Province, I am enabled to lay here before the reader twelve stone implements discovered by this gentleman in the vicinity of his station. They had been loaned by him to the Royal Scottish Museum of Edinburgh, and Mr. Walter Clark, Curator of the Museum, by request of Mr. Couling, has shown me the courtesy of forwarding these specimens to me for investigation. I avail myself of this opportunity to herewith express my thanks also publicly to both Mr. Couling and Mr. Clark for their generous liberality, to which a considerable advance in our scanty knowledge of stone implements from China is due. Mr. Couling, who deserves the honor of being credited with the discovery, wrote to me on September 22, 1905, from Ts'ing-chou fu in regard to these finds: "These specimens, with the exception of one (Plate XII, Fig. 5) the origin of which is unknown, have been found in this immediate neighborhood, say within a radius of ten miles from the city during the last few years. Most of them have been obtained through my schoolboys. On knowing of what I wanted some remembered to have seen such things, some knew neighbors who had them; others went out searching and found a few. The finds are made in ploughed fields or in river beds [2] or in the loess cliffs not far down. The Chinese pay no heed to them, only sometimes troubling to keep one as being a somewhat curious stone. I should say there must be plenty more, though it is nearly a year since I obtained the last, but they are not easy to collect, as the people do not recognize their value."

Jade does not occur in any of these specimens, for the apparent reason that this mineral is not found *in situ* in Shantung; they are all made of easily procured common local stones of the character of talco-hem-

---

[1] The reader may be referred to Chapter VIII of DE GROOT, The Religious System of China, Vol. VI (Leiden, 1910), where a full and able discussion of this subject is given.

[2] This statement is in full accord with that given by Chinese authors (see below).

STONE CELTS FROM SHANTUNG PROVINCE.

atite schist, with the exception of the grooved hammer, which is diorite, but all of them are highly polished.

In glancing over the eleven objects represented on the three Plates X–XII, it will be noticed that all of them lack that one characteristic feature of the Shensi implements, the perforation. Mr. Couling has, however, succeeded in finding at a later date a perforated chisel, reproduced after a sketch of his in Fig. 7, of a grayish white hard marble-like stone with slightly convex lateral edges and with a perforation not far above the centre of the surface. The borings have been effected from each face, meeting inexactly at the middle, as shown by the dotted lines in the sketch. This piece perfectly agrees in shape with the corresponding types of Shensi and has probably been used as a mattock.

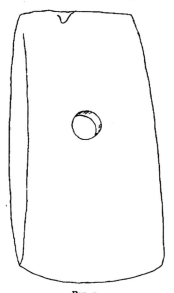

FIG. 7.
Perforated Stone Chisel from Shantung (after Sketch furnished by Mr. Couling).

Perforations in stone implements have had a significance and a purpose; if they are large enough to allow the insertion of a wooden or bone haft, we shall not fail to conclude that such has actually been the case. If the perforations are so small in diameter that such a contrivance seems out of the question, we are led to the belief that they served only for the passage of a thong or cord from which the implement was suspended and perhaps fastened to the girdle, or that some kind of ceremonial usage was involved; we may infer in this latter case that the perforation has the function of a conventional survival in remembrance of its former more intense utilization. As GIGLIOLI correctly supposed, the broad rectangular type of stone chisel with large eye near the back seems to have served as a mattock in husbandry. I am not inclined to think that pieces of precious jade in beautiful colors have ever been turned to such a purpose; but if we realize that such types as represented on Plate II, Fig. 2, and Plate IV were simply made of ordinary stone with all necessary adaptations, we can recognize in them the agricultural implement in question. And in the specimen from Shantung (Fig. 7) a real mattock of common stone has come down to us.

In certain parts of northern China, such mattocks are still actually in use. W. W. ROCKHILL (Diary of a Journey through Mongolia and Tibet, p. 46) made the following observation on the bank of the Yellow

River in the north-eastern part of Kansu Province: "There is now going on a curious process of agriculture which shows how little the Chinese understand saving of labor. The farmers dig up a large patch of the surface of each field, cart it back to their farm-yards and there let the clods of earth dry, when they take a mallet (or a stone hammer with an eye drilled through it in which to fix a long handle), and reduce it all to powder; with this is then mixed what manure they have been able to collect on the road, and this top dressing is laboriously carted back and spread over the field from which nine-tenths of its component parts were a few days before quite as laboriously taken away."

Even the modern iron mattock has clearly preserved in shape and perforation its relationship to its stone predecessor. It is of the same rectangular form with straight edge, and a wooden handle standing vertically against its surface is stuck through the hole.

This subject is of great significance for the history of agriculture. Everywhere in Eastern Asia we can observe two principal and distinct methods in the cultivation of cereals, which are often employed side by side in the same geographical area, but then as a rule by representatives of different tribes differentiated as to the degree of their culture. The one method bears a close resemblance to our process of gardening, except that broadcast sowing obtains, and the hoe or mattock is almost the only tool utilized in it (hoe-culture). The other method identical with true agriculture is based on the principle of the plough drawn by cattle, on the laying-out of fields in terraces and the appliance of artificial irrigation. There is a sharp line of demarcation between hoe-culture and plough-culture, each being a well-defined sphere in itself, the latter not having developed from the former. The aspect of the development of the two stages is of a purely historical character, as far as Eastern Asia is concerned. There, the Chinese are the representatives of plough-culture, and so are the great groups of Shan and Burmese tribes, in short the entire stock comprised under the name Indochinese because of their affinity in language; the aboriginal tribes gradually pushed back by the Chinese in their onward march towards the south and designated by them with the generic name *Man*, as well as the Mon-Khmer or South-east-Asiatic group (SCHMIDT'S Austronesians), were originally only representatives of hoe-culture. In many localities, they received the plough from their more powerful conquerors and adopted with it their methods of tilling; in others, they have still preserved their original state, as may be seen from numerous reports,[1]

[1] This subject deserves a special monograph. Many intricate problems, as the domestication of cattle, the history of the wheeled cart which appears only in the stage of plough-culture and is absent in hoe-culture, the history of rice-cultivation and terraced fields, are here involved, which could be discussed only at great length. I can make here only these brief allusions, in order to define the historical position of the stone mattock.

1   2
3   4

STONE HAMMERS FROM SHANTUNG PROVINCE.

STONE PESTLES FROM SHANTUNG PROVINCE.

1
2

5

3
4

from which I may be allowed to select one. "The most common form of cultivation (among the Kachin of Upper Burma) is the wasteful process of hill-clearing. The method employed is to select an untouched hill slope, fell the jungle about March and let it lie on the ground till it is thoroughly dry. This is set fire to in June or July, and the surface of the earth is broken up with a rude hoe, so as to mix in the wood ashes. The sowing is of the roughest description. The worker dibbles away with the hoe in his right hand and throws in a grain or two with his left. The crop is left to take care of itself till it is about a foot high, when it is weeded, and again weeded before the crop gets ripe. The crop is usually reaped about October. The same field cannot be reaped two years running. Usually it has to lie fallow from seven to ten years where the jungle does not grow rapidly, and from four to seven years where the growth is quicker." [1]

At the present day, the tribal differences which once prevailed between hoe and plough culture have disappeared to a large extent, though not so much as to escape the eye of a keen observer, and the difference now chiefly rests on economic grounds, as can be seen from the example of Siam where rice is grown in hoe-culture on the rude hills, and by the methods of agriculture on the fertile plains by the same population; the poor hill-people being simply forced to their mode of life by sheer economic necessity. Thus, also, numerous low-class Chinese who took refuge among the wild tribes of the mountains descended to their style of husbandry as an easier manipulation, and the poor colonists on the banks of the Yellow River met with by Mr. Rockhill must have been in a similar condition of wretchedness, for what they did was nothing but a relapse into hoe-culture. From their use of the stone mattock as observed by Mr. Rockhill,— and we shall see later on in another connection that stone mattocks have been in use for this purpose throughout the south-east of Asia,— we are justified in concluding that the stone mattock is *the* mattock employed in hoe-culture, and further that also the ancient stone mattocks found in Shensi and Shantung must have been associated with hoe-culture. It follows from an historical consideration of this subject that these stone mattocks are to be attributed to a non-Chinese population which lived there before the invasion of the Chinese and was gradually absorbed by them, rather than to the Chinese themselves.

All the specimens from Shantung have been apparently turned to practical purposes; nearly all of them show traces of having been used. The two rectangular chisels illustrated in Plate X fairly agree in shape

[1] Gazetteer of Upper Burma and the Shan States, Vol. I, Part I, p. 424 (Rangoon, 1900). This is one of their methods, but in other regions wherever they learned from their Shan and Chinese neighbors, wet paddy cultivation, *i.e.* agriculture, has been introduced among them.

with those of Shensi and show the same straight cutting edge, with slightly rounded, beveled corners in Fig. 1, where we find also the convex lateral edges.

Figure 2 on Plate XI exhibits a similar type, only that the blade is much broader here.   The hammer in Fig. 1 of Plate XI is particularly interesting as revealing the stone prototype of the carpenter's iron hammer common all over China.   The blade starts from about the middle of the stone and gradually slopes on both faces towards the cutting edge.   In Fig. 3 of the same Plate the back is oval-shaped, and the blade does not occupy the entire front part, but is cut out in the way of an arch; the cutting edge is round.   Fig. 4, Plate XI, evidently belongs to the same type, but the blade is much weathered out.

The five stone objects united on Plate XII seem to be pestles for pounding grain, with exception of Fig. 5.   This is a rough fragment of black pebble flattened on one side and marked with a small circle on the top; it was not found like the other stones near Ts'ing-chou fu, but it came into the possession of Mr. Couling by purchase, its origin not being known; I do not venture any theory in regard to its possible use.

The greatest surprise among the stone implements of Shantung is afforded by the find of a grooved axe or hammer of diorite (Plate XIII, Fig. 1), 9 cm long and 6 cm wide, with a deeply furrowed groove running all around, about 2 cm wide.   It is the first and the only known type of this kind from China, and of particular interest to us, because it is a type very widely spread in North America.[1]   The Chinese specimen is better worked than any from America known to me and exhibits a remarkable regularity and proportion of form, that same sense for dimension which elicits our admiration in their most ancient productions of pottery, metal or stone.   There is a ridge-like projection over one side of the groove.

As this type stands alone in the Chinese field, according to our present state of knowledge, and is generally of greatest rarity in Asia, it will be appropriate to determine its position by calling attention to finds of a related character in other regions.   Only one of this type, as far as I know, has become known from India.   It was found at Alwara, two miles north of the Jumna, and thirty-seven miles southwest of Allahabad by Mr. J. Cockburn, placed together with a number of other stones under a sacred tree.   It was figured and described by

---

[1] The Chinese specimen comes nearest to that figured by TH. WILSON in *Report of National Museum*, 1888, p. 647, No. 72. — Different from this type are the grooved globular clubs of Scandinavia as described by SOPHUS MÜLLER (Nordische Altertumskunde, Vol. I, p. 144), which, however, seem to be plummets or sinkers (compare WILSON, *l. c.*, p. 653, Nos. 107, 108).

EXPLANATION OF PL. XIII.

Fig. 1. Grooved Diorite Axe from Shantung Province, in British Museum, London.

Fig. 2. Grooved Quartz Axe from India, in British Museum, London. After Rivett-Carnac.

Fig. 3. Grooved Stone Hammer from Saghalin. After Iijima.

Fig. 4. Grooved Stone Hammer of the Chukchi. After Bogoras.

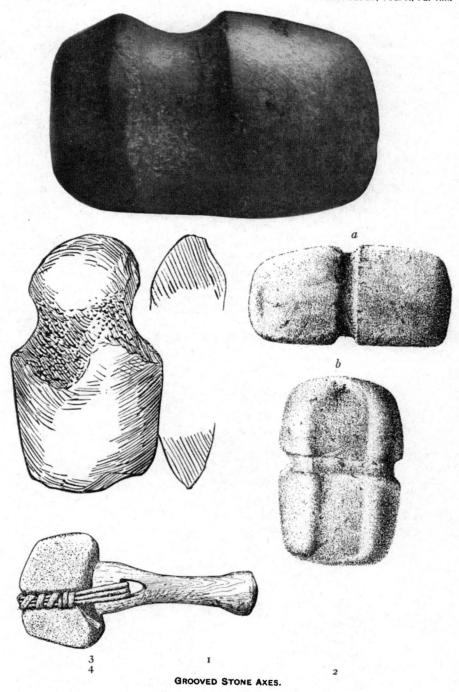

a

b

3
4

I

2

GROOVED STONE AXES.

J. H. RIVETT-CARNAC,[1] after whom it is here reproduced on Plate XIII, Fig. 2, *a* and *b* (upper and lower faces).   The original is now preserved in the British Museum.   It is 13 cm long and 6.5 cm wide.   RIVETT-CARNAC describes it as made of a tough, grayish quartzite, somewhat resembling a modern hammer in form, being flat at the ends and slightly curved on the upper surface.   A groove has been carefully carried round the centre.   The base has been hollowed out with equal care in a gouge-like form.   The whole arrangement suggests that the hammer was attached by a ligature to a wooden or withy handle, the ligature being kept in its place by the upper groove, while the lower groove held the hammer in position on the rounded haft.   Mr. Cockburn has pointed out certain minute marks, especially on the lower groove, which suggest the possibility of metal implements having been used in the fashioning of the hammer, and it may be that this implement belongs to the transition stage from stone to metal, when metal, though available, was scarce.   This specimen is believed, concludes the author, to be the first of this description found in India; he adds that his collection contains several other grooved hammers of a less perfect form, bearing no trace of metallic tooling, which appear to be water-worn pebbles grooved to admit of being attached to a withy handle.

Figure 3 on Plate XIII[2] shows a grooved stone hammer found in a shell-mound north-west of Korsakovsk on the southern shore of Saghalin Island by Dr. Iijima.

N. G. MUNRO (Prehistoric Japan, p. 140, No. 1, Yokohama, 1908) has figured a similar type from Japan, without defining the locality of the find, but he diagnoses it as a sinker,[3] with the remark: "This stone is sometimes described as a hammer, but those that I have seen are made of rather friable lava and would not stand much concussion. I have more than once seen these objects placed on tomb-stones, in fishing localities."   I do not believe that this supposition is correct, but think that Mr. EDWARD S. MORSE (Shell Mounds of Omori, p. 15, and Plate XVII, 1-2, Tokyo, 1879) who has discovered two grooved hammers in these shell-mounds of Ainu origin is perfectly right in identifying them as hammers, and in saying "it is hardly probable that they were intended for net sinkers."

Farther north-east, we find the grooved stone hammer in modern times as a common household utensil among the Chukchi (Plate

[1] On Stone Implements from the North Western Provinces of India, p. 6 (Calcutta, 1883).

[2] Derived from the *Journal of the Anthrop. Soc. of Tokyo*, Vol. XXI, No. 247, 1906.

[3] Doubtless prompted by J. EVANS, Ancient Stone Implements of Great Britain, p. 236 (Second ed., London, 1897).

XIII, Fig. 4). It is a maul oblong in shape with a groove all around in about the lower third of the stone, to which a short wooden or bone perforated haft is tied by means of strong leashings. It is especially used for splitting marrow-bones to extract the marrow, also for breaking all kinds of bones from which tallow is to be extracted, and for crushing frozen meat, fish and blubber. The crushing is done over a large flat round stone.[1]

GERARD FOWKE[2] has devoted a chapter to the description of grooved axes in North America which, according to him, seem to be of general distribution throughout the United States, being, as far as can be learned from various writers, much more numerous east of the Mississippi River than west of it. But he is inclined to think that no deductions can be made concerning their relative abundance or scarcity, as collectors have more diligently searched in the east than in the west.

The grooved stone hammer does not survive in any other object in modern China; its unique occurrence in one specimen in the times of antiquity seems to show that it takes a rather exceptional position. The finds of this type in shell-mounds of Saghalin and Japan which must be connected with the culture of the Ainu inhabiting this region, in addition to the live tradition of the Chukchi, prove that it belongs to the Palæ-asiatic, or as I prefer to say, North-Pacific culture-area. And the Chuckchi on Bering Strait present the natural stepping-stone linking it with the American continent. Coincidences in material objects as well as in ideas underlying myths and traditions finally rest on historical causes. Nobody competent to judge will deny at present that there have been mutual historical influences between Asia and America revealed by numerous indications, steadily growing as our knowledge advances. I am not an advocate of the theory that American cultures in their whole range are derived from Asia; there was a continuous undisturbed indigenous development going on for ages on this continent with a keynote of striking originality which cannot be explained by Asiatic ideas. On the other hand, human ideas have never been stationary, but mobile and constantly on the path of migration. Ideas have poured in from Asia into America, and from America into Asia, in a process of mutual fertilization. The grooved stone axe may well be claimed as an autochthonous product of North America; there it occurs in greatest abundance and in a number of variations. It occurs

---

[1] A. E. v. NORDENSKIÖLD, Die Umsegelung Asiens und Europas, Vol. II, p. 111 (Leipzig, 1882) and W. BOGORAS, The Chukchee I, Material Culture, p. 187 (Jesup North Pacific Expedition, Vol. VII, 1904), from whom our illustration is borrowed.

[2] Stone Art in XIIIth *Annual Report of the Bureau of Ethnology*, pp. 62–72. Compare also MOOREHEAD, The Stone Age of North America, Vol. I, pp. 222 *et seq.*, 287 *et seq.* (Boston, 1910).

on this side óf[1] and beyond Bering Strait as an every day household object, but otherwise sporadically only in Eastern Asia. Each case must be pursued individually, and attempts at premature generalization be suppressed, as very well outlined by O. T. MASON.[2] In other archæological types, America may have borrowed from Asia as *e. g.* the stone weights from Lower Columbia Valley[3] whose shapes and peculiar handles agree with the Chinese bronze weights of the Ts'in dynasty (B. C. 246–207).

It would be a premature venture to attempt to set a date for the stone implements discovered in Ts'ing-chou fu. The material is too scanty to allow of far-reaching conclusions. It is clearly distinguished in its character as surface-finds from the mortuary specimens of Shensi. Internal evidence might lead one to attribute a comparatively higher age to the Shantung stone implements, but such evidence based on considerations of this kind is often fallacious. There is no doubt that the types represented by them are older, but even this granted, the actual specimens under view may notwithstanding come down from a later period, because we have as yet no clue as to the time when the manufacture of such implements ceased in Shantung. The question as to the identification of these stone implements with a Chinese or a non-Chinese culture, though it cannot be definitely solved at present, yet may be approached on the supposition that they are much more likely to have been produced by a non-Chinese tribe than by the Chinese.

As regards the type of the mattock, culture-historical considerations switched us on the same track. The history of Shantung furnishes proof that the Chinese settlers struck there, as in other territories, an aboriginal population of whose culture we can unfortunately form no clear idea from the ancient meagre records. The region of Ts'ing-chou, from which our stone implements are derived, is said to have been inhabited by a tribe called Shuang-kiu under the Emperor Shao-hao, whose time is dated traditionally at the 26th century B. C. Subsequently, the tribes Ki-shê and P'u-ku take their place,— whether these names simply denote a change of the former name or a new current of immigration, we do not know. The Ki-shê belong to the time of the Emperor Shun and the Hia dynasty (approximately B. C. 23d–19th century). At the end of the Shang dynasty (B. C. 1154), the P'u ku were counted among the feudal states of China;

[1] For Alaska see, *e. g.*, A. P. NIBLACK, The Coast Indians of Southern Alaska (*Report of U. S. Nat. Mus.*, 1888, Plate XXI).

[2] Migration and the Food Quest (*Smithsonian Report for* 1894, pp. 538-539).

[3] HARLAN I. SMITH in *American Anthropologist*, 1906, p. 305.

Ch'êng Wang, the second ruler of the Chou dynasty (B. C. 1115–1078) put an end to their rule, when they rebelled against him. At the time of the Emperor Yü (alleged to have lived about B. C. 2200), two barbarous tribes are mentioned in the eastern part of Shantung, the Yü living around the Shantung promontory, and the Lai who subsisted on cattle-rearing and left their name in the present Lai-chou fu.[1]

There is no record to the effect that any of these tribes availed itself of stone implements, but there is little on record regarding them anyway. The choice is only between the Chinese and the primeval population, and the fact that the latter has existed there before the arrival of the Chinese cannot well be doubted. As the Chinese, when settling and spreading in Shantung, were in the possession of metal and bronze implements and are silent about the use of stone implements on their part, it may be assumed with a tolerable degree of certainty that the stone implements of Ts'ing-chou have emanated from the hands of aboriginal man.

In reviewing the whole material as presented here, we may draw from it the following conclusions:

(1) All stone implements so far found in China are polished, many of them elaborately and elegantly polished. Therefore, they belong to that class which, as far as prehistoric Europe, Egypt, India and America are concerned, has been styled neolithic. No stone of palæolithic and eolithic character has as yet come to light in China.

(2) These implements are found scattered in certain parts of the country and are generally scarce. There are two groups as to the character of the finds noticeable,— finds on or immediately beneath the surface or in river-beds, and grave-finds. The former are more primitive and rougher in technique, the latter of much superior workmanship. Whether a chronological difference exists between the two, it is hard for the present to say; they may have been contemporaneous, after all, the one for the practical use of the living generation, the others for ceremonial and funeral purposes. Local and tribal differentiations have to be equally taken into account in this connection.

(3) Chisels, hammer-shaped axes, and mattocks are the prevailing types thus far discovered.

(4) No deposits of stone implements, so-called work-shops, have as yet been found anywhere in China which would allow the conclusion that man, without the aid of any metal, depended solely on stone utensils at any time, or that a stone industry for the benefit of a large

---

[1] *Shu king*, Tribute of Yü 6, 7. CHAVANNES, Se-ma Ts'ien, Vol. I, p. 113. I do not wish to refer my readers to F. v. RICHTHOFEN's Schantung (pp. 87 *et seq.*) on account of his obvious errors, as known to the initiated (compare HIRTH, Schantung und Kiau-tschou in *Beilage zur Allgemeinen Zeitung*, 1898, Nos. 218 and 219, where an able history of eastern Shantung is given, utilized for the above notes).

local population was carried on to any extent. It is therefore, in the present state of our knowledge, not justifiable to speak of a stone age of China, and still less, as we shall see from a consideration of native records, of a stone age of the Chinese.

(5) The stone implements thus far found need not be credited with any exaggerated age, nor is the term "prehistoric" applicable to them. This term is not absolute, but denotes a certain space of time in a relative sense requiring a particular definition for each culture area, and varying according to the extent in time of historical monuments and records. The burial of jade implements was much practised during the historical period of the Chou dynasty (B. C. 1122–249) and continued down to the epoch of the two Han dynasties (B. C. 206–221 A. D.). While the jade implements in our collection come down from the Chou period, though in regard to some it may be fairly admitted that they are comparatively older, this does not certainly mean that jade or stone implements sprang up at just that time. Their forms and conventional make-up undeniably show that they are traceable to older forms of a more realistic and less artistic character. This primeval age of stone implements, however, can only be reconstructed artificially on the basis of internal evidence furnished by objects of a more recent epoch, or in other words, it remains an hypothesis, an assumption evolved from logical conclusions of our mind, pure and simple. It is not a fact, by any means, but an idea. The substantial, tangible facts have not yet come to the fore.

Turning now to what the Chinese themselves have to say regarding the subject of stone implements, we meet with some allusions to them in the traditions relative to the culture-heroes of the legendary epoch. Shên-nung is credited with having made weapons of stone, and Huang-ti some of jade. This is simply a construction conceived of in later times, without any historical value. In the Tribute of Yü (*Yü kung*) embodied in the *Shu king*, one of the oldest documents of Chinese literature, the composition of which may be roughly dated at about B. C. 800, we read twice of stone arrow-heads offered as tribute to the Emperor Yü (alleged about B. C. 2200). The tribes residing in the territory of the present province of Hupeh brought among other objects, metals of three qualities, mill-stones, whetstones, and stones from which to make arrow-heads; and the tribute of the inhabitants of the province of Liang (in Shensi) consisted in jade for resonant stones, iron, silver, steel, stones from which to make arrow-heads, and ordinary resonant stones.[1] It appears from this account that these two groups of tribes

---

[1] Compare *Shu king* ed. LEGGE, p. 121; ed. COUVREUR, pp. 73, 77. CHAVANNES, Se-ma Ts'ien, Vol. I, pp. 123, 129. G. SCHLEGEL, Uranographie chinoise, p. 758.

must have been acquainted with and in the possession of metals[1] which they offered as tribute; they cannot have lived, therefore, in a true stone period, and the stone arrow-heads must then have been rare and precious objects, otherwise they would not figure in the tribute-list.

It is worthy of note that the name for the flint arrow-head appearing for the first time in those two passages is a single word *nu* (GILES No. 8394); the written symbol expressing it is composed of the classifier *stone* and a phonetic element reading *nu*. The latter element, again, denotes also "a slave, a servant," so that the original meaning "stone of the slaves" *i. e.* stone of the subjugated tribes may have been instrumental in the formation of this character.[2] There is further another word *nu*,— having like the word *nu* "flint arrow-head" the third tone and therefore perfectly identical with it in sound,— with the meaning of "crossbow," the character being composed of the classifier *bow* and the same word *nu* "slave" as phonetic complement. Here, we have accordingly "the bow of the slaves." Now, in the language of the Lolo, an independent aboriginal group of tribes in the mountain-fastnesses of southwestern Sze-ch'uan, the crossbow is called *nu*,[3] and the crossbow is the national weapon not only of the Lolo, but of the whole *Man* family, the remnants of which are now scattered throughout southern China. I am under the impression that the Chinese derived the crossbow with many other items of culture

[1] The metals of three qualities are supposed to be gold, silver and copper. The mention of iron, I believe, is not an anachronism as supposed by HIRTH (The Ancient History of China, p. 237); the ancient Chinese certainly knew iron ore and meteoric iron; what they received and learned from the Turks was not simply iron, but a specific method of working iron.

[2] I am well aware of how deceitful such dissections of characters are, and how cautiously any historical conclusions based on such analysis must be taken. The present forms of Chinese characters represent a recent stage of development teeming with alterations and simplifications in comparison with the older forms; many of these changes are due to subsequent reflections on, or modified interpretations of, the ideas associated with the word which they symbolize. Thus, the present way of writing the word *nu* is possibly only the outcome of an afterthought, but not the original form. The sinological reader may be referred to K'ang-hi's Dictionary where two old forms of this character are given which evidently show no connection with the modern form. I can hardly hope to discuss this question here without the use of Chinese types.

[3] Compare PAUL VIAL, Les Lolos, p. 71 (Shanghai, 1898), who remarks: "*Nou*, arbalète, ce mot si singulier, si anti-chinois, unique comme son, vient du lolo *nou*, arbalète, d'autant que cette arme elle-même n'est pas d'origine chinoise." In his Dictionnaire français-lolo, p. 27 (Hongkong, 1909), Father VIAL gives the word for crossbow in the form *nă*, which is a dialectic variant based on a regular phonetic alternation between the vowels *a* and *u* in the Lolo group of languages. The same vowel-change takes place also in ancient Chinese and in modern Kin-ch'uan (*Jya-rung*) as compared with Central Tibetan, and plays such an important rôle in Indo-chinese languages in general that we can speak of *A*-groups and *U*-groups.

as *e. g.* the reed pipe,[1] several kinds of dances and songs,[2] the well-known bronze drums, from this once powerful and highly organized stock of peoples.   In opposition to the prevalent opinion of the day, it cannot be emphasized strongly enough on every occasion that Chinese civilization, as it appears now, is not a unit and not the exclusive production of the Chinese, but the final result of the cultural efforts of a vast conglomeration of the most varied tribes, an amalgamation of ideas accumulated from manifold quarters and widely differentiated in space and time; briefly stated, this means, China is not a nation, but an empire, a political, but not an ethnical unit.   No graver error can hence be committed than to attribute any culture idea at the outset to the Chinese for no other reason than because it appears within the precincts of their empire.

At all events, whenever Chinese authors speak of flint arrow-heads, these generally refer to foreign non-Chinese tribes.   Especially the Su-shên, a Tungusian tribe akin to the later Niüchi and Manchu, are looked upon by Chinese tradition as typical makers and owners of such arrow-heads.   When the Chou dynasty rose in power and extended its influence into the far north-east, the chief of the Su-shên offered as tribute arrows provided with stone heads and wooden shafts, one foot eight inches long.   The Emperor Wu (B. C. 1122–1116) caused the words "Arrow of the tribe Su-shên" to be engraved on the shaft to transmit the matter to posterity.   Another tradition tells that, "when Confucius was in the kingdom of Ch'ên in B. C. 495, the king took one day his meal on a terrace of his garden, and suddenly a bird pierced by a stone arrow fell down in front of him.   Confucius when consulted as to this arrow replied: 'The bird is a sort of sparrow-hawk originating from the land of the Su-shên, and the arrow-head resembles that of the Emperor Wu which he bestowed as an emblem of rank on the prince in whose favor he raised the country of Ch'ên into a kingdom.'   Thereupon a search was made in the arsenal of the king where in fact the stone arrow-head was found which the princes

---

[1] A musical instrument consisting of a windchest made of gourd with a mouth-piece attached to it, and a series of tubes or pipes, five of which vary in length.   An interesting article, with illustrations, on the wind-pipes of the Miao-tse by Ryuzo Torii will be found in No. 169 of the *Kokka* (June, 1904).

[2] The Chou emperors had a special master of ceremonies called *mao jên*. *Mao* is a flag made from the tail of the wild ox which the dancers held in their hands as signals.   It was the task of the *mao jên* to teach the foreign dances with the music accompanying them.   All people from the four directions of the compass who had duties at court as dancers were under his command, and in solemn sacrifices and on the occasion of visits of foreign ambassadors, representations of these dances were given.   The ancient Chinese furnish the naïve explanation that they were adopted to show that all peoples under Heaven form only one empire or family.

of Ch'ên had religiously kept."[1] Though this is no more than an anecdote *ben trovato*, it may reveal several important points,— that at the time of Confucius flint arrow-heads were no longer generally known, that they were precious rarities preserved in the royal treasury, and that as early as the twelfth century B. C. they had sunk into a mere emblematic significance and served as insignia of authority,[2] and that the Su-shên, a Tungusian tribe, are made responsible for their origin.

Chavannes (Se-ma Ts'ien, Vol. V, p. 341) quotes a passage from the *San kuo chi* to the effect that in 262 A. D. the governor of Liao-tung informed the court of the Wei dynasty that the country of the Su-shên had sent as tribute thirty bows, each three feet and five inches long, and three hundred arrows with a point of stone and shaft of the tree *hu*, one foot and eight inches long. But there is a still later reference to the use of flint arrows on the part of this tribe.

As late as the middle of the fifth century A. D. we hear again of the same Su-shên as being in possession of flint arrows, as attested by a passage in the *Wei shu*, the Annals of the Wei dynasty (386–532 A. D.). Under the year 459 A. D. it is there recorded that the country of Su-shên offered as tribute to the court arrows with wooden shafts and stone heads, and the same tribute is ascribed for the year 488 A. D. to the territory of Ki in the present P'ing-yang fu, Shansi Province. This account offers a twofold interest in showing that flint arrows were then still held in reverence by the Chinese and regarded as valuable objects, and in affording evidence of the long-continued use of flint arrow-heads among the Su-shên for whom we can thus establish a period spent on their manufacture lasting over a millennium and a half.[3]

---

[1] G. Schlegel, Uranographie chinoise, pp. 758, 759. This story is derived from the *Kuo yü* and reproduced in the Annals of *Se-ma Ts'ien* (see Chavannes' translation, Vol. V, p. 340) where a fuller version of it is given.

[2] Bow, arrows and quiver were conferred upon the vassal princes by the emperor as sign of investiture.

[3] Palladius discovered in 1870 a stone hatchet near the bay of Vladivostok. He was under the impression that it was made of nephrite; microscopical investigation, however, proved that the substance was diorite-aphanite (H. Fischer, Nephrit und Jadeit, pp. 283–284). Palladius drew from this find a somewhat hazarded conclusion; he believed that "it would decide the question regarding the famous stone arrow-heads made by the aboriginal inhabitants of Manchuria, the Su-shên, and their direct descendants and successors, the I-lou, Ugi and Mo-ho, from oldest times down to the twelfth century." Palladius evidently labored under the error that the arrow-heads of the Su-shên were of nephrite. There is, however, no account to this effect. All Chinese accounts are unanimous in speaking of these arrow-heads as being of plain stone, and never use the word for jade (*yü*) in connection with them. The stone hatchet of Vladivostok certainly has no bearing on the whole question, and the further conclusions of Palladius in regard to alleged sites in Manchuria and at the mouth of the Amur where, according to Chinese sources, nephrite should have been found, which is not at all correct, are not valid. I am quite familiar with the Amur region, and having seen a good number of stone implements from there, can positively state that no implements whatever of nephrite have been found there.

The Su-shên seem to have been a warlike nation at that time and fought two wars with the Japanese in 658 and 660 A. D., after they had already settled, in 544 A. D., on the island of Sado, west from Hondo where they subsisted on fish-catching. Their relations with Japan are described in the Japanese annals, *Nihongi* (see ASTON, Nihongi, Vol. II, pp. 58, 257, 260, 263, 264). I here allude to them because they contain a passage from which it may be inferred that the Su-shên did not possess iron at that time. In 660 A. D. an expedition of two hundred Japanese ships under Abe no Omi with some Ainu on board was despatched against twenty ships of the Su-shên. The Japanese commander sent messengers to summon them, but they refused to come. Then he heaped up on the beach colored silk stuffs, weapons, *iron*, etc., to excite their cupidity. Two old men sent forth by the Su-shên took these articles away. In the ensuing battle they were defeated, and when they saw during the fight that they could not resist the power of their enemies, put to death their wives and children.

"In the country of the *I-lou*,[1] they have bows four feet long. For arrows they use the wood of the tree *hu*,[2] and make them one foot eight inches long. Of a dark (or green) stone they make the arrow-heads, which are all poisoned and cause the death of a man when they hit him."[3]

In a small treatise on mineralogy *Yün lin shih p'u*, written by TU WAN in 1133 A. D.,[4] is the following note under the heading "Stones for arrow-heads:" "In the district Sin-kan in the prefecture of Lin-kiang in Kiang-si Province, there is a small place called *Pai yang kio* ("Horn of the white sheep") ten *li* from the district-town. There is a mountain called *Ling-yün ling*, on the summit of which a plain stretches level like a palm. A military out-post was stationed there in ancient times, and everywhere in the land occupied by it, ancient arrows with sharp-pointed blades and knives have been preserved; examining the blades of these knives, it is still possible to cut with them. The material of these arrows and knives consists of stone;

---

[1] A Tungusic or Korean tribe located between the Fu-yü and the Wo-tsü, peoples inhabiting Korea. They are described as resembling in their appearance the Fu-yü, but speaking a different language; they were agriculturists without cattle and sheep, and pig-cultivators; they were not acquainted with iron; they wore armor of skin covered with bone. Compare PLATH, Die Völker der Mandschurei, pp. 75–77 (Göttingen, 1830).

[2] BRETSCHNEIDER, Botanicon Sinicum, Part II, No. 543: an unidentified tree. Legge's translation "arrow-thorn" is based on the error that Confucius on one occasion referred to the famous *hu* arrows; his reference is made only to the shafts being of this wood. Many Chinese editors, from not understanding this word, have changed it in the few passages where it occurs into *k'u* "decayed tree;" but it is not plausible that rotten wood was ever used for arrow-shafts.

[3] *Hou Han shu, Tung I chuan*, quoted in *P'ei wên yün fu*, Ch. 100 A, p. 18.

[4] Reprinted in the collection *Chih pu tsu chai ts'ung shu*, Section 28; Ch. 2, p. 8 b.

they are over three and four inches in length.   There are among them also short ones like those which Confucius under the name 'stone arrow-heads with wooden shafts' made out as objects of the tribe of Su-shên. [Then follow the quotations from the *Yü kung* above referred to.]   In the Ch'un Ts'iu period (B. C. 722–481) they were collected in the palace of the state of Ch'ên in such a way that the wooden shafts were perforated and strung; the stone arrow-head of these was one foot and eight inches long.   There are, further, stone coats-of-mail consisting of scales ("leaves") like tortoise-shells,[1] but somewhat thicker.   There are stone axes as big as a palm, the wooden hafts of which have been pierced to enable convenient carrying.   They are all dark (or green) colored and hard; when struck, they emit sounds."

CH'ANG K'ü, the author of ancient records relating to Sze-ch'uan (*Hua yang kuo chi*) written at the time of the Tsin dynasty (265–313 A. D.) mentions stone arrow-heads in the district of T'ai-têng (Sze-ch'uan Province), on a mountain situated on a lake *Ma hu* ("Horse Lake") into which a small river *Sun shui* or *Pai sha kiang* ("White-sand River") falls.   "When these arrow-heads are burnt by fire, they will harden like iron."[2]   JOHAN NEUHOFF (Die Gesantschaft der Ost-Indischen Gesellschaft, p. 318, Amsterdam, 1669) tells of a peculiar kind of stone found on the mountain *Tiexe* near K'ien-kiang in Sze-ch'uan; when heated by fire, iron pours out of them well suited for sabres and swords.

In the great archæological work edited by the two brothers FÊNG in 1822, the *Kin-shih so* (*kin so*, Vol. 2), three stone arrow-points with inscriptions are published (reproduced in Fig. 8).   In the upper one on the left an ancient form of the character *yu* "right" is engraved, in the one on the right the symbols for the two numerals "eight" and "thousand;" in the lower arrow-head the character *t'ung* "together" appears in relief,[3] while the lower face of the same specimen figured beside it is without character.   The Chinese editors do not express an opinion in regard to the meaning of these symbols; maybe they merely take the place of property marks, if they are not, which is even worse, collectors' marks only.   I hardly believe that they originated contem-

[1] The *Shih i ki* written by WANG KIA in the fourth century records: "In the first year of the period *T'ai shih* (265 A. D.) of the Tsin dynasty, men of the country of Pin-se came to court with clothing adorned with jade in five colors, in the style of the present coats-of-mail (*i. e.* jade plaques were fastened to the coat as metal plaques in an armor).   Further the country of Po-ti presented a ring of black jade in color resembling lacquer."

[2] Quoted in *P'ei wên yün fu*, Ch. 100 A, p. 29 b.

[3] As shown by its black color in the reproduction, whereas the two others are white.   The illustrations were made from rubbings of the specimens.

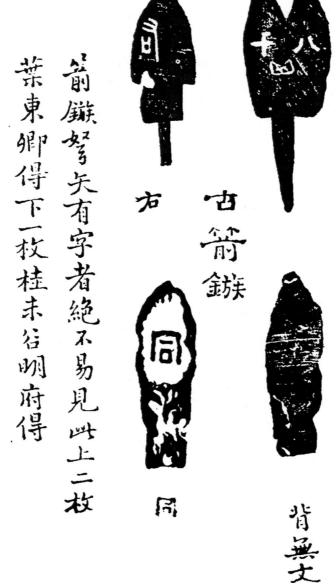

箭鏃弩矢有字者絶不易見此上二枚
葉東卿得下一枚桂未谷明府得

古箭鏃

右

同

同

背無文

FIG. 8.
Three Stone Arrow-Heads (from *Kin-shih so*).

poraneously with these arrow-heads. There is no testimony that the Chinese themselves ever made practical use of flint arrows, and if we want to suppose such a period in their culture, it must certainly be far back in times antedating the invention of writing. It will therefore be more reasonable to argue that these flint arrow-heads were incidentally found by Chinese, and that some one with antiquarian tastes had these characters engraved in them. This would not be an unprecedented case, for there are numerous examples for such procedures. Especially in the K'ien-lung period (1736–1795) when a mania for antique scripts broke out, numberless genuine ancient objects fell victims to this fashion and were covered with date-marks and other inscriptions of archaic style. Bronze swords and other weapons, tiles, and jades were the favorite objects of such improvement, so that this period is apt to become the *crux* of the archæologist. The text of the editors says: "Arrow-heads with points of stone and inscribed. They are certainly objects of great rarity (*lit.* not easily seen or found). The two upper specimens were obtained by Yeh Tung-k'ing; the one below by Kuei Wei-ku in Ming-fu (ancient name for Ning-po)." As shown by the black ink, the reproductions were made by means of rubbings and accordingly teach nothing about the character of the surface of the specimens; we only receive a glimpse of their outlines which are presumably correct. We are surprised at the gracefully elegant shapes of the two upper specimens, as are rarely, if at all, found in flint arrow-heads; they convey the impression that they are imitations of bronze arrow-heads, to which also the long tangs seem to point. Indeed, when glancing over the pages of the *Kin-shih so*, I had many times taken these illustrations for bronze arrow-heads, until the reading of the editorial comment convinced me of my error. The twin tips in the upper specimen on the right are also striking, and I am not aware of any analogous example to this phenomenon in other flint arrows. The lower specimen seems to be a lance-head rather than an arrow-head. The two brothers *Fêng* were immensely capable and ingenious archæologists, and it would be unfair to suspect that they became the victims of a mystification in this case. While I am inclined to regard the characters as epigone additions,[1] I think of the objects themselves as authentic, but as having been made in a bronze period as reproductions of bronze arrow-heads presumably for use as amulets, as far as the first two are concerned.

As regards stone hatchets, we saw them mentioned by TU WAN in 1133 A. D. But there are some earlier records of such finds.

[1] These are not forgeries in Chinese estimation, but improvements or embellishments.

In the fourth year of the period Ta-ming (460 A. D.) the governor of Sü-chou,[1] Liu Tao, descended into the river Pien and found a hatchet of white jade which he presented to the Emperor.[2]

During the reign of the Emperor Su-tsung (756–762 A. D.) of the T'ang dynasty, a Buddhist priest, Ni-chên-ju by name, made a present to the Emperor of eight precious objects which he alleged to have received from Heaven for transmission to the Son of Heaven. The sixth of these was styled "Stones of the God of Thunder." It consisted of two pieces having the shapes of hatchets, about four inches long and over an inch wide; they were not perforated and hard like green jade.[3]

TUAN CH'ÊNG-SHIH, the author of the *Yu yang tsa tsu*, who died in 863 A. D., mentions stone axes occurring in a river in the district of I-tao which is the ancient name for I-tu in King-chou fu, Hupeh; some of them as big as ordinary (copper) axes, others small like a peck (*tou*).[4]

It seems that actual use of primitive jade axes was still made at the court of the Mongol emperors in Peking; for, as PALLADIUS[5] pointed out, T'AO TSUNG-I who wrote the interesting work *Ch'o kêng lu* at the close of the Yüan dynasty, mentions two life-guards standing next to the Khan who held in their hands "natural" axes of jade. PALLADIUS adds that they were axes found fortuitously in the ground, probably primitive weapons.

LI SHIH-CHÊN, the great naturalist of the sixteenth century, summed up the knowledge of his time regarding ancient stone implements in his *Pên ts'ao kang mu* (Section on Stones, Ch. 10) as follows. He comprises them under the generic term *p'i-li chên* (or *ts'ên*) which means "stones[6] originating from the crash of thunder." Before giving his own notes, he quotes CH'ÊN TSANG-K'I, the author of a Materia medica under the T'ang dynasty in the first half of the eighth century as saying: "Suchlike objects have been found by people who explored a locality over which a thunderstorm had swept, and dug three feet in the ground. They are of various shapes. There are those resembling choppers and others like files. There are some pierced with two

---

[1] In Kiang-su Province.

[2] *Sung shu, fu jui chi* (quoted in *P'ei wên yün fu*, Ch. 100 A, p. 213 a).

[3] G. SCHLEGEL, *l. c.*, p. 760.

[4] Quoted in *P'ei wên yün fu*, Ch. 100 A, p. 30 b.

[5] Elucidations of Marco Polo's Travels in North-China (*Journal China Branch R. Asiatic Society*, Vol. X, 1876, p. 43).

[6] The word *chên* (GILES No. 626) is properly a flat smooth stone block as occurring on the bank of a river or brook used by women to beat clothes on when washing them. Li Shih-chên remarks that of old this word was written with the character *chên* meaning "needle" (GILES No. 615) "which is an insignificant mistake;" but maybe stone needles really existed in ancient times.

holes. Some say that they come from *Lei-chou* in Kuang-tung[1] and from *Tsê-chou fu* in Shansi (*Ho tung shan*) where they have been found after a storm with lightning and thunder. Many resemble an axe. They are dark (or green) in color with black streaks and hard like jade. It is stated by several that these are stone implements made by man and presented to the celestial deities, a matter the truth of which I ignore."

Now LI SHIH-CHÊN himself takes the word: "The Book on Lightning (*Lei shu*) says: 'The so-called thunder-axes are like ordinary axes made of copper or iron. The thunder washing-blocks (*lei chên*) resemble those of stone in real use; they are purple and black in color. The thunder hammers weigh several catties. The thunder gimlets are over a foot in length, and are all like steel. They have been used by the God of Thunder in splitting things open or in striking objects. The thunder rings are like jade rings; these have been worn as girdle-ornaments by the God of Thunder and have subsequently fallen down. The thunder beads are those which the divine dragon (*shên lung*) had held in its mouth and dropped. They light the entire house at night.' — In the *Po wu chi* (a work by CHANG HUA, 232–300 A. D.) it is said: 'Fine stones in the shape of small axes are frequently seen among the people. They are styled axes of the crash of thunder (*p'i-li fu*) or wedges of the crash of thunder (*p'i-li hieh*).'— In the *Hiüan chung ki* (by KUO-SHIH of the fifth century) it is narrated: 'West of *Yü-mên* (near Tun-huang, Kansu) there is a district with a mountain on which a temple is erected. There the people of the country annually turn out gimlets to offer to the God of Thunder as a charm against lightning. This is a false practice, for thunder partakes of the two forces of *Yin* and *Yang* (the female and male power) and has accordingly a loud and a low voice, so that it can produce in fact divine objects (*shên wu*). Thus, numerous objects come to light out of hidden places, like axes, gimlets, washing-stones, hammers, which are all real things.[2] If it is said that in Heaven conceptions arise, and that on earth forms arise,[3] we have an example in stars falling down on earth and being stones there. And so it happens that it rains metal and stone, millet and wheat, hair and blood, and other queer things assuming shape on earth. There are certainly in the universe (*lit.* the great void) divine

---

[1] The name *Lei-chou* means Thunder-City. Whether it received this name from the finds of thunder-stones, or whether it is credited with the latter for the sake of its name, I cannot decide. The God of Thunder is much worshipped in that prefecture (HIRTH, Chinesische Studien, p. 140; CL. MADROLLE, Hai-nan et la cote continentale voisine, p. 79, Paris, 1900).

[2] Compare *Mémoires concernant les Chinois*, Vol. IV, p. 474.

[3] Quotation from *Yi king*.

objects which can be utilized. There was, *e. g.*, Su Shao at the time of the Ch'ên dynasty (557–587 A. D.) who had a thunder hammer weighing nine catties. At the time of the Sung (960-1278 A. D.), there lived Shên Kua[1] who found during a thunder-storm under a tree a thunder wedge resembling an axe, but not perforated. The actions of the spirits are dark and cannot indeed be fully investigated."

It will be seen that Li Shih-chên does not divulge his own opinion on the subject, but is content with citing his predecessors. We notice that the almost universal belief in thunderbolts presumably suggested by falls of meteors and shooting-stars prevails also in China.[2] Fig. 9 is reproduced from the *Pên ts'ao* and exhibits six sketches in outline of implements mentioned in the article,[3] — from left to right explained as wedge, axe, gimlet, inkcake, pellet, and washing-stone, the latter rather looking like a club than an anvil. Regarding the

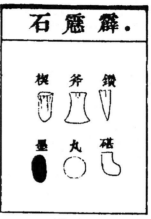

FIG. 9.
Stone Implements
(from *Pên ts'ao kang mu*).

so-called inkcake, LI SHIH-CHÊN has the following additional remark: "The Book on Lightning (*Lei shu*) says: 'Every lightning writes in wood and stone which are then called wooden writing-tablets (*mu cha*). The writing is two or three-tenths of an inch deep of dark-yellow hue. Others say that flowers of sulphur, indigo-blue and vermilion are combined in the writing of the documents of lightning. Again, others say that it is grease from the stones of Mount P'êng-lai to furnish this

[1] The author of the *Mêng ki pi t'an* who lived from 1030 to 1093 (GILES, Biographical Dictionary, No. 1691). He expresses himself in this book as follows: "The people of the present time have found numerous thunder hatchets and thunder wedges, in all cases, after a thunderstorm. Of the thunder hatchets, many are made of iron and copper; the wedges are of stone and resemble hatchets, but are not perforated."

[2] R. ANDREE (Ethnographische Parallelen, Neue Folge, pp. 30–41) offers a series of notes on the propagation of this idea.— The term thunderbolt is not limited to stone implements, but is also applied to those of bronze. In 1902 I obtained two bronze spear-heads and three bronze chisels excavated near the ancient city of Shao-hing in Chekiang Province through Mr. Gilbert Walshe (see *Journal China Branch R. Asiatic Society*, Vol. XXXIII, p. 92). In a letter, dated Shanghai, June 31, 1902, Mr. Walshe then remarked: "The so-called thunderbolts are, I imagine, really bronze chisels of a bronze age, — I will not say *the* bronze age, — and are said to be found buried in the earth some three feet beneath the spot where a man has been struck by 'thunder' (according to the Chinese ideas)."

[3] Reproduced also by F. DE MÉLY, Les pierres de foudre chez les Chinois et les Japonais (*Revue archéologique*, 1895, p. 5 of the reprint) and Les lapidaires chinois, p. LV.

writing-material.   In Lei-chou every thunderstorm is connected with
a big downpour of rain in which large objects like sandstones fly down
and small ones like bits of stones, others in the shape of fingers, hard
like stone, of black color, lustrous and very heavy.'— LIU SÜN says in
his book *Ling piao lu* (T'ang dynasty): 'In Lei-chou, after a rain,
men have frequently found in desert places stones like jet[1] which they
call inkcakes of the God of Thunder; when struck, they yield a sound
like metal; they are of bright lustre and nice.'— LI CHAO narrates in
his book *Kuo shih pu* (beginning of ninth century): 'Thunderstorms
abound in Lei-chou.   At the advent of the autumn they cease, and it
looks as if thunder would then hibernate in the soil.   The people dig
it up, take and eat it, and in view of this fact, such objects appear as
the fruits of thunder.' "[2]

These "inkcakes" are doubtless natural productions, and not
wrought stone.   Soft stones available for writing with are frequently
mentioned.   JOHAN NEUHOF (Die Gesantschaft der Ost-Indischen
Gesellschaft, p. 317, Amsterdam, 1669) alludes to a stone found near
Nan-hing in Kuang-tung Province "very similar in color to Chinese
ink, by means of which the Chinese write on white polished boards in
the same manner as Europeans with chalk."

The thunder stones have also an artistic or ornamental function
in a curious wood-engraving inserted in the *Fang-shih mo p'u* (published
in 1588) and here reproduced in Fig. 10.   Six so-called "precious
objects of good omen" are there united in a circular zone on a background of cloud-ornaments.   On the right side, two hatchets will be
recognised which are explained as "thunder stones."   The other
objects are: below, two hooks by means of which the empress cuts
the leaves of the mulberry-tree,[3] a jade seal, beads of the white coral
called *lang-kan*,[4] the red jewel called *mo-ho*, and the precious pearl
granting every wish.

LI SHIH-CHÊN refers in another passage also to acupuncture needles
having been made of stone in ancient times.   They are called *pien
shih* (GILES No. 9194) or *chên shih* (No. 610).   His note runs as follows:

---

[1] See LAUFER, Historical Jottings on Amber in Asia, pp. 222–225.

[2] In the *Wu li siao shih* by FANG I-CHIH (HIRTH, *T'oung Pao*, Vol. VI, p. 428)
it is said that thunder assumes the shape of a swine in the ground.   HIRTH (Chinesische Studien, p. 158) thinks that a plant like a truffle is understood.

[3] This is an ingenious instrument combining the two functions of a hook and a
cutter.   Those which I have seen in central China were in the shape of a small
scythe to the back of which a projecting hook is attached.   The instrument is
provided with a socket into which a long bamboo stick is inserted.   The farmer uses
the hook to seize and bend down towards himself the branches of the mulberry-tree,
and when they are within easy reach, he cuts them off with the scythe-like knife.

[4] See CHAVANNES, *T'oung Pao*, 1907, p. 182; FORKE, *Mitteilungen des Seminars*,
Vol. VII, 1904, p. 148.

"The *Tung shan king* says: 'The mountains of *Kao-shih* and *Fu-li* abound in the stone for acupuncture needles.' KUO P'O (276–324 A. D.) remarks in his commentary to the dictionary *Êrh-ya* that instead of *pien* also *chên* can be written.— The medical work *Su wên i fa fang i lun* says: 'In the regions of the eastern quarter there is a place where salt is fished from the sea. The water along the beach of the sea is

FIG. 10.
Stone Hatchets in Ornamental Composition (from *Fang-shih mo p'u*).

wholesome in the cure of sores and ulcers and conveys its beneficial effects to the acupuncture stones; hence the latter come also from the eastern region.— WANG PING, a physician of the eighth century, states in his commentary (to the *Huang-ti su wên*): 'The stones called *pien* are like jade and may be called also needles (*chên*). The ancients made the acupuncture needles of stone; the more recent generations substituted iron for the stone. The people of the present time use a porcelain needle to perform acupuncture on the same principle derived from the stone needles. Only we do not know any longer the stone used in acupuncture (no specimens having come down to us), but it may be supposed that it belonged to the class of stone from which arrow-heads were made.' "[1]

In some localities, "stones of the Thunder-God" of special fame are pointed out. Thus, *e. g.*, according to the Imperial Geography (*Ta Ts'ing i t'ung chi*), there is in the north-west of the city of Nan-fêng in Kiang-si Province a summit in the shape of a lotus-flower (*Lien hua fêng*), on the top of which there is a stone of the Thunder-God. Also the Emperor K'ANG-HI shows himself familiar with thunder-bolts in his Jottings on Natural Science (translated in *Mémoires concernant les Chinois*, Vol. IV, p. 474, Paris, 1779); he says that their shapes and materials vary according to the localities, and that the nomadic Mongols, whereby he understands Mongols proper and Tungusians on the eastern sea-coasts, avail themselves of such implements in the manner of copper and steel; some are shaped like hatchets, others like knives, and some like mallets, some of blackish, and others of greenish color.

It is a favorite idea inferred *a priori* that stone implements must be infinitely old and called prehistoric. For stone implements found on Chinese soil at least, this is merely illusory, as we have trustworthy historical accounts relating to the manufacture of such implements in comparatively recent time. Thus, *e. g.*, it is recorded in the geographical work *Huan yü ki* written by LO SHI at the end of the tenth century that the people in the present locality of *T'êng hien* (in *Wu-chou fu*, Kuang-si Province) manufacture knives and swords of a dark-colored (or green) stone (*ts'ing shih*) of which their women turn out armlets and rings; with the former, it takes the place of iron and copper, with the latter of pearls and gems.[2] This notice would become one of importance, if stone objects of this description would ever turn up in that district which, without any additional evidence, would have to be dated in the tenth century A. D.

There are even still more recent accounts of stone implements actually manufactured. JOHAN NEUHOF (Die Gesantschaft der Ost-Indischen Gesellschaft, p. 317, Amsterdam, 1669) mentions the occurrence near the city of Nan-hing in Kuang-tung Province of a kind of stones so hard that the inhabitants can make from them hatchets and

---

[1] Then follows a note concerning flint arrow-heads which it is not necessary to reproduce, as we are familiar with its contents. It is, however, interesting to see that to Li Shih-chên the flint arrow-heads come only from the country of the Su-shên.

[2] The same account is given also in the *Pên ts'ao kang mu* (Section on Stones, Ch. 10, p. 36 b) where the clause is added: "The people of that district, in bringing a field under cultivation, use a knife (*i. e.* mattock) made of stone and over a foot long."

knives. The statement of the same tenor made by GROSIER (Description générale de la Chine, Vol. I, p. 191, Paris, 1818) seems to go back to that source.[1] There is, in my opinion, not the faintest reason to connect these modern manufactures with the idea of a stone age or even to consider them as survivals; they are merely the outcome of chance and convenience. Hundreds of utensils are turned out of stone by the Chinese, so that there is no wonder that occasionally and sporadically also a knife or a hatchet is listed among these objects, when a suitable material offers.

After having surveyed the existing material and the records of the Chinese, it may be well to go back to the assertion which several authors have made in regard to a stone age of China, some very positively, others more guardedly by merely pointing to the possibility of this case. In view of the scanty material before them, there is certainly occasion to admire the courage of such writers. As early as 1870, EDWARD T. STEVENS (Flint Chips, p. 116) who knew of just one stone adze from China exhibited in the Christy Collection, London, wrote: "St. Julien has extracted passages from different Chinese works which prove the existence of a stone age in China. Not only are arrow-heads and hatchets of stone noticed, but also agricultural implements made of the same material." JULIEN can hardly be made responsible for these notes which consist of four brief and incomplete references; they were communicated by him to CHEVREUL who published them under the title "Note historique sur l'âge de pierre à la Chine" (Comptes rendus de l'Académie des Sciences, Vol. LXIII, pp. 281–285, Paris, 1866). But as one swallow does not make a summer, one stone adze does not yet go to make a stone age, and four literary allusions of recent date do not help much to support it.

R. ANDREE (Die Metalle bei den Naturvölkern, p. 103, Leipzig, 1884) says: "However early and highly developed the knowledge of metals appears among the Chinese, yet this people does not make an exception and has had like all other peoples a stone period; it even seems that in some provinces stone implements were used in comparatively recent times." ANDREE justly calls attention to the aborigines in the south and south-west among whom stone implements may have been longest in use.

Sir JOHN LUBBOCK (Prehistoric Times, fifth ed., p. 3) is inclined to assume that the use of iron was in China also preceded by bronze, and bronze by stone; and M. HOERNES (Urgeschichte der Menschheit, p. 92) strikes the same note by saying: "The remains of a stone age

---

[1] The same author (Vol. I, p. 439) asserts: For the rest none of those ancient cutting stones wrought to supplant the use of iron are found in China; at least, the present literati have never heard of such.

which has passed long ago are preserved in the soil of China; in this great empire, there are provinces where not so long ago axes and cutting instruments were made of hard stone," etc.

Giglioli (*l. c.*), as shown by the very title of his article, and C. Puini[1] have taken the same stand.

Prof. Hirth has adopted a special platform (The Ancient History of China, p. 236), and his argument deserves a hearing, as it is based on a discourse of the philosopher Kuan-tse. Hirth believes that "this philosopher was fully conscious of the extent and sequence of cultural periods in high antiquity, knowledge of which, as the result of scientific reasoning, is a comparatively recent acquisition with westerners." The words of the philosopher are then construed to mean that the time of the primeval emperors (about b. c. 3000) was a stone age in which weapons were made of stone and were used for splitting wooden blocks for the construction of dwellings, and that this first period is followed by a second age extending from about b. c. 2700–2000 in which jade was used for similar purposes. "This may be compared," adds Hirth, "to our neolithic period, when hatchets and arrow-heads were made of polished stone, either jade or flint." All Chinese philosophers evince a great predilection for evolutionary theorizing which appears as the mere outcome of subjective speculation and cannot stand comparison with the methods and results of our inductive science; deduction there, and induction on this side, make all the difference. It is impossible to assume that the Chinese speculators of later days should have preserved the memory of cultural events and developments which must lie back, not centuries, but millenniums before their time. Just the intentional interpretation of an evolution read into the past which looks so pleasant on the surface is the strongest evidence for the fact that this is a purely personal and arbitrary construction or invention, not better than the legend of the golden, silver and iron ages. Thus, I cannot agree either with Hirth (*l. c.*, pp. 13, 14) in regarding the traditions clustering around the ancient emperors as symbolizing "the principal phases of Chinese civilization" or their names as "representatives of the preparatory periods of culture." They are, in my opinion, culture-heroes (*Heilbringer*) of the same type as found among a large number of primitive peoples, downright mythical creations which have no relation whatever to the objective facts of cultural development. Reality and tradition are two different things, and the thread connecting reality with tradition is usually very slender. Nowhere has the history of reality been so evolved as traditional or

---

[1] Le origini della civiltà secondo la tradizione e la storia nell' Estremo Oriente, p. 163 (*Pubblicazioni del R. Istituto di Studi Superiori*, Firenze, 1891).

recorded history will make us believe, for it is not only traditions themselves which in the course of time change and deteriorate, but above
all their interpretations and constant re-interpretations in the mind
of man. A custom, *e. g.*, may survive at the present time and be
practised in exactly the same or a similar manner as thousands of years
ago; but another reason for it may be given, another significance
attributed to it by modern man. And these explanations of customs,
of rites, of traditions, have possibly nothing to do with the objective
development of the matter in the world of reality. They are certainly
important, but more as folklore or psychological material, while their
historical value is small and only relative in that they may be apt to
furnish the clue to the correct scientific explanation. Applied to the
case under consideration, this means: Kuan-tse's argumentation is
certainly interesting as characterizing the intellectual sphere of the
man, the trend of his thoughts, and his manner of reasoning, and as
furnishing a good example of this mode of Chinese philosophizing;
but to make use of it as the foundation of far-reaching conclusions
regarding the existence of certain cultural periods is, in my estimation
at least, out of the question.[1] Such conclusions must be reached
by other methods.

In weighing the records of the Chinese in the balance of our critique,
we are, above all, confronted with the fact that, throughout Chinese
literature, there is not one single instance on record in which the Chinese
would admit that stone implements like arrow-heads, knives or hatchets
have ever been made and used by them in ancient times. In attempting to account for the occasional finds of stone implements, the mere
thought that these might have originated from their forefathers, did
not even enter their minds. They were strange to them and looked
upon with superstitious awe. As far as Chinese history can be traced
back, we find the Chinese as a nation familiar and fully equipped with
metals, copper or bronze, or — copper and bronze, the beating and

---

[1] The same holds good for the culture-periods established by Hirth in his paper
Chinesische Ansichten über Bronzetrommeln, pp. 18–19, on the ground of a passage
in the *Yüeh tsüeh shu* compiled in 52 A. D. and possibly containing views attributable
to the fifth century B. C. Also here no historical source is involved from which
inferences could be drawn in regard to historical events, but only the theorizing
opinion of a philosopher couched in the style of a biblical sermon. According to
him, in oldest times, weapons were made of stone to cleave timber for making
palaces and houses; the dead were buried by dragons, for God the Lord had so intended; up to the time of Huang-ti, weapons were made of jade to fell trees for building houses, and to bore into the soil, for jade was also a divine substance; and as the
Lord still intended so, the dead were buried by dragons. In the same stilted language, with reference to Providence, bronze and iron are treated to conclude the
evolutionary series. For lack of all palpable sources, this philosopher was, of course,
entirely ignorant of any facts relating to the periods of which he speaks. His utterances are philosophy of history, not history.

casting of which was perfectly understood. The jade implements of
the Chou period are not only contemporaneous with the Chinese
bronze age, but also from an epoch when the bronze age after an existence of several millenniums was soon nearing its end and iron gradually
began to make its way; *i. e.* from an archæological viewpoint, they
are recent products. They are not the index of a stone age, and the
literary records are in full agreement with this state of affairs. At
the time of the Chou, the Chinese lived surrounded by numerous
foreign peoples who partially made use of flint arrows and possibly
other stone weapons; but also the stock of *Man* tribes was acquainted
with copper and employed copper utensils. The stone implements
of their neighbors were a source of wonder, mere curiosities, to the
Chinese.

Another notable fact to be gleaned from the references above
given is that the association of worked stone with the God of
Thunder is a rather late idea and sets in only from times long after the
beginning of our era; in all probability, it is not earlier than the T'ang
dynasty, for Ch'ên Tsang-k'i who lived in the beginning of the eighth
century is the first in whom this idea has crystallized (p. 63). We
observe that the thunderbolts are not found anywhere and everywhere, but that they are restricted to certain localities, among which
Lei-chou is prominent. As the Tungusic Su-shên, from the days of
antiquity till the present time, are, in the minds of the Chinese, the
typical representatives of flint arrow makers, so the notion of thunderbolts centers around Lei-chou, the Thunder City. This cannot be
accidental. We know that the Chinese have been conquerors and
colonists in this territory, and that it was inhabited before their arrival
by an aboriginal tribe, the *Li*, the remnants of which are to be found
nowadays in the interior of the island of Hainan.[1] During the Sung
period (960–1278 A. D.) they were still settled in the prefecture of *Lei-
chou*, as at that time their language is mentioned as one spoken in that
locality (HIRTH, Chinesische Studien, p. 169). We may safely assume
that the stone implements there discovered and not understood by the
Chinese must be credited to the *Li*. And if other regions like Kiang-
si and Sze-ch'uan are involved, we have the same state of affairs in
that these too were and are still inhabited by non-Chinese tribes; in
regard to Yün-nan, I expressed the same opinion above (p. 32).
Generally speaking, wherever in southern China, the land south of the
Yangtse, stone implements turn up, there is the greatest probability
of their origin being non-Chinese.

[1] Regarding this tribe compare HIRTH, Die Insel Hainan (Reprint from *Bastian-
Festschrift*, pp. 24 et seq.).

The fact that stone implements were once more widely distributed in China than the actual finds hitherto made will allow us to conclude, may be traced from some survivals existing in other forms. We shall meet a number of such survivals in the group of jade symbols which, during the Chou period, were emblems of rank and dignity; part of these are traceable to former implements, as *e. g.* the imperial emblem of sovereignty to an original hammer. These types will be discussed in the following chapter. By comparing a jade chisel with one of bronze (p. 39), we have made the acquaintance of another kind of survivals,— of stone forms in bronze. Many types of bronze chisels and hatchets bear indeed such a close resemblance to corresponding jade objects that the assumption of an historical connection between the two groups is forcibly impressed upon our minds. As the number of such bronze implements in our collection, however, is too large, and this subject would require a long digression into the bronze age, I must leave it here and come back to it in a future monograph on these bronze objects.

I wish to call attention in this connection only to one type of a stone-form survival in bronze which thus far has become known to us in China only in this material, but whose origin most probably goes back to an older form in stone. This type has been rather unfortunately termed shoulder-headed celt; I prefer to adhere to the term spade-shaped celt familiar to us in America where this stone implement widely occurs,[1] because it is more appropriate to the matter, for in all likelihood this implement was once really a spade.

The Chinese admit that in ancient times coinage was unknown and only barter practised, or as one Chinese author puts it: "In ancient times they carried on trade merely by using what they possessed in exchange for what they did not possess."[2] Lumps of metal, metal implements, cloth and silk, also shells seem to have taken the place of money. This primary exchange of actual implements may have led to the practice of casting miniature tools and inscribing them with a fixed valuation. The word *ts'ien* which long ago assumed the meaning of money, once occurs in the *Shi king* (but pronounced *tsien*) in the sense of a hoe; also the *Shuo wên* attributes this former meaning to the word and defines it as an agricultural implement.[3] A coin current during the Chou dynasty under the name "spade-money" (*ch'an pi*) reveals the form of a spade or shovel[4] and may have been derived from

[1] Compare MOOREHEAD, The Stone-Age of North America, Vol. I, p. 335, pp. 418 *et seq.*

[2] L. C. HOPKINS in *Journal Royal Asiatic Society*, 1895, p. 329.

[3] G. SCHLEGEL (Uranographie chinoise, p. 273) takes it in the sense of "sickle."

[4] HOPKINS, *l. c.*, p. 324.

an agricultural implement of this type (Fig. 11). T. DE LACOUPERIE [1] looked upon it as a survival of an implement of the stone age and

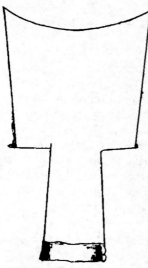

FIG. 11.
Spade-Shaped Bronze Coin (from Specimen in Field Museum).

likened it to "the shouldered celt proper to southeastern Asia, which has hitherto been found only in Pegu, Cambodia (Tonlé-Sap), and Central India (Chhōtā-Nāgpur)." This identification, however, is by no means perfect and only partially justified, as there is a remarkable difference in the curves of the edges which are convex in the stone celts and concave in the Chinese spades. Figure 11 is drawn after an ancient specimen of a bronze spade-shaped coin in our collection which is covered with a fine russet, blue and green patina. The edge is curved in, of almost half-circular shape, terminating in two lateral tips; the blade is 1 mm thick. The shoulders are symmetrical, each 1 cm wide. The handle, the sides of which are in the shape of triangles, is a hollow cast forming a socket betraying the application of a wooden handle; there is a small triangular opening [2] in the upper part of one of the broad faces of the handle to admit of the passage of a cord for closer attachment to the wooden hilt. Related types may be seen also in the jade dance-hatchets of the Han period, as illustrated in Figs. 1–4. These are far advanced products of a higher art, and an attempt to trace them back to their primeval ancestral forms will probably lead to a reconstruction closely related to the type of the spade-shaped celt of stone.

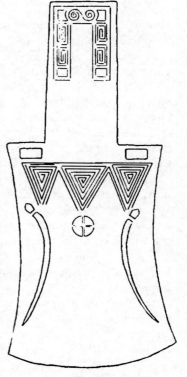

FIG. 12.
Bronze Hatchet of Han Period.

[1] Catalogue of Chinese Coins, p. 4 (London, 1892).
[2] This is not made in the cast, but cut out after casting. Also the corresponding pieces figured in the *Kin-shih so* show such apertures of various irregular forms.

I here add two other bronze hatchets of the Han period which offer a still more striking analogy to the stone spades. Both are in my collection in the American Museum, New York, and were obtained at Si-ngan fu in 1903. The one shown in Fig. 12 (15.3 cm × 6.5 cm) comes nearest of all to the supposed ancestral form and has, of course, assumed under the clever hands of the Chinese bronze-caster a more

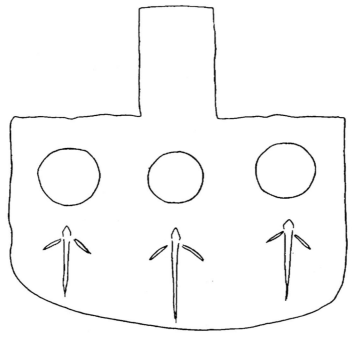

FIG. 13.
Bronze Hatchet of Han Period.

regular and graceful appearance. The two shoulders and the head-piece have remained, and the socket is wanting. The headpiece was stuck into the cleft of the wooden hilt, the section decorated with meanders being left uncovered and projecting freely. Cords or leashes passing through the two rectangular apertures in the butt were fastened around the hilt. The blade is covered with three triangles filled, so to speak, with triangular convolute spirals; below these, two crescents surround a circle in slight relief. The significance of the whole ornamental composition is beyond our knowledge. The bronze axe in Fig. 13 (18 cm × 18 cm) shows the same head-piece without socket, and in the butt three large round perforations (each about 3 cm in diameter), an axe-shaped ornament in relief being between each circle and the blade. Shafting was perhaps done differently in this case, the hilt enclosing the head-piece being set vertically on the back of the axe, attachment being strengthened by leashes passing through the three apertures.

The spade-shaped implement of stone is a peculiar characteristic of the Colarian tribes of Central India and the Mon-Khmer group in Farther India who speak related languages (Southeast-Asiatic or Austronesian stock). Figure 14 shows three of these celts after A. GRÜNWEDEL.[1] Up to 1873, this type was known only from Pegu and the Malayan Peninsula. In that year, two specimens of the same type were discovered also in Chhōtā Nāgpur in Central India and described by V. BALL,[2]—one made of a darkgreen hard quartzite, the other of a black igneous rock. Ball was able to show that the material from which they are made occurs *in situ* within the district of Singbhum where the finds had been made, and which belongs to the habitats of the Colarian group. The identity of these two types with those from Pegu was at once recognized, particularly by I. F. S. FORBES[3] who hailed this discovery as a welcome confirmation of the results of comparative philology. These finds have recently much increased, and NOETLING obtained two of these spade-shaped celts in Burma, which

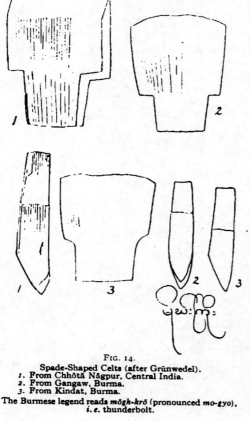

FIG. 14.
Spade-Shaped Celts (after Grünwedel).
1. From Chhōtā Nāgpur, Central India.
2. From Gangaw, Burma.
3. From Kindat, Burma.
The Burmese legend reads *mōgh-krō* (pronounced *mo-gyo*).
i.e. thunderbolt.

[1] Prähistorisches aus Birma (*Globus*, Vol. 58, 1895, p. 15).

[2] On some Stone Implements of the Barmese Type, found in Pargana Dalbhum (*Proceedings of the Asiatic Society of Bengal*, 1875, No. VI, pp. 118–120, 1 Plate).

[3] Comparative Grammar of the Languages of Further India, p. 142 (London, 1881).

are now in the Berlin Museum, and have been compared by GRÜN-
WEDEL with the Indian types in the article quoted. P. O. BODDING[1]
brought together more material from the region of the Santal, a branch
of the Colarian or Munda group, and combats the view meanwhile
set forth by S. E. PEAL that these implements have been used as hoes.
PEAL,[2] when in 1893 at Ledo and Tikak, villages of the Naga tribe,
east of Makum in Assam, secured two small iron hoes used by women
in weeding the hill paddy. They are full-sized instruments, yet the
blade measures only two inches square, and the shoulder less than one
inch; they have handles of split cane a foot long, the cane being firmly
bound round the shoulder. PEAL assumes that these hoes are simply
the Kol-Mon "shoulder-headed celts" made in iron, and that hence
we see not only the meaning of the peculiar "shoulder," but the office
of the complete implement as a miniature hoe. BODDING[3] objects to
this conclusion on the ground that, if these celts should originally have
belonged to the ancestors of the Mon-Khmer and Munda peoples,
one would expect, if Peal's deductions are correct, to find an iron hoe
of the same shape used by these peoples also; but no such implement
is found, at least not among the Santal, who, of agricultural tools,
know only a stick with a flat piece of iron attached to the end for the
purpose of digging roots, or making small holes in the ground. This
objection is, in my estimation, not very weighty. We must always
be mindful of the overwhelming sway of history and historical events.
The Colarian group, who we are bound to suppose migrated in a remote
period from Farther India, where it was in close touch with the Mon-
Khmer peoples, into their present habitats, have had a long and varied
history. Their own traditions carry their migrations back into a period
when they were settled in the country on both sides of the Ganges.
Starting from the north-east, they gradually worked their way up the
valley of the Ganges, until we find them in the neighborhood of Benares
with their headquarters near Mirzāpur. Here the main body which
had occupied the northern bank of the river, crossed and, heading
southward, struck the Vindhya hills, until they at length reached the
tableland of Chhōtā Nāgpur.[4] These events, however, present only

---

[1] Ancient Stone Implements in the Santal Parganas (*Journal Asiatic Society of Bengal*, Vol. 70, Part III, No. 1, 1901, pp. 17–22, 4 Plates) and Shoulder-headed and other Forms of Stone Implements in the Santal Parganas (*Ibid.*, Vol. 73, Part III, No. 2, 1904, pp. 27–31).

[2] On some Traces of the Kol-Mon-Annam in the Eastern Naga Hills (*Ibid.*, Vol. 75, Part III, No. 1, 1896, pp. 20–24).

[3] *L. c.*, p. 29.

[4] Compare L. A. WADDELL, The Traditional Migration of the Santal Tribe (*Indian Antiquary*, Vol. XXII, 1893, pp. 294–296) and A. CAMPBELL, Traditional Migration of the Santal Tribes (*Ibid.*, Vol. XXIII, 1894, pp. 103–104).

the last landmark in the long migration history of the Colarians. There are reasons to believe that in still earlier times, chronologically not definable, the Colarian-Mon stock when it formed a coherent unbroken ethnic body must have lived along the southern ranges of the eastern Himalaya, extending into the territories of Bengal and Assam. In Tibetan literature and even in the modern Tibetan colloquial language, the word *Mon* still appears as a generic designation for all non-Tibetan tribes living southward of the Himalaya and is particularly used in composition with the names of those kinds of cereals and pulse early received by the Tibetans from India.[1] Consequently, the Tibetan name *Mon* originally referred to *the* Mon tribes, and to those exclusively; while, at a later period, after a disintegration of this group resulting in a migration of the Mon southward into Farther India and of the Colarians into a southwesterly direction, this name was retained by the Tibetans and transferred to new-coming tribes occupying the place of the former emigrants, and then to Northern India in a generalized way. In view of such historical events, to come back to our proposition above, it is quite conceivable that Mon or Colarians, or both as a prehistoric undivided unit, once covered also the territory of the Naga and left there these peculiar stone celts which could have subsequently given the incentive for their imitation in iron. The Colarians preserved them faithfully, until they reached their new home where they gradually dropped into oblivion, as they received iron from their more cultivated neighbors. This consideration is also apt to prove that the spade-shaped stone celt must be of considerable antiquity, as also indicated by the extent of the area over which it has been found. It has been a long-lived implement, too, and seems to have still been in actual use during the bronze age. J. DENIKER (The Races of Man, p. 364, New York, 1906) figures one polished spade-shaped celt excavated with several others in Cambodja, side by side with objects of bronze.[2]

All the tribes among which this spade has been found once practised hoe-culture and still partially practise it (see p. 48). Hence it is evident that this implement was the mattock which they used for this

---

[1] LAUFER in *Mémoires de la Société finno-ougrienne*, Vol. XI, pp. 94–101 (Helsingfors, 1898), where all the evidence in this question is brought together. The conclusions as formed above are the same as those at which I arrived thirteen years ago, and after renewed examination, I see no reason to modify them.

[2] Curiously enough, he does not refer to the corresponding types of Central India, and is quite unaware of the great importance which this trifling object bears on a chapter in the primeval history in Eastern Asia. His note that the Naga have still at the present day axes of precisely the same form which they use as hoes deserves correction, for the article of Peal to which he refers speaks of hoes made of *iron* in a similar shape, and which are *possibly*, but not positively, connected with those ancient stone celts.

purpose, and it thence follows that not the Chinese, full-fledged agriculturists from the beginning of their history, invented it, but the hoe-culturists adjoining them on the south. This aspect of things will account also for the absence of this stone celt on Chinese soil, as the Chinese had no use for it. They imitated it in a miniature bronze form adapted to the purpose of barter, and if it occurs again in the ceremonial dance-axes of bronze and jade, the reason for this derivation is not far to seek, if we remember that the Chinese, according to their own accounts, derived many of their pantomimic dances from their southern barbarian neighbors, the *Man*.[1] If they derived the dances, there is good reason to believe that they derived simultaneously also the paraphernalia belonging to them. The spade-shaped form of implement in its bronze derivate, accordingly, is one of the numerous objects and ideas which the Chinese took up from the culture-sphere of the South-east at a period when these two great cultural provinces were still separated. The gradual welding of these two into one finally resulted in that culture unit which we now call simply China.

[1] And most probably, all their ancient dances come from that source. The Chinese have never been a dancing nation, as is easily seen in modern China where no man and no woman is given to dancing; but with the Tibetans, the Man and all Southeast-Asiatic tribes including the Malayan, dancing is popular and national.

II. JADE SYMBOLS OF SOVEREIGN POWER

Among the numerous offices of the Chou dynasty (B. C. 1122–255) there was a steward of the treasury (*t'ien fu*) charged with the superintendence and preservation of the Hall of the Ancestor (*Hou-tsi*) of the imperial house, in which all precious objects transmitted from generation to generation were hoarded. He was responsible for the regulations governing the treasury and took care of the emblems of jade and all valuables belonging to the dynastic family. On the occasion of the celebration of a great sacrifice or of a great funeral service, he brought the desired objects out to the place of ceremony, and at the termination of the rites, locked them up again (BIOT, Vol. I, p. 480). In the beginning of the spring, he sprinkled with the blood of the victims the precious objects and the jewels of the imperial costume. At the end of the winter, he arranged the pieces of jade used in the ceremony which was observed to determine whether a favorable or unfavorable new year would ensue (*Ibid.*, p. 482). In case the emperor transferred his residence and his treasury, he handed all valuables over to the chief of the office in the new place (p. 483).

There were, further, at the court of the Chou dynasty, special artisans to execute works in jade (*yü jên*, "jade men"), in particular the official insignia of jade the care of which was placed in the hands of an officer, called *Tien-jui* (BIOT, Vol. II, p. 519; Vol. I, p. 483). *Jui* is the general name for the jade tablets conferred by the emperor on the five classes of feudal princes (*wu jui*) as a mark of investiture and a symbol of their rank, and held by them in their hands, when they had audience in court. The *tien-jui* official was obliged to distinguish their kinds and names, and to define the ceremonies where they come into action. There were four great audiences, one in each season, and occasional and combined visits of the fief-holders.

Different from this office is that of the *yü fu* (BIOT, Vol. I, p. 124). The *tien jui* was a master of ceremonies in direct connection with the rites in which the treasures in his charge were involved, and taking an active part in the proceedings. The *yü fu*, however, was an executive official in superintendence of the manufacture of jades and other valuable objects touching on the ceremonial life of the imperial family. The sphere of his competency is not clearly defined from that of the *nei-fu* who received all precious objects offered to the emperor by the great dignitaries like gold, jade, ivory, furs, weapons, etc., a duty

80

ascribed also to the *yü fu*. It seems that the former had the mere function of a collector, the latter that of a preparator who supplied jewelry and ornaments for actual use down to such banalities as an imperial sanitary vessel.

In the official hierarchy of the Chou, everything was defined and regulated according to a well devised scheme which found its expression in a series of jade insignia of power and rank. This is, for several reasons, one of the most difficult subjects of Chinese archæology. In the *Chou li* the names and utilizations of these insignia have been handed down without a full description of them which was only supplied by the commentators of the Han period. Most of these insignia were then lost, and the commentators seem to speak of them merely on the ground of traditions. The drawings of these insignia added to the later editions of the Rituals like the *San li t'u* of NIEH TSUNG-I of the Sung period[1] are not made from real specimens of the Chou dynasty, but are imaginary reconstructions based on the statements of the commentators to the *Chou li*, and are therefore worthless, in my opinion, for archæological purposes. The same judgment holds good for the numerous illustrations of these insignia embodied in the *Ku yü t'u p'u* which, aside from the spurious inscriptions carved in them, are suspicious because of their striking similarity to the reconstructive drawings of Nieh Tsung-i[2] and because of an abundance of decorative designs which plainly betray the pictorial style of the Sung period and cannot have existed at the time of the Chou dynasty. The ingenious investigations of WU TA-CH'ÊNG release me from the task of pursuing this criticism, and I propose to supersede all the doubtful material of imaginative Chinese draughtsmen by his positive results in the shape of a series of genuine jade tokens of the Chou period.

I first give a brief review of what the *Chou li* and its commentators have to say in regard to these insignia, and then proceed to lay before the reader the material of Wu Ta-ch'êng which bears all external and internal evidence of representing the objects spoken of in the Ritual.

The emperor was, according to the *Chou li*, entitled to several jade tablets. Prominent among these are two, "the large tablet" (*ta kuei*, BIOT, Vol. II, p. 522) with hammer-shaped head, three feet long, which he fixed in his girdle; and "the tablet of power" (*chên kuei, l. c.* p. 519) being one foot two inches long and held by the sovereign in

---

[1] I availed myself of the Japanese edition printed in Tokyo, 1761, with a preface by LAN CH'ÊNG-TÊ, dated 1676. The date of the original is 962 A. D.

[2] It is also noteworthy that the Sung Catalogue of Jades quotes his work throughout, but not the *Chou li* or *Li ki*. I am under the impression that the compilers of the catalogue made it their object to reconstruct the material described or figured in the *San li t'u*.

his hands.   It was adorned with bands embroidered in five colors, and the emperor having the *ta kuei* in his girdle and the *chên kuei* in his hands offered, during the spring, the sacrifice to the Sun in the morning (BIOT, Vol. I, p. 484).[1]   According to the opinion of the commentators of the Han period, designs of hills were engraved on the symbol of sovereign power (*chên kuei*).   This view doubtless arose from the fact that the word *chên* in the designation of this tablet means not only pacification, submission, power, but is also the name given to the four protecting mountains of the frontier; hence the subsequent illustrators represented this tablet with a conventional design of four hills simply based on this misunderstanding.   There can be no doubt that the meaning of *chên kuei* is plainly tablet of power or emblem of sovereignty, and that it has no reference to the four mountains in whose worship it serves no function.[2]   We shall see from the actual specimen of WU that the *chên kuei* was unadorned indeed, and this is quite in harmony with the spirit of the Chou time, all the jade objects of which are of extreme simplicity.   It is entirely out of the question that mountain scenery, as the epigones will make us believe, was carved on these jade implements which are connected with most primitive and primeval ideas.   These mountain drawings are downright inventions of the Sung period, and suspicion must increase, as different conceptions of them exist.

One may be viewed, *e. g.*, in the book of GINGELL, p. 33, where the tablet ends above in a pointed angle, and where the four hills are arranged in one vertical row, one placed above the other; also the silken band is here added.   This illustration is identical with that in the *K'ien-lung* edition of the Rituals.   Another cut is inserted in the Dictionary of COUVREUR, p. 433, in which the tablet is surmounted by a rounded knob, and where two hills are placed side by side in the upper part and two others in the same way at the foot.

The fact that these imperial emblems were not ornamented is plainly borne out by the wording of the *Li ki*, for "acts of the greatest reverence admit of no ornament" (LEGGE, Vol. I, p. 400; COUVREUR, Vol. I, p. 549), and for this reason, the *ta kuei* of the sovereign was not carved with any ornaments; as added in another passage (COUVREUR, p. 600), because it was only the simplicity of the material which was appreciated.[3]

[1] Compare DE GROOT, The Religious System of China, Vol. VI, p. 1172.

[2] This is expressly stated also by the K'ien-lung editors of the *Chou li*: by means of the *chên kuei*, the sovereign rules (*chên*) and pacifies the empire.

[3] This entire disquisition of the *Li ki* is highly instructive and of primary importance.   In some ceremonial usages the multitude of things formed the mark of distinction, in others the paucity of things formed the mark of distinction; in others

During the Han dynasty the custom obtained that the jade emblem *kuei*, of a length of one foot four inches, was interred with the sovereign; it was presented with a piece of red cloth three inches square and hemmed on all sides with scarlet silk of red lining (DE GROOT, The Religious System of China, Vol. II, p. 404).

There was a round jade tablet (*yüan kuei*, GILES No. 13724) nine inches long, fastened with a silk band, used, as the *Chou li* says, "to regulate virtue" (BIOT, Vol. II, p. 523), or in another passage, "to call forth virtue and to perfect good sentiments" (Vol. I, p. 491). The commentary explains that this tablet is entrusted to the delegates of the emperor; when a feudal prince shows himself virtuous, this tablet is conferred upon him by imperial order as a reward. The tablet is round, another commentary remarks, having no points, which seems to mean an "all-round" perfection.

In opposition to the round tablet of perfect virtue, there was the "pointed tablet" (*yen kuei*, GILES No. 13073) serving "to change conduct, to destroy depravity" (BIOT, Vol. I, p. 491). The projecting point, remarks the commentary, is the emblem of wrongs and offences, of the attack on and appeal to duty, of blame and punishment; when the emperor orders a dignitary to abandon his bad behavior and to reform, he sends this tablet to reprimand and to warn him. According to another commentary, it is also a tablet of credence for the delegates of the emperor and of the princes; when a prince despatches a prefect to obtain instructions from the emperor, he enjoins on him to take this tablet, and thus to indicate his mission. COUVREUR (p. 433) has figured this tablet with a spiral-shaped cloud-ornament in the upper triangular part and two others placed side by side at the foot. In the K'ien-lung edition of the Rituals, a continuous cloud-ornament covers the body of the *kuei*, while the triangular point is blank. That the ideas of the Chinese regarding this instrument are much confused, is evident from the confounded descriptions given by the two commentators translated by BIOT (Vol. II, p. 524).

A jade tablet called *ku kuei*, "tablet with grains," seven inches long, is offered by the emperor to the woman whom he marries (BIOT, Vol. II, p. 525), *i. e.* it accompanied the bridal presents, and designs of grains were engraved on it. This was not, however, a realistic plant-

greatness of size formed the mark, in others smallness of size formed the mark; in others, the height formed the mark of distinction, in others lowness formed the mark; in others ornament formed the mark, in others plainness formed the mark. This lesson should be taken to heart by our school of evolutionists who construct the development of all human thoughts by means of artificial and illogical evolutionary and classificatory schemes and know everything with dogmatic peremptoriness about thought evolutions, as if they had rendered themselves actual midwifery services at the birth of every thought.

design, but "grains" (*ku*) was merely the name of a geometric orna-ment consisting of rows of small raised dots or knobs which from a supposed resemblance to grains received this name. It occurs also on sacrificial bronze bowls of the Chou period, one of which is in our collection.

The tablets called *chang* (GILES No. 400) will be discussed below in connection with the actual specimens (p. 100).

When the sovereign received the feudal lords in audience, he availed himself of a jade tablet called "cap" (*mao kuei*, BIOT, Vol. II, p. 520). It is described by the commentaries as a sort of cube of jade, each side being four inches in length, an arch-shaped section being cut out on the lower face, in order to indicate "that the emperor's virtue can cover and protect the empire." The feudal lords were supposed to hold their jade insignia of rank in their hands, while the sovereign placed the *mao* over them ("capped" them) to ascertain whether they were genuine. It is mentioned as early as in the *Shu king* (*Ku ming* 23, ed. COUVREUR, p. 356).

The feudal prince of the first rank (*kung*)[1] is invested with the jade tablet called *huan kuei* "pillar tablet" of a prescribed length of nine inches. The traditional representations figure it either as a pointed *kuei* with two vertical lines inside running parallel with the lateral sides, or with a top consisting of a horizontal line with two adjoining slanting lines, as may be seen in Fig. 15.[2] The commentary adds that the feudal lords of the first rank are the great councillors of the emperor and the descendants of the two first sovereigns; the two "pillars" are emblematic of the palace and support it, as the princes support the emperor.

The feudal prince of the second rank (*hou*) is distinguished by the jade tablet *sin kuei*, seven inches long. Figure 16 shows the traditional representation of it, with flat top, while again in the illustrations to the *Chou li* a pointed roof-shaped top appears. But it is noteworthy that in both cases the tablet is unornamented, so that also the K'ien-lung editors had lost confidence in the artificial picture given in the *San li t'u* and identical with that of Couvreur to be mentioned presently. The word *sin* "faith" should be read here *shên* "body," explain the

---

[1] The five degrees of feudal rank (*wu kio*) alleged to have been instituted by the mythical emperors Yao and Shun are called *kung, hou, po, tse, nan*, commonly ren-dered into English as duke, marquis, earl, viscount, baron (W. F. MAYERS, Chinese Reader's Manual, p. 320); but as our own political institutions fundamentally differ from those of the Chinese, such translations are misleading, and I therefore prefer to adhere to the plain terminology introduced by BIOT: prince or feudal prince of first, second, etc. rank.

[2] Derived from the Palace edition of the *Li ki* (1748); in the illustrations of the same edition to the *Chou li*, the same tablet is represented with a pointed top.

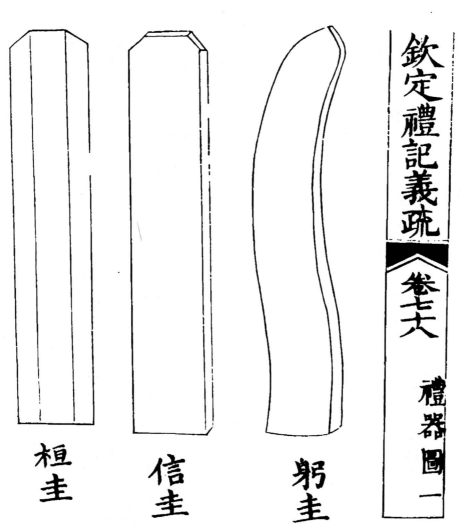

FIG. 15. FIG. 16. FIG. 17.
Reconstructions of the Jade Tablets *huan kuei* (Fig. 15), *sin kuei* (Fig. 16), and *kung kuei* (Fig. 17)
(from the Palace Edition of the *Li ki*).

commentaries, so that BIOT translates *tablette au corps incliné* and COUVREUR *tablette du corps droit*.   But what does that mean?   COUV-REUR gives a reproduction of this tablet on which is engraved the figure of a man with long sleeves standing and holding a tablet in his hands.   If the proposed reading *shên* should refer to this figure, it is just a subsequent and secondary reflection as this figure itself, which certainly cannot be a production of the Chou period, but is also a comment and the outcome of a misled imagination of the epigones.

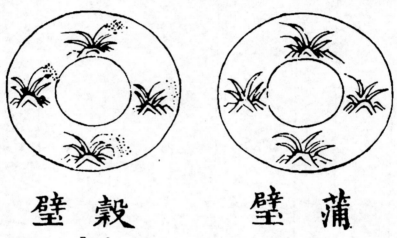

FIG. 18.
Jade Disk with "Grain" Pattern.

FIG. 19.
Jade Disk with "Rush" Pattern.

According to the Notions of the Sung Period (from the Palace Edition of the *Li ki*).

I think the wording of the text simply says what the written symbol implies,— *sin kuei*, a tablet of credence, a badge of trust and confidence.

The feudal prince of the third rank (*po*) is honored with a jade tablet seven inches long, called *kung kuei* "curved tablet."[1]   The word *kung* seems to imply also the idea of submission or subordinance.   Figure 17 shows the conception of this tablet in the K'ien-lung period which seems to come nearer to reality than the figure of the *San li t'u* or *Leu king t'u* reproduced by Couvreur.

While the tablets of the three first feudal ranks belong to the class of *kuei*, *i. e.* oblong, flat, angular jade plaques, those assigned to the fourth and fifth ranks are jade disks or perforated circular plaques

[1] Translated by BIOT (Vol. I, p. 485): *kuei du corps penché* or *tablette au corps droit* (Vol. II, p. 520) with reference to the engraved figure of a man holding a tablet (in COUVREUR, p. 433) which is, of course, a late invention of the Sung period conceived of in justification of this commentatorial explanation.

(*pi*), the one intended for the lord of the fourth rank (*tse*) decorated with a pattern of grain *ku* (Fig. 18), and the other for the lord of the fifth rank (*nan*) ornamented with the emblem of rushes *p'u* (Fig.

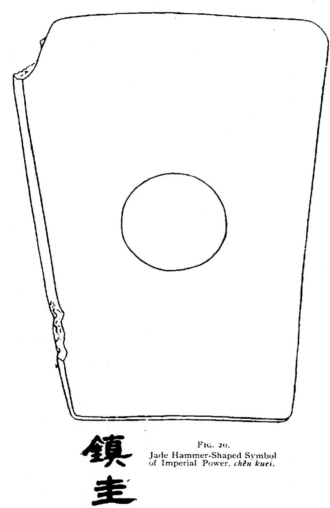

FIG. 20.
Jade Hammer-Shaped Symbol
of Imperial Power, *chên kuei*.

19).[1] The Chinese illustrators of the Sung time represented the former with a naturalistic design of four bundles of grain and the latter with four naturalistic rushes. We explained the grain pattern above; the "rush" pattern was in reality a geometric design consisting of rows of hexagons, as we shall see later on (Fig. 72). We do not treat in this

[1] Also figured in GINGELL, p. 34, and COUVREUR, p. 620.

chapter of these two emblems, but in the section on the jade symbols
of Heaven, as they belong typologically to the series of *pi*.

The main differences of the tablets of rank consisted not only in
their length and in their shape, but also in the quality of the material.
The Son of Heaven alone was entitled to the privilege of using pure
white jade of uniform color, while the princes from the first to the
third rank were restricted to the use of jades of mixed colors (Biot,
Vol. II, p. 521).[1]

Jade tablets were also to be sent along with presents of which six
kinds were distinguished (*leu pi*): these were horses accompanied by
the tablet *kuei;* furs presented with the tablet *chang;* plain silks with
the jade ring *pi;* variegated silks with the jade tube *ts'ung;* embroidered
silks with the jade carving in shape of a tiger (*hu*); and silks embroidered in black and white with the *huang,* a semicircular jade piece.[2]

We have seen that the *chên kuei* or tablet of power was the symbol
of imperial sovereignty. Wu Ta-ch'êng has discovered a specimen
which he believes he is justified in identifying with this object, and which
appears as the first illustration in his book, reduced to $\frac{1}{10}$ of the original.
It is here reproduced in Fig. 20. It is of dark-green jade mottled in
various colors. The small essay dwelling on this interesting object
displays a great deal of acumen on the part of the author and does
much credit to his critical faculty which dares to oppose the sanctioned
interpretations of the past. He takes the statement of the *Chou li*
(Biot, Vol. II, p. 522) as his starting-point and discusses the meaning
of the sentence that the tablet (*kuei*) of the Son of Heaven had a *pi*
in the centre. This word is written in the ancient text in a purely
phonetic way, — as then so often occurred, — with the character *pi*
meaning "to be necessary" (Giles No. 8922), *i. e.* the classifier has
evidently been omitted. The commentator Chêng had proposed to
read *pi* (Giles No. 9001) with the meaning of "fringe, cord." Thus,
also Biot joining this opinion translated that the tablet carried by the
emperor has in the middle a cord called *pi.* Wu rejects this view on
the ground that in the present tablet the perforation has a circumference
of three inches, that from the upper periphery of this circle to the upper
edge there is an interval of four inches and a half, and not quite so much
from the lower periphery down to the lower edge; in other words, the
perforation is situated almost in the centre of the tablet and is much
too large to make it possible to suppose that it might have merely served

[1] The interpretation of this passage by the commentary of Chêng Ngo seems
to me an arbitrary opinion which the wording of the text does not bear out.

[2] The conventional Chinese designs of the first four tablets see in Gingell, p. 38.

for the passage of a cord. In such cases simple small holes are drilled somewhere very near to the lower edge as seen in the following specimens. The present one, however, was not provided with a cord, but seized by the hand. Wu therefore suggests the combining of the word *pi* in the text of the *Chou li* with the classifier 75 for tree or wood, a character (not in Giles) reading also *pi*. This word means, according to the dictionary *Shuo wên*, "a palisade erected by means of bamboo poles," but also a socket, *e. g.* in a spear-head, for the insertion of a wooden handle. It may therefore be conjectured with tolerable certainty that this perforation served the purpose of wedging in a handle. If this was really the case, *i. e.* if this perforation presents an axe-hole, we are further justified in concluding that this instrument either is itself an axe or hammer or at least derives its shape and essential features from the model of an axe, in the manner, so to speak, of a ceremonial survival.

This conclusion of mine is corroborated by the further investigation of Wu. He refers to the definition given in the *Chou li* of "the great tablet" (*ta kuei*) as being three feet long, made sloping (*chu*, GILES No. 2611) in the upper part so as to form a hammer's head (*chung k'uei shou*, see GILES No. 6491, last item); the same definition is repeated in the *Shuo wên* where the *ta kuei* is named also *t'ing*. Wu correctly points out that this condition is fulfilled in the implement under consideration where the upper left corner is chamfered, and that this chamfered portion is identical with the so-called hammer's head.[1] Wu dilates on a discussion of this term which may be safely omitted here, being of philological, not archæological interest, and which would require the reproduction of the text in Chinese characters to become intelligible. He finally reminds us of the so-called medicine-spades (*yao ch'an*) in connection with this hammer. We may safely adopt the sober and judicious result of Wu's investigation which is based on and in full harmony with the statements of the *Chou li*, and which furnishes a satisfactory explanation of the peculiar features of this instrument. While it is not necessary to go so far as to regard it as a real hammer, it shares the essential characteristics of a stone hammer, perhaps with some modifications growing out of its ritualistic purpose. Its identification with the *chên kuei* of the Chou period is further justified in that it agrees in regard to its length with the measure

---

[1] BIOT (Vol. II, p. 522) was unable to render the passage correctly, as he had no knowledge of what this object really was, and as the Chinese drawing, the result of an imaginary attempt at reconstruction, was naturally apt to lead him astray. This figure shows a rectangular wedge with a square knob at the end. It is figured also in DE GROOT (The Religious System of China, Vol. VI, p. 1172) who is inclined to regard this implement as an exerciser of demons in connection with solar worship.

of twelve inches given in the *Chou li* when recalculated on the foot-measure of that period.   The so-called *chên kuei*, the tablet or symbol of imperial power, was accordingly a hammer-shaped implement of jade.

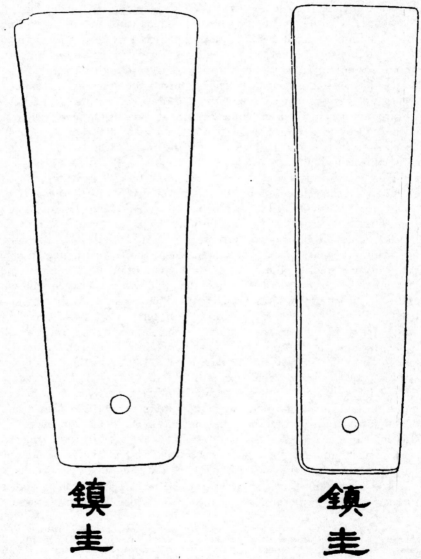

FIG. 21.   FIG. 22.
Jade Hammer-Shaped Symbols of Imperial Power, *chên kuei*.

Wu Ta-ch'êng figures three further specimens of a similar type (Figs. 21–23), also designated as *chên kuei*, but in the latter cases, this identification seems rather doubtful and somewhat hazarded, for the chief characteristics of the former implement are lacking here. None of these objects is provided with the large central perforation, but they have only small holes drilled through the base near the edge, and these could have been utilized for no other purpose than for the passage of a cord. One of them, reproduced in Fig. 22, has on its lower side two pairs of shallow cavities communicating through a passage below the surface, so that a wire or a thread could be drawn through. The same piece entirely lacks the chamfered hammer-head-like portion which is rather weakly developed in Fig. 21; if there it is not due to a mere accidental cause, as the two rounded notches would seem to indicate, while it is plainly brought out with manifest intention in Fig. 23. Whereas it is apparent that these three pieces belong to the type of *kuei*, they cannot be traced back to that of the *chên kuei*. *Wu* evinces a feeling of a similar kind, because also their measures deviate from the standard type of twelve inches. The piece in Fig. 21 is described as being of dark-green jade, that in Fig. 22 of the same color mixed with black spots, and that in Fig. 23 as uniformly red.

The jade object[1] in Fig. 24 is identified by Wu with the *ta kuei* or great tablet which, according to the *Chou li*, was carried by the emperor in his girdle (Biot, Vol. I, p. 484). We have alluded above to the other passage which says that this instrument was to be three feet long, made sloping above so as to form a hammer's head. But the specimen in question is only one foot nine inches long. Wu conjectures that the datum of three feet in the *Chou li* may be a mistake for two feet. Such an error may have, of course, crept into the text, but the attempted amendment will never rise above the degree of a conjecture. In its favor it might be said that three feet seems to represent a considerable length for an object to be placed in the girdle, and that one only two feet long would do much nicer in view of this purpose. Whether this specimen may be justly identified with the *ta kuei* or not, it loses nothing of its great value. It is most interesting in that it shows the four corners chamfered or, according to the Chinese idea, hammer-shaped; that is to say, in this specimen, this feature has developed into a mere ornamental form, void of any practical use. The lower end is apparently shaped into a handle, and two hammer-shaped faces are here part

[1] It is described under the heading: "It has also the name *t'ing;* of dark-green jade with black designs under which two figures like a dragon and phenix seem to be hidden [a natural phenomenon in the stone causing this impression]; below the perforation, there is to the extent of three or four inches a zone of yellow color. The illustration is by $\frac{5}{10}$ smaller than the implement (*i. e.* half of the original size)."

of the handle. This piece, accordingly, represents a much advanced stage of conventionalization compared with the ceremonial hammer in Fig. 20. Its development is conceivable only when it is referred to the latter, and if it is supposed to have been derived from it.

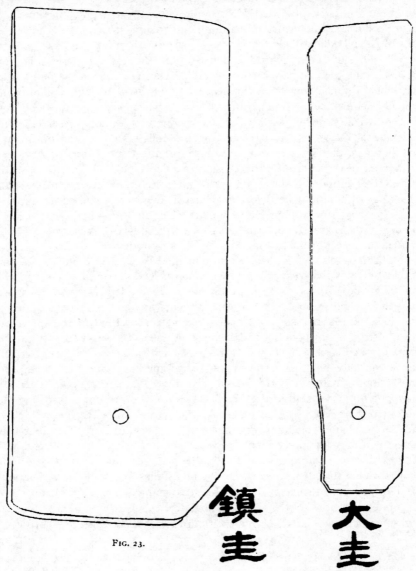

FIG. 23.

FIG. 24.

Jade Hammer and Knife-Shaped Symbols of Imperial Power, *chên kuei* and *ta kuei.*

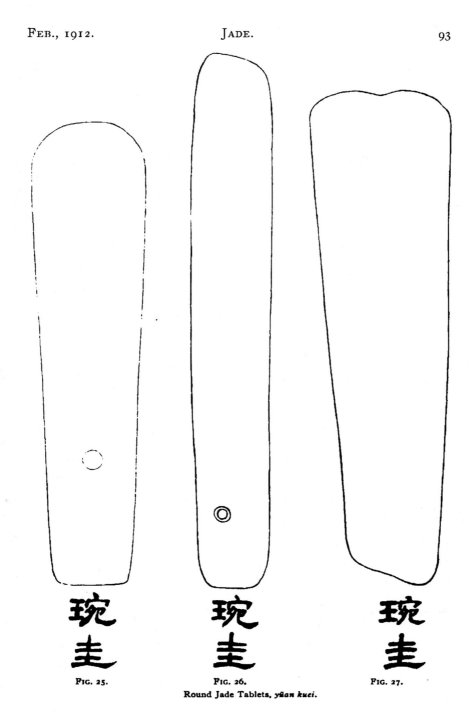

FIG. 25.   FIG. 26.   FIG. 27.
                    Round Jade Tablets, *yüan kuei*.

Viewed in this light, the specimen in Fig. 23 becomes more intelligible; here too the hammer form has sunk into a purely decorative trait, as is obvious from the rectangular shape of the object which is unfit for any pounding. The perforation bears further testimony to its ornamental utilization. While the primeval form in Fig. 20 shows in its outlines the design of a hammer or axe, the pieces in Figs. 21–24 exhibit the rectangular shapes of knives identical with the jade knives described above on pp. 38, 39. These original knife-forms have then been modified in Figs. 23 and 24 under the influence of the imperial hammer-symbol which lent to them the peculiar feature of a chamfered edge or edges. The development then terminated in pieces like Figs. 21 and 22 where this trait has disappeared. Their apparent relation to the imperial hammer is conclusive evidence also of this series having formed symbols reserved to the emperor. This group represented by Figs. 21–24 may therefore be defined as knife-shaped imperial emblems of jade developed by a process of adaptation and conventionalization from the hammer-shaped symbol of sovereign power.

Figures 25–27 represent three specimens of the so-called "round tablets" (*yüan kuei*) which, as we saw from the report of the *Chou li* (p. 83) symbolize virtue by their rounded shape and are bestowed by the emperor upon virtuous vassal princes. The piece in Fig. 25 is of dark-green jade, and twelve inches (modern) long corresponding to nine inches of the Chou time which is the required measure for this tablet. Wu says he acquired it at the bazar of the city of Ts'i-ning in Shantung, and has no doubt of its being an ancient *yüan kuei*. The word *yüan* is explained after TUAN as a mound consisting of two superposed hills (*k'iu*) and identical with the notion of a hillock; this hillock is seen in the rounded top. Wu quotes, besides the commentaries mentioned by us, a passage from TAI TÊH, the author of a Ritual known under the name *Ta Tai li* ("Ritual of the senior Tai") as saying "that the points in all tablets called *kuei* are one inch and a half long; if this point is made level by angular measurement to form just a true right angle, the *yüan kuei* rises with its lofty height." This quotation shows how fond the Chinese mind of the Chou period was of geometrical constructions and geometric symbolism, as the round line could be symbolic of the perfection of qualities. The virtue-emblem in Fig. 26 is of similar shape, of dark-green jade with earth spots, reduced to $\frac{7}{10}$ by Wu, and therefore appears somewhat larger than the other; that in Fig. 27 is of red jade, a piece being broken off below, but otherwise a large *yüan* of twelve inches.

In Fig. 28, Wu has illustrated, without further explanation, a type

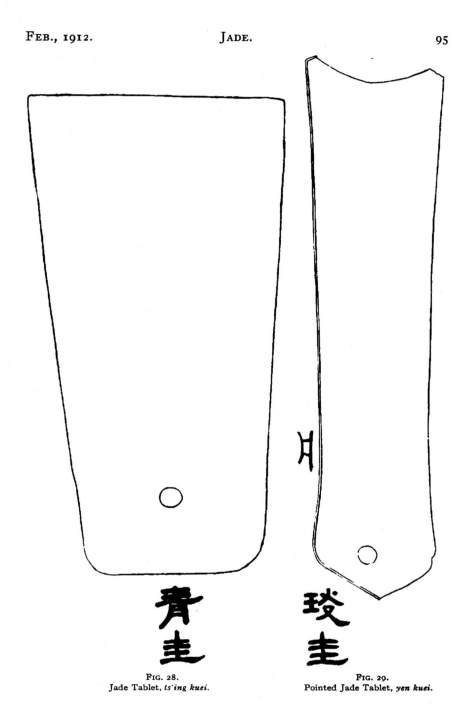

FIG. 28.
Jade Tablet, *ts'ing kuei*.

FIG. 29.
Pointed Jade Tablet, *yen kuei*.

called *ts'ing kuei*, "dark or green tablet," of dark-green jade, reduced to $\frac{9}{16}$; its shape is evidently derived from that of an axe.[1]

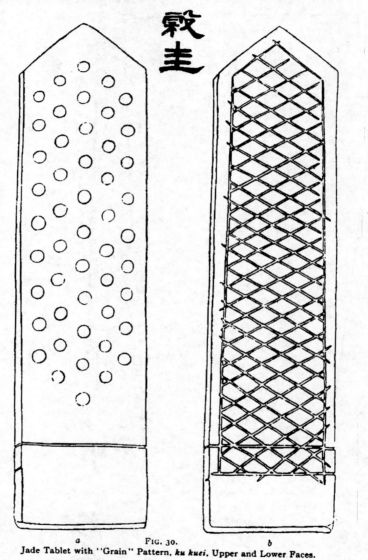

a   FIG. 30.   *b*
Jade Tablet with "Grain" Pattern, *ku kuei*, Upper and Lower Faces.

In Fig. 29 the *yen kuei* or pointed tablet, characterized above (p. 83) according to the *Chou li*, is reproduced. "It is of black jade, $\frac{4}{5}$ of the

[1] Compare the chapter Jade in Religious Worship, p. 172.

original size (in Wu's reproduction), above made into the shape of a half-circle, the two horns a bit broken." The author opens the explanatory text by saying that the color of jade is uniformly black, *i. e.* not mixed with any other color, which is not identical with what is usually designated as saturated with mercury. *Wu* develops a peculiar view in regard to the origin of the crescent-shaped upper edge which he sets in relation to the ancient form of a written symbol for a lance-head (reproduced to the left of Fig. 29), and remarks, that, while all other *kuei* derive their form from a hammer-head, this is the only one to derive it from a lance-head. This assumption, however, is not forcible and finds no echo in any ancient document; while, as the other *kuei*, this one also goes back to the same original form of a weapon, the peculiar features of its shape can be sufficiently explained as a geometrical construction based on the curious symbolism connected with it.

FIG. 31.
Jade Tablet *kuei* on a Han Bas-Relief (from *Kin-shih so*).

In the jade tablet of Fig. 30 (of green jade with black stripes) Wu believes to recognize the "tablet with grain-pattern," mentioned above after the *Chou li*, which the emperor bestowed on his bride elect. The upper face of this tablet (Fig. 30 *a*) is decorated with five vertical rows of raised knobs, — five being the number of the earth, — arranged in the numbers 9, 10, 11, 10, 9, yielding the sum of 49, by which also a symbolism was presumably expressed. The lower face of the tablet

(Fig. 30 *b*) is ornamented with a diapered pattern of lozenges.   It will
be noticed that the shape of this tablet *kuei*, — an oblong rectangle

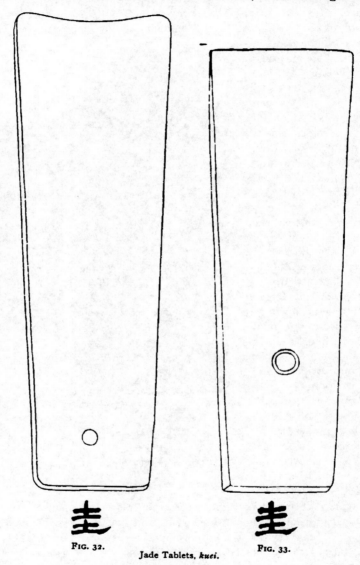

FIG. 32.   FIG. 33.

Jade Tablets, *kuei*.

surmounted by a roof, — agrees with the familiar traditional form
under which the *kuei* has been handed down in the later illustrations
of the Rituals and as a frequent emblem in the hands of Taoist deities.

The best idea of the authentic shape of this jade tablet *kuei* will be conveyed by the engraving on stone preserved on the Han bas-reliefs of Wu-liang (Fig. 31, reproduced after the *Kin-shih so*) where it appears among "the marvellous objects of good omen" (*fu jui*) accompanied by the inscription, "The black tablet: when the rivers and sources flow and reach the four oceans in such a way that all waters communicate, then it will appear." This sentence is doubtless prompted by the alleged black jade tablet of the Emperor Yü (*Shu king, Yü kung* 38, ed. COUVREUR, p. 89), by means of which he is said to have obtained control over the waters of the flood. The extraordinary feature of this tablet is only the black-colored jade and perhaps also its ornamentation, — a network of lozenges with dots in the centres. The editors of the *Kin-shih so* refer to its identity with that on the *pi-liu-li* (Fig. 39). They call it *sie tou* "slanting bushels" (*tou*, a measure of capacity). The shape of this tablet is presumably derived from that of a spear.

Pan Ku, the author of the *Po hu t'ung*, who died in 92 A. D., remarks that the tablet *kuei* is pointed above and angular below, and that the tapering part means the male principle *yang*, and the lower square portion the female principle *yin*. This may hint at a possible original phallic significance of this emblem, and such a supposition would be corroborated by the two facts that the *kuei* with the grain emblem is conferred by the emperor upon his consort, thus perhaps alluding to their progeny, and that "the green kuei" is utilized as a symbol in the worship of the East corresponding to the spring (BIOT, Vol. I, p. 434).[1]

There were also other shapes of *kuei*, as we see from the definition of the *Shuo wên* that they are round above and angular below, and from the two specimens of WU in Figs. 32 and 33, both simply designated as *kuei*, no further explanations being given. In Fig. 32, the upper edge is concave, and in Fig. 33 it forms a straight line. It seems likely that also these tablets are derived from knife-shaped implements.

This type *kuei* is doubtless very ancient and may go as far back as the Shang dynasty. At least, we hear in one report of jade tablets buried in the grave of Pi Kan (twelfth century B. C.).

In the *Mo chuang man lu*[2] it is on record: "In the period Chêng-ho

---

[1] On traces of phallicism in China see E. C. BABER, Travels and Researches in Western China, p. 19 (*Royal Geographical Society, Supplementary Papers*, Vol. I, London, 1886).

[2] In ten chapters, written by CHANG PANG-KI about the middle of the twelfth century. It contains a large collection of facts, supplementary to the nat'onal records; and although some incredible marvels occasionally find a place in the course of the work, there is much to establish the author's reputation for depth of research and penetration (A. WYLIE, Notes on Chinese Literature, p. 164). The work is reprinted in the collection *Shuo ling* (30 Vols., 1800) a copy of which was procured by me for the John Crerar Library of Chicago (No. 119).

(1111–1117 A. D.),[1] the Imperial Court was anxious to hunt up bronze vessels of the San-tai period (the three dynasties Hia, Shang, and Chou). Ch'êng T'ang who was Collector of Taxes on Tea and Horses (*T'i tien ch'a ma*) of Shensi, and Li Ch'ao-ju who was Intendant of Grain (*Chuan yün*) of Shensi, despatched a man to Fêng-siang fu (Shensi Province) to break up the grave of Pi Kan of the Shang dynasty[2] where he found a bronze plate of more than two feet in diameter, with an inscription consisting of sixteen characters on it. He further found jade slips, forty-three pieces, over three inches in length, in the upper part rounded and pointed (*i. e.* shaped like a Gothic arch), in the lower part broad and angular; they were half a finger in thickness. The color of the jade was bright and lustrous. These objects were all widely different from those buried as substitutes for human sacrifices (*sün tsang*)."

Under the name *chang* (GILES No. 400) a series of jade tablets is comprised which are explained as representing the half of the tablet *kuei* divided in its length from top to bottom.

The tablet *chang* is mentioned as early as in the *Shi king* (ed. LEGGE, Vol. II, p. 306; ed. COUVREUR, p. 223): "On the birth of a boy, a jade tablet will be given him to play with," as an emblem of the dignity with which he is hoped to be invested when grown up. Thus, the word *lung chang* "one playing with the *chang*" has come to assume the meaning of a new-born son. The girl, according to the same song, will receive a tile, "the emblem of her future employment when, on a tile upon her knee, she will have to twist the threads of hemp" (LEGGE). Prof. GILES, alluding to this passage in his valuable notes on "Jade" (*Adversaria Sinica*, No. 9, p. 312, Shanghai, 1911) justly comments on its sense as follows: "It has been too hastily inferred [from this passage] that the Chinese have themselves admitted their absolute contempt for women in general. Yet this idea never really entered into the mind of the writer. The jade tablet, it is true, was a symbol of rule; but the tile, so far from being a mere potsherd implying discourtesy, was really an honorable symbol of domesticity, being used in ancient times as a weight for the spindle."

In the *Chou li* (BIOT, Vol. II, p. 525), three kinds of *chang* are distinguished regulated according to length at nine, nine and seven inches, respectively, and called great (*ta*), middle (*chung*), and side (*pien*). According to K'ung Ying-ta, as quoted by BIOT, these three *chang* are

---

[1] The art-loving Emperor Hui Tsung reigned at that time (1101–1125).

[2] Put to death by the last emperor of that dynasty. See CHAVANNES, Se-ma Ts'ien, Vol. I, pp. 199, 203, 206–7. *Se-ma Ts'ien* (also *Shu-king*, Ch. *Wu-ch'êng*, 3) informs us that a tumulus was placed over the grave of Pi Kan, so that it was very possible to identify his burial-place. A photograph of this tomb is reproduced in the journal *Kuo suei hio pao*, Vol. V, No. 1.

in reality basins to hold aromatic wine, but another opinion is expressed by Chêng K'ang-ch'êng who says: "For the great mountains and rivers, the great *chang* is used, with the addition of ornaments; for the middle mountains and rivers, the middle *chang* is used, with proportionately smaller ornaments; for the small mountains and rivers, the *pien chang* was used, with only a half decoration, — which may be recognized from the fact that only the upper half of the *chang* is carved with ornaments."

The jade tablet in Fig. 34 has been identified by Wu with this tablet *chang*, and it is possibly a *pien chang*. The linear ornaments in the triangular portion are engraved only on the upper face, the lower face being plain. It is stated to be of green jade with russet speckles.

The *Chou li* (Biot, Vol. II, p. 527) mentions a jade tablet called *ya chang*, *i. e.* the tablet *chang* with a tooth which serves to mobilize troops, and to administrate the military posts. Hence, says Wu Ta-ch'êng, its form has been chosen from among military weapons, and its head resembles a knife, while its long sides have no blade; it usually goes under the name of "jade knife" (*yü tao*); in all *kuei* and *chang*, the two sides are straight whereas this one is the only type with sides curved like teeth, and hence the name. Figure 35 illustrates the specimen thus described which is of dark-green and white jade; it has the shape of a knife with sloping upper

Fig. 34.
Jade Tablet, *chang.*

and lower edges and a pierced handle clearly set off from the body. The commentator Chêng Se-nung remarks: "The *ya chang* is carved into the shape of a tooth; teeth symbolize warfare, and hence troops are levied by means of this instrument, for which purpose bronze tallies in the shape of a tiger are used at present."[1]

In the *Chou li*, two kinds of these *chang* are distinguished, the one being the *ya chang* mentioned, the other being called *chung chang*, *i. e.* middle *chang*, both of the same length of nine inches (seven for the body and two for the point or tooth). The present specimen in Fig. 35, however, is seventeen inches and a half long, so that Wu, because of this discrepancy, suspects that it is an object of the time posterior to the Eastern Chou, *i. e.* of the time in the latter part of the Chou dynasty reckoned from B. C. 722. The Chinese illustrators like Nieh Tsung-i of the Sung period, the author of the *San li t'u*, have depicted this instrument in the form of a rectangle with a sloping row of seven saw-like teeth in the upper edge, taken in by the explanations of commentaries speaking of a plurality of teeth which, as we noted, exists in our specimen, but which must not be imagined in a continuous row. Wu's specimen is a palpable and living reality, NIEH's drawing a fantasy corresponding to no real or possible object.

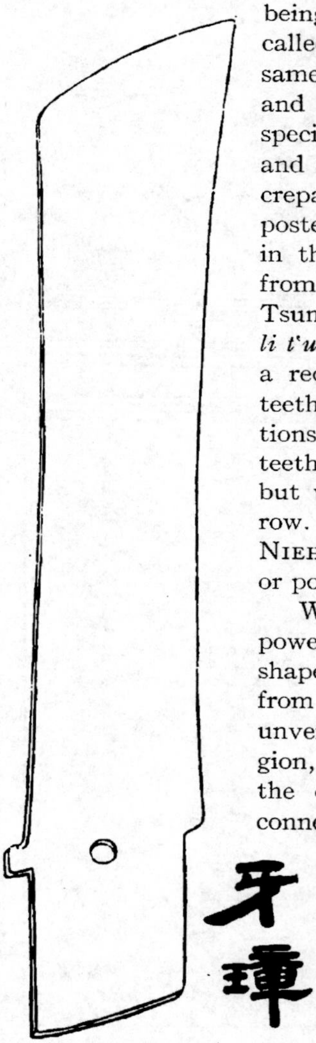

We see that all these jade symbols of sovereign power are imitations of implements and derive their shapes from hammers and knives, possibly also from lance and spear-heads. This investigation unveils a new aspect of most ancient Chinese religion, for I concur with Prof. DE GROOT (*l. c.*) in the opinion that these emblems were originally connected with some form of solar worship. In the Chou period, only slight vestiges of this cult appear on the surface; the emblems themselves had then sunk into a mere conventional and traditional, nay, an hieroglyphic use within the boundaries of the rigid official system. The geometric forms and lines of these emblems were curiously explained in the sense of sober moral household maxims as they befitted a patriarchal government based on an ethical state policy. These rational reflections growing out of Confucian

FIG. 35.
Jade Tablet, *ya chang*.

[1] *I. e.* in the Han period.

state wisdom can certainly lay no claim to representing an original or primeval stage of symbolism. On the contrary, they indicate a recasting of primeval ideas of a remote antiquity into the particular mould of the spirit of the Chou time. In the same manner, as these ceremonial insignia point back to primitive implements from which they were developed, so also the ideas associated with them in the age of the Chou point to a more rudimentary and elementary form of symbolism and worship. The Chou emperor worshipped the sun by holding in his hands the hammer-shaped jade symbol of sovereignty. This means, in my opinion, that at a prehistoric age a jade (or perhaps common stone) hammer was regarded as the actual image of the solar deity worshipped by the sovereign, and I believe that the burial of jade implements in the Chou period, discussed in the previous chapter, was as a last survival also connected with this ancient cult of the sun. We thus find in prehistoric China the same condition of religious beliefs as is pointed out for prehistoric Europe.

Sophus Müller (Urgeschichte Europas, p. 151, Strassburg, 1905) sums up as follows: "On Crete, the axe was worshipped; it was not a symbol, but the direct image of the deity; supernatural power resided in it. The same ideas must have obtained also in other parts of Greece and the rest of Europe; in Italy and Scandinavia, there are stone and bronze hatchets, either too small or too large for real use; in Scandinavia, there are large and small hatchets of amber; in the French stone chambers or on stones freely exposed, axe-blades and shafted axes are carved in. Everywhere, the axe appears, but not a deity holding it, as in the subsequent mythologies. The sun was worshipped as a deity; the round disk appears carved on stone slabs in Scandinavia, England and Ireland, as it plays a rôle in religious representations also in the south and in the orient. The personal solar god, however, arises in Greece but late in the last millennium B. C., and from the northern regions of Europe we are ignorant of an image of him. The fact that originally an impersonal solar deity was adored is best confirmed by the bronze solar disk on a chariot drawn by a horse from Nord-Seeland (Denmark)."

III. ASTRONOMICAL INSTRUMENTS OF JADE

In the second chapter of the *Shu king* (*Shun tien,* 5), it is said with regard to the mythical Emperor *Shun* that he examined an instrument called *süan ki yü hêng* of which he availed himself to regulate the Seven Governors (*i. e.* the sun, the moon, and the five planets).[1] This passage has therefore been understood by Chinese and foreign commentators in the sense that Shun employed a kind of astronomical instrument manufactured, as inference from the name allows, of jade; for the first word in the compound *süan* (GILES No. 4813) is interpreted as designating a kind of fine jade; *ki yü hêng* is, according to WU TA-CH'ÊNG, the designation for the astronomical instrument, so that the translation of the name would be "the astronomical instrument *ki yü hêng* made of the jade *süan.*" Others, however, present the opinion that only the word *ki* signifies an instrument, and that the term *yü hêng* means a part of this instrument itself, taken literally in the sense of a piece or tube of jade placed crosswise over the machine. It is undoubtedly this literal interpretation which has given impetus to the later conception of this instrument as of a regular armillary sphere, and which has resulted in the reconstruction of an elaborate figure repeated in many Chinese books on astronomy reproduced in COUVREUR'S edition of the *Shu king,* p. 15. This is a complex apparatus of spheroid shape representing the celestial sphere with the equator. We need hardly insist on the fact that this Chinese illustration is simply a reproduction of the armillary sphere constructed as late as in the Mongol period of the thirteenth century,[2] and that it cannot be adduced as evidence for the supposed astronomical instrument of the ancient legendary Emperor Shun. At the outset, it is most unlikely that such a complicated machine should have been constructed at that mythical age.

Our Wu Ta-ch'êng is doubtless more fortunate in identifying the instrument *süan-ki* with a perforated disk of jade described by him as "white interspersed with russet spots." As will be seen from Fig. 36 reproducing this piece, the outer edge of the ring is very curiously shaped and divided into three sections of equal length marked off by a deep incision forming a pointed angle on the inner and a pointed projec-

[1] Compare SCHLEGEL, Uranographie chinoise, p. 504.

[2] Compare A. WYLIE, The Mongol Astronomical Instruments in Peking, pp. 5 *et seq.* (in his *Chinese Researches*) and Plate A.— Chang Hêng (78–139 A. D.), an eminent astronomer and mathematician of the Han period, is said to have constructed an armillary sphere (GILES, Biographical Dictionary, No. 55).

104

tion at the outer side. Each of these three divisions is indented in such
a way that six small teeth of irregular shape[1] project over the edge leaving five slightly curved notches in their interstices. It will further be
noticed that these three divisions are treated much alike in their measurements, that the protruding teeth and their interstices bear the same

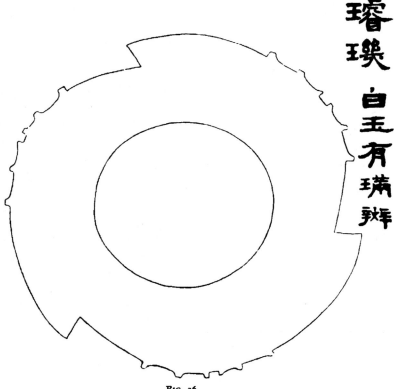

FIG. 36.
Jade Astronomical Instrument *süan-ki*.

shape and measurements in each section, and that the total of teeth is
arranged in the same distance from the ends of each section which on
the one side is double that of the other side. As number and measurement plays such a significant part in all these ancient jade pieces, we
are justified in concluding that also here a well meditated symbolism
is intended, that the main division of the circle into three parts, the
six teeth and the five notches as well as their regular formations and the
regular proportions of all parts must have a peculiar meaning. Each

[1] WU TA-CH'ÊNG calls them *kua* (GILES No. 6288) "nocks of an arrow."

division, it will be seen, consists of altogether seven lines which could be interpreted as symbolical of the Seven Governors, *i. e.* the sun, the moon, and the five planets, to which the instrument employed by Shun, according to the passage in the *Shu king*, was devoted. In this case, I should like to regard the longest line as the symbol of the sun and the line opposite this one, on the other side of the section, being half the length of the sun-line, as emblematic of the moon, while the five indentations between might denote the five planets. The question would now be, — what is the triple repetition of this design to mean? But not being an astronomer, I do not feel like embarking on this problem and must leave its solution to specialists in this field; perhaps M. L. de Saussure will find here a welcome task for the exercise of his acumen.

Wu Ta-ch'êng has illustrated also the lower side of this instrument (Fig. 37) which is altogether identical with its upper side, except that there is a fourfold division of the circle by lines into four parts of different size. The author has made no statement regarding this feature; it can hardly be otherwise than that these lines are not accidental or natural veins in the stone, but saw-marks originating from cutting the disk out of the living stone into its present shape.

From CHAVANNES' translation of Se-ma Ts'ien's Annals (Vol. I, p. 58) it will be seen that SE-MA TS'IEN has developed a fundamentally different view of the above passage of the *Shu king*. He thinks that the seven stars of the Great Bear are here involved, and is no doubt prompted to this view by the fact that the expression *yü-hêng* denotes also the star Alioth ε or the three stars *piao* in the Great Bear; he would mean to say that Shun observed the seven stars to determine the seven domains on which they exert their influence, *i.e.* the four seasons, the movements of the astral bodies, the configuration of the earth, and the conduct of man. This explanation is far-fetched and artificial, and besides, contradicts Se-ma Ts'ien's own view of the *Ts'i chêng* or Seven Governors which are, as expounded in his Chapter XXVII (CHAVANNES, Vol. III, p. 339) the sun, the moon, and the five planets.

I should add that the symbolic interpretation of the instrument as given above is my own, and not that of Wu Ta-ch'êng, and certainly remains hypothetical; it pretends to be nothing more than a suggestion. The Chinese scholar goes only so far as to assert that, owing to its peculiar numerical divisions, this jade disk could have been utilized for celestial observations, and that he supposes it was an astronomical instrument (*hun t'ien i*), of whose proper significance, however, we are ignorant, as the tradition concerning it has been lost. Neither Wu Ta-ch'êng is, nor am I, of course, naïve enough to believe that the specimen in question is identical with just the one employed by the Emperor Shun, and the quotation from the *Shu king* is to us merely a vehicle of interpretation. Wu Ta-ch'êng remarks in regard to the age of this piece: "Although it is not an object of the Hia dynasty, it is, as

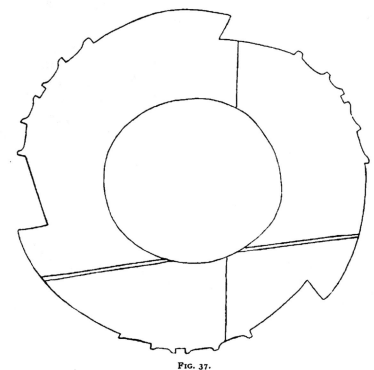

FIG. 37.
Lower Face of Astronomical Instrument in Fig. 36.

shown by an examination of its make-up, not far off from the days of antiquity."

In this connection we should consider also the next object depicted by Wu Ta-ch'êng and here reproduced in Fig. 38, entitled by him *I yü*, *i. e.* Jade of the Tribes called *I*; this heading is followed by the words: "Some call it *pi-liu-li;* in its make-up, it is identical with the *süan-ki.*"

As the discussion of Wu Ta-ch'êng added to this specimen is of particular interest, I let a literal translation of it follow:

"The piece in question is a ring (*huan*); the color of the jade is yellow like gold, and bright like amber. It is not met with in present collections, but a treasure of greatest rarity; judging from its make-up, it is also of an extraordinary age. In the convex and concave parts (*i. e.* the indentations) of the edge, clayish spots are still preserved,[1] so that it is decidedly not an object posterior to the period of the *San tai* (Hia, Shang and Chou dynasties). It is an ancient *sün-yü-k'i* (GILES No. 4873). The chapter *Ku ming* of the *Chou shu* in the *Shu king* (ed. COUVREUR, p. 352) contains the two terms *ta yü* 'large jade' and *I yü* 'jade of the tribes *I*', commenting on which WANG SU says:

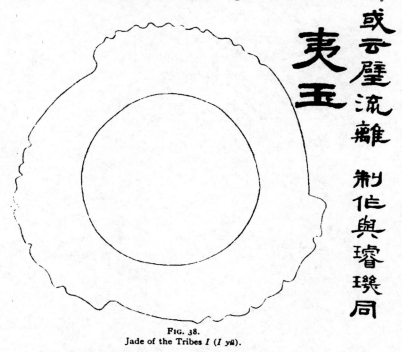

FIG. 38.
Jade of the Tribes *I* (*I yü*).

'*I yü* is the finest jade of the *I* tribes of the east.' CHÊNG K'ANG-CH'ÊNG says: '*Ta yü* 'the large jade' is the precious stone (*k'iu*, GILES No. 2321) of the Hua shan (the sacred mountain in Shensi); *I yü* 'the jade of the *I* tribes' is identical with the stone *sün-yü-k'i* of the north-east.' The dictionary *Êrh ya* explains it as the most beautiful stone of the eastern region, and speaks of the *sün-yü-k'i* of the *I-wu-lü*, to which the commentary of KUO adds that *I-wu-lü* is the name of a mountain which is situated in what is called at present *Liao tung*, and that *sün-yü-k'i* belongs to the class of jade. The dictionary *Shuo wên* notes under the heading *sün* the expression '*sün-yü-k'i* of *I-wu-lü*.' The jade of the *I* tribes mentioned in the *Chou shu* is called by some also a jade vessel (or implement, *k'i*).

[1] See Introduction, p. 27.

"When, following an imperial edict, I, Ta-Ch'êng, proceeded to the province of Kirin,[1] and passed through *Kuang-ning hien* in *Kin chou* in *Fêng t'ien*, I obtained a piece of jade produced in the *I-wu-lü* mountains.[2] It was cut and polished into the shape of a girdle pendant, in size not exceeding an inch. I confess I have not yet seen such big ones. The common name is 'stone of *Kin chou*.' It is not very expensive or esteemed. The jade substance in the ring under consideration is similar to the *Kin chou* stone. There are especially differences between the old and the modern ones: if it has lain underground for a long time, the color receives a moist gloss and reflects under the light. Truly it is an unusual kind of jade.

"Others assert the identity of this specimen with the ancient *pi-liu-li*. The dictionary *Shuo wên* says: 'Those of the *ya-shih* (*ya* stones) possessed with a lustre are called *pi ya;* they are produced in the country of the Western Turks (*Si Hu*).'[3] The Commentary of TUAN remarks: '*Pi-ya* is identical with *pi-liu-li*.' The *Ti li chi* says: 'The bright beads for sale at the sea-ports are *pi-liu-li*.'[4] The Account of the Western Regions (*Si yü chuan* in the *Ts'ien Han shu*) says: 'The country of Ki-pin (Cashmir at the time of the Han) produces *pi-liu-li*.' On the stone bas-reliefs of the ancestral hall of Wu-liang of the Han dynasty, the motive of the *pi-liu-li* appears, with the inscription: "When the sovereign does not commit secret faults, it will arrive."

Then follows the well-known passage from the *Wei lio* regarding the *liu-li* of ten different colors of *Ta Ts'in*. The author continues: "TUAN (the commentator of the *Shuo wên*, who lived 1735–1815) says: 'The three characters composing the word *pi-liu-li* form the name which is derived from the (or a) language of the *Hu*, in the same way as the word *sün-yü-k'i* comes from the language of the *I* tribes.[5] The people of the present time have, in their provincial speech, altered the

---

[1] Probably in 1884, when he was sent to Corea as commissioner. See Introduction, p. 12.

[2] A range of mountains stretching west of Mukden (see D. POSDNEYEV, Description of Manchuria, Vol. I, pp. 132–133, St. Pet., 1897, in Russian; and CHAVANNES, Voyageurs chinois, *Journal asiatique*, 1898, p. 408, Note).

[3] *Pi-ya-se* (GILES No. 9009, a kind of cornelian) is rendered in the "Dictionary of Four Languages" by the Emperor K'IEN-LUNG (Ch. 22, p. 66) into Manchu *langca*, Tibetan *nal*, and Mongol *nal ärdäni*. ABEL-RÉMUSAT (Histoire de la ville de Khotan, p. 168) translates this word "rubis balais" (balas ruby) and derives the Chinese name from *balash* or *badakhsh*, to be traced to the name of the country of Badakshan. It is apparently a Turkish word. *Nal* and *langca* go back to Persian *lal*, the balas ruby. The stone *pi-ya-se* is used on the sable caps of all ordinary imperial concubines, while those from the first to the fourth rank are privileged to wear a Japanese pearl (*tung chu*), as is recorded in the "Institutes of the Present Dynasty" (*Ta Ts'ing hui tien t'u*, Ch. 43, p. 5).

[4] *I. e.* the stone *vaidūrya* (and not precious rings of glass, nor glass, as has been translated; see farther below).

[5] Probably a Tungusian language.

two characters *liu-li* (GILES No. 7248, p. 908 b) into *liu-li* (GILES No. 7244). Thus, the ancients pronounced in their provincial speech *pi-ya*. At present, they are seldom seen in China, though in Turkistan (*Si yü*) they may be still constantly met with. Hence, at the time of the Han, they were considered as 'marvellous objects of good omen.' This is, however, a subject not easily settled, but one awaiting the investigation of a widely read scholar."

FIG. 39.
The Ring *pi-liu-li* (from *Kin-shih so*).

This discussion opens a wide perspective into many archæological questions. It is evident that our author is right in pointing to the similarity in the construction of this piece with the preceding specimen *süan-ki;* we have here the same division into three sections and a similar arrangement of notches in each, but they are irregular and differ in number in each section, so that this object may be different after all, or may have been employed for another purpose.

The identification with the *I yü* rests on the ground that the jade of the mountain I-wu-lü was recognized to be of a substance similar to that in this ancient jade ring; we have no means to check this statement, though it may very well be so. At all events, this is more plausible than to regard the object as an ancient *pi-liu-li*. The only representation of this ring appears on the Han bas-reliefs of Wu-liang (Fig. 39), where it is a plain ring without notches, identical in shape with the common jade rings called *pi*, and engraved with dotted squares (*lo wên* "net-pattern"). Wu Ta-ch'êng's disquisition, however, is apt to solve another problem. We see that he takes this word as a unit derived from a foreign language, and as denoting a precious stone, but not as meaning glass, as believed heretofore by foreign writers. The translation of *pi-liu-li* "the precious ring *liu-li*" (the latter taken in the sense of "glass") [1] which, for the rest, would be rendered in Chinese as *liu-li pi*, cannot, therefore, be accepted; the word as a whole is apparently a derivation from and phonetic transcription of Sanskrit *vaiḍūrya* "beryl" or "lapis lazuli" (EITEL, Handbook of Chinese Buddhism, p. 191; JULIEN, Méthode, No. 1374; F. PORTER SMITH, Contributions towards the Materia Medica of China, p. 129). The Buddhist transcription as given by EITEL and JULIEN goes to show that the writing of the syllable *pi* was arbitrary and originally conveyed no meaning; if the word *pi* "jade ring" was substituted for it, this was a process of adaptation of which there are many other examples.

[1] Given by CHAVANNES, La sculpture sur pierre en Chine, p. 34. It seems to me that it is impossible to assume that the first objects in glass known to the Chinese were shaped like rings; for these rings, as represented on the bas-relief and handed down to us in specimens of jade, are things essentially Chinese, of an ancient indigenous form, which does not occur in the west. If the first objects of glass were imported into China from the west, how should it have happened that they were shaped into a Chinese form? This militates against the opinion that the *pi-liu-li* on the Han bas-relief is supposed to be of glass; it is a ring either of a highly prized kind of jade or of beryl. It is gratifying to see that CHAVANNES in his study Les pays d'occident d'après le Heou Han Chou (*T'oung Pao*, 1907, p. 182) has now adopted also the view that the word *pi-liu-li* is to be regarded as a unit and to be traced back to Sanskrit *vaiḍūrya;* but I do not believe that the latter designated the stone *cat's eye*, at least not in this early period. This opinion goes back to the small treatise of Narahari (edited by R. GARBE, Die indischen Mineralien, p. 85); there, the word *vaiḍūrya* appears as one of the many designations or attributes of the cat's-eye, but not as the one exclusive name of it. Further, this work is by no means authoritative, but contains a good many errors, and above all, it represents a recent production, not written earlier than the beginning of the fifteenth century, as Prof. Garbe has been good enough to write me on May 22, 1911; his former calculation dating this work between 1235 and 1250, remarks Prof. Garbe, was due to an error. Hence, the book of Narahari cannot be quoted as an authority holding good for matters relative to the first centuries of our era. The cat's eye is always called in Chinese *mao-tsing* 'cat's essence' (see *Ko chih king yüan*, Ch. 33, p. 3), and there was no reason to adopt the word *pi-liu-li* with this meaning. It would lead us too far to demonstrate here that it is in many cases just the Chinese and Tibetan terms of precious stones and other minerals which are apt to shed light on the definitions of the corresponding Sanskrit words. How could the Tibetan authors distinguish blue, green, white and yellow *vaiḍūrya*, if the word should denote the cat's-eye?—The mystery why this word has been referred also to the cat's-eye is easily solved by consulting the mineralogists on this subject. Cloudy and opaque specimens of chrysoberyl often exhibit in certain directions a peculiar chatoyant or opalescent sheen similar to that of cat's-eye (quartz-cat's-eye), only usually much finer. This variety of chrysoberyl is known to mineralogists as cymophane, and to jewelers as chrysoberyl-cat's-eye, oriental cat's-eye, Ceylonese cat's-eye, more briefly as opalescent or chatoyant chrysoberyl, or simply as cat's-eye (MAX BAUER, Precious Stones, p. 302). And on p. 304: As these stones are frequently referred to in descriptions of the precious stones of Ceylon simply as cat's-eye, it is often impossible to decide whether chrysoberyl or the variety of quartz, also known as cat's-eye, is meant. See also E. W. STREETER, Precious Stones and Gems, Part II, p. 40.

I fail to see that the word *liu-li* has generally had the meaning of glass which was and is called *po-li*. Special inquiries were made by me in regard to this subject in the glass factories of *Po shan* in Shantung Province where the word *liu-li* with reference to glass, whether translucid or opaque, is entirely unknown; glass is called *po-li*, and strass or colored glass *liao* (GILES No. 7070), while *liu-li* refers only to ceramic glazes, as *e. g. liu-li wa* means glazed pottery, and the use and distinction of these three terms is uniform all over northern China. [1]

I concur with HIRTH (Chinesische Studien, p. 63) in the view that the word *po-li* is traceable to Turkish *bolor*. But then it is impossible to identify the word *pi-liu-li* also with *bolor*, for *liu-li* and *po-li* are two different words, and, as admitted also by HIRTH, two different articles or substances; consequently, the two words cannot be credited with the same etymological origin. This subject cannot be pursued any further in this connection, as lying outside of the pale of this publication; I must be content with these indications which will possibly lead to a revision of the history of glass in China. Nothing could induce me to the belief fostered by Prof. Hirth that the Chinese with their ever vital instinct for the value of natural products and with their keen sense of trade should ever have been so unsophisticated as to mistake colored glass beads for precious stones, and to honor them with exorbitant prices; no child in China could be enticed into such a game, and the most confiding and optimistic mind can see or feel the difference between glass and stone. Certainly, the *liu-li* looked upon by the Chinese as precious stones have been so indeed; they were *vaiḍūrya*, whatever species of precious stone this word may have served to denote. Wu Ta-ch'êng's discussion goes to prove that it is also the judgment of modern Chinese scholars that the *pi-liu-li* of old was not glass or strass, but a precious stone.

Another astronomical instrument is mentioned in the *Chou li*, but no specimens of it have survived from which we could form any definite ideas. This is the *tu kuei, i. e.* a jade piece for measuring, which, according to the *Chou li* (BIOT, Vol. II, p. 522), should be one foot five inches in length and serve "to determine the point where the sun comes, and to measure the earth," *i. e.* it was an instrument to measure the length of the solar shadow which is said to have been one foot and a half at the summer solstice and thirteen feet at the winter solstice, the gnomon having eight feet (compare BIOT, Vol. I, pp. 200–204, 488).

I do not mean to deal in this connection with the jade figure known

---

[1] Also BUSHELL (Chinese Art, Vol. I, p. 61) understands, and I believe correctly, that what the artisans from the kingdom of the Indoscythians taught in China early in the fifth century was the art of making different kinds of colored glazes (*liu-li*).

as the "south-pointing chariot," as this subject has been amply treated by such scholars as Hirth, Giles and Bushell. The following references may be useful to readers interested in this topic. Illustrations of the jade figure in question are given in the *Ku yü t'u*, Ch. 1, p. 2, and *Ku yü t'u p'u*, Ch. 47, p. 1, the former being reproduced in the *T'u shu tsi ch'êng* and after this one by GILES (*Adversaria Sinica*, No. 4, p. 114); the latter has been reproduced by BUSHELL in BISHOP's work, Vol. I, p. 31. A corresponding contrivance of bronze is shown in the *Kin-shih so* (*kin so*, Vol. 2) reproduced by HIRTH in *T'oung Pao*, Vol. VII, 1896, p. 501, and after HIRTH by FELDHAUS (Ruhmesblätter der Technik, p. 432, Leipzig, 1910). Of modern authors, HIRTH (*l. c.*, pp. 498–501) was the first to call attention to this curious object; see also his article "Origin of the Mariner's Compass in China" in *The Monist* (Vol. XVI, 1906, pp. 321–330), the same reprinted in his "The Ancient History of China" (New York, 1908, pp. 126–136). Dr. BUSHELL (*l. c.*) has translated the text of the *Ku yü t'u p'u*, and Prof. GILES (The Mariner's Compass, *l. c.*, pp. 107–115) has elucidated the whole subject with very valuable comments. It is interesting to note that Hirth, Bushell, and Giles pursued their studies independently one of another, merely from Chinese accounts, without referring the one to the other, and that all three were unaware of the fact that this theme had been taken up twice in our literature long ago. There is indeed nothing new under the sun,— not even in sinological research. Old Father DE MAILLA (Histoire générale de la Chine, Vol. XIII, Paris, 1785, p. 296) had already described the south-pointing chariot of Ch'êng Wang after the *T'ung kien kang mu*, and EDOUARD BIOT (Note sur la direction de l'aiguille aimantée en Chine, *Comptes rendus des séances de l'Académie des sciences*, Vol. XIX, 1844, pp. 1–8) has studied this problem with a great amount of ingenuity and acumen. If I am allowed to express an opinion on this subject, it seems to me that it is not worth while wasting energy on the explanation of this so-called chariot. The specimens figured in the Chinese books mentioned are, as so many other antiquities of recent date, reconstructions or restorations based on misconceptions and misunderstandings of the wording of the ancient texts, — misunderstandings easily fostered by the loss of the ancient originals so that unlimited play was allowed the imagination. That the jade or bronze figure itself was not magnetic, goes without saying and, if a magnetic needle was employed in this case, it was certainly somewhere suspended freely and not in direct contact with this figure which itself is, in my opinion, an afterthought or purely imaginative creation of the Sung period. However this may be, it has no importance whatever for the archæology of jade and may be duly dismissed with these remarks.

IV. JADE AS WRITING–MATERIAL

Jade cut into polished slabs was used as material to write upon in ancient times, but its use was reserved to the emperor. It was a tablet called *t'ing* of rectangular shape to symbolize that the emperor should give a "straight and square deal" to all affairs of the empire, and he carried it in his girdle. The material for the writing-tablets of the feudal princes and of the great prefects (called *shu*) was ivory; that of the former was rounded at the top and straight at the bottom (*i. e.* rectangular) to symbolize that they should obey the Son of Heaven, that of the latter was rounded at the top and bottom to express that they had only superiors to obey (COUVREUR, Li Ki, Vol. I, pp. 682, 685). The general name for such tablets serving as official records was *hu* (GILES No. 4926), apparently, as the composition of the character shows, made of bamboo originally; they were worn suspended from the girdle belonging to the outfit of any young gentleman (*Li ki, Nei-tse*, I, 2) and used as memoranda for jotting down any notes. In general use during times of antiquity, they were reserved, at a later epoch, for the organs of government and became at the same time insignia of dignity. When an official had audience at court, he had inscribed on the tablet what he had to say, and added what the emperor replied or commanded. [1] From Yen Shih-ku we learn that the tablets *hu*, at least in his time, were made also of wood; [2] it is evident that they differed in shape, in their mode of use and official significance from the contemporaneous bamboo slips and wooden boards used for writing.

The notebook of jade was, accordingly, a prerogative of the emperor, but was used only during an abundant season; if the year was bad, he abandoned it and adopted the common bamboo tablet of the ordinary officials, wearing at the same time linen clothes; in the same way he did not partake of full meals nor indulge in music, when no rain had fallen in the eighth month (*Li ki, Yü tsao* I, 11). Jade, therefore, indicated also in this case a symbol of plenty and luck.

The fact that the emperor's memorandum was really made of jade is expressly stated in the same chapter of the *Li ki* (*Yü tsao* II, 16).

---

[1] "When the great prefect (*tai-fu*) had washed his head and bathed, his secretary brought him the ivory tablet to write down his thoughts, his replies, and the orders of the prince." *Li Ki, Yü tsao* I, 16.

[2] CHAVANNES, Les livres chinois avant l'invention du papier (*Journal asiatique*, 1905, p. 26, Note). It will be useful to add here jade and ivory to bamboo, wood, and silk as writing materials.

114

"For his memorandum-tablet, the Son of Heaven used a piece of sonorous jade;[1] the prince of a state, a piece of ivory; a great officer, a piece of bamboo, ornamented with fishbone;[2] ordinary officers might use bamboo, adorned with ivory at the bottom" (LEGGE, Li Ki, Vol. II, p. 12). The great importance attached to these tablets appears from the following paragraph: "When appearing before the Son of Heaven, and at trials of archery, there was no such thing as being without this tablet. It was contrary to rule to enter the grand ancestral temple (*ta miao*) without it. During the five months' mourning, it was not laid aside. When the prince, bare-headed, performed a funerary ceremony,[3] he laid it aside. When he put it in his girdle again, he was obliged to wash his hands; but afterwards, though he might have had a function to fulfill at court, it was not necessary to wash the hands. Whoever had something he desired to call to the attention of the ruler or to illustrate before him, used the tablet. Whoever went before him and received his orders, wrote them down on the tablet. For all these purposes the tablet was used, and therefore it was ornamented (with reference to the rank of the bearer). The tablet was two feet and six inches (52 cm) long; its width at the middle was three inches (6 cm), and it tapered at the ends to two inches and a half (5 cm)." These tablets had the shape of a shuttle and apparently were different from those described above.

Also in K'ANG-HI's Dictionary we find that the dictionary *Kuang yün* (T'ang period) defines the word *t'ing* as "a designation for a jade (*yü ming*)," the dictionary *Po ya* (or *Kuang ya*) of the third century as a *hu*, and that the Commentary to the *Tso chuan* gives the full definition of "a jade writing-tablet" (*t'ing yü hu ye*). Although nowhere expressly mentioned, we may infer that the writing on these memoranda of jade, ivory, and bamboo could easily be erased to make place for other notes, and that this constituted their principal difference from the bamboo or wooden documents and books which were to be permanent. From other facts known to us, we are justified in concluding

[1] *K'iu yü.* LEGGE's translation is based on the explanation of the *Shuo wên* (*yü k'ing ye*); COUVREUR's translation "fine jade" is justified by the *Kuang yün* (*mei yü ye*). This means that it is not known what variety of jade was understood by this name.

[2] LEGGE's rendering is correct, and there is no reason, in the light of archæological facts, to conjecture with K'ung Ying-ta (COUVREUR, Li Ki, Vol. I, p. 698, Note) that it was an ornament made from the barb of a crocodile or made in shape of a barb. We now have a number of pieces of ancient pottery unglazed and red-burnt called by the Chinese archæologists *yü ku kuan, i. e.* "fishbone jars" in which small pieces of white gypsum are inlaid in the surface, doubtless for the purpose of giving it a glittering aspect, and perhaps for some symbolical reason still unknown to us. In the same way as pottery, I believe that also the bamboo tablets *hu* were inlaid with gypsum.

[3] I here deviate from Legge's translation and follow Couvreur.

that the instrument for writing was a wooden or bamboo stylus[1] (of the same kind as still used in Tibet) placed in a tube which was carried in the girdle on the right side,[2] and that the ink was a kind of black varnish.

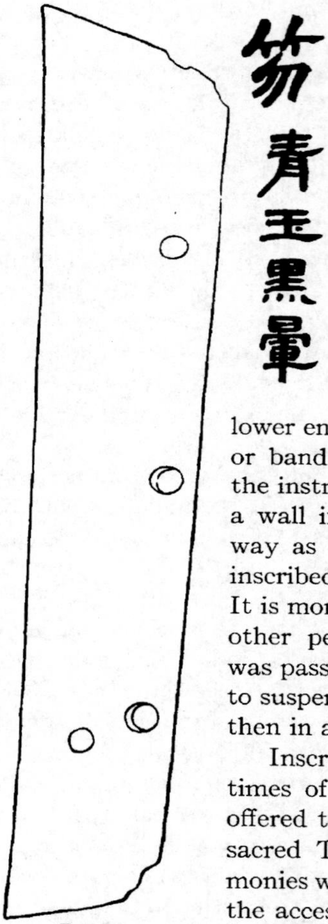

笋青玉黑暈

WU TA-CH'ÊNG has illustrated a specimen of jade (here reproduced in Fig. 40) which he thinks he is justified in identifying with the imperial jade tablet *hu* for writing. It is of green jade with a "black mist," and in its outward shape, resembles the other imperial tablets, *i. e.* it is constructed in the shape of an implement, seemingly a knife. There are three perforations arranged in a vertical line near one of the lateral edges, and a smaller perforation outside of this row near the lower end; the latter served for the passage of a cord or band, apparently for two purposes, — to fasten the instrument to the girdle, and to suspend it from a wall if writing on it was required. In the same way as the ancient bamboo slips, these were also inscribed with just one vertical line of characters. It is more difficult to see for what purpose the three other perforations were made, — possibly a band was passed through these and fastened in the middle to suspend the tablet from the girdle, so that it was then in an horizontal position.

Inscribed jade tablets played a prominent rôle in times of antiquity for the sacrifices *fêng* and *shan* offered to Heaven and Earth on the summit of the sacred T'ai-shan. The main object of these ceremonies was to announce to these two powerful deities the accession to the throne and the prosperity of a new dynasty: the emperor, recalling the merits of his predecessors and ascribing to their blessings his own virtues and success, returned his thanks to the two deities for the support which they had rendered to him and his line. This announcement was made by means of a document written and engraved on tablets of jade; five in number, one foot two inches in length, five inches in width and one inch in thickness. The slabs were

FIG. 40.
Imperial Jade Tablet
hu for writing.

[1] See CHAVANNES, *l. c.*, pp. 65 *et seq.*
[2] *Li ki, Nei tse* I, 2.

piled one above the other and thus reached a total thickness of five inches equalling their width. The whole package was protected on the upper and lower face by a jade slab two inches thick and as long and wide as the inscribed tablets. A golden cord was fastened five times around and held by a seal placed in a notch. Thus, the package was laid in a box of jade in which it fitted exactly, and the box was introduced into a stone coffer.[1]

Dr. BUSHELL informs us that the first sovereign of the Han dynasty, the Emperor Kao-tsu (B. C. 206–195), announced his accession to the throne by sacrificing to Heaven on a jade tablet engraved with one hundred and seventy characters. The jade was of a bright white color spotted and with moss-markings, shining in colors of red, blue, vermilion, and black. The writing was in the *li shu* of the Han, and the style was clear and strong.

The question of varicolored jade was brought on the tapis when the Emperor Kuang-wu (25–57 A. D.) made his preparations for the sacrifices on the T'ai-shan and gave instructions to search for a blue stone without blemish, but it should not be necessary to have varicolored stones.[2] At this time when the seal-makers were not capable of engraving the jade tablets, the emperor decided to avail himself of red varnish to write on the slabs; but this plan was not carried out, as a man was found, able to do the work of engraving. We here notice an important difference between the ancient writing-tablets and these sacrificial tablets of jade; the former were memoranda to be inscribed only; in the latter, the writing had to be carved as a permanent document. Probably for this reason, the engraved characters under the T'ang dynasty were incrusted with gold.[3] In a decree of the Emperor T'ai-tsung (627–649 A. D.) the reasons for the employment of jade tablets as the essential feature in these ceremonies are accounted for by the firmness and solidity of the material, its density and perfect supernatural qualities "which are transmitted indefinitely, for ever preserved and unalterable."[4]

These tablets, therefore, do not present a continuation of the writing-tablets *hu*, but are developed, as observed also by CHAVANNES,[5] from the ancient bamboo slips or wooden splints which served as writing-material before the invention of paper. These were inscribed with only one line of writing by means of a bamboo stylus dipped into black varnish, and fastened together with a silk cord or leather strip

---

[1]According to the investigations of CHAVANNES, Le T'ai Chan, pp. 22 *et seq.*, where also three Chinese illustrations and a description of the stone coffer will be found.

[2]CHAVANNES, *l. c.*, pp. 162, 163.　　[3]*Ibid.*, p. 226.

[4]*Ibid.*, p. 173.　　[5]*L. c.*, p. 174, Note.

to form a coherent book. In the same manner, also the sacrificial jade slips were covered with only one vertical line of characters and tied together into the appearance of a book. In the year 1747, two boxes of jade were discovered on the summit of the T'ai-shan; one of them was opened and contained seventeen jade slabs dating from the year 1008 when the Emperor Chên-tsung celebrated the sacrifice *fêng*, each slab having but one line of writing carved in. A similar find had been made at an earlier date in 1482 (*l. c.*, pp. 55, 56). I do not know if any of these ancient jade slips have been preserved to the present day. The "jade books" in vogue among the emperors of the Manchu dynasty appear as their natural offshoot, with the only distinction that more than one line is engraved on one tablet in which the page of a book printed on paper is imitated.

Under the present dynasty it was still customary to engrave important state documents and poetical productions of the emperors on jade slabs of book-size, and to unite these into a so-called jade book (*yü shu*). The Bishop collection contains such a book consisting of four oblong slabs of nephrite framed in sandal-wood in which a eulogy on the Seven Buddhas composed by the Emperor K'ien-lung (1736–1795) is inscribed (BISHOP, Vol. II, p. 173, not illustrated). F. W. K. MÜLLER (*Zeitschrift für Ethnologie*, Vol. XXXV, 1903, p. 484) mentions a jade book preserved in the Museum für Völkerkunde, Berlin, coming down also from the time of K'ien-lung in which the emperor himself is said to have carved a few lines in the summer residence of Jehol. I am in a position to illustrate on Plate XIV the oldest work of this kind from the epoch of the reigning dynasty, originating in the year 1648, four years after the Manchu had taken possession of China. Being written in Manchu and Chinese, this document presents at the same time one of the earliest specimens of the Manchu language in existence. The first literary attempts of the Manchu are the political essays of Nurhaci of the year 1616; their earliest epigraphical record is an inscription of 1639 erected in Sam-jön-do, Corea, in commemoration of the subjugation of this country in 1637; their earliest print comes down from the year 1646, being a translation of the Sacred Edicts of the Emperor Hung-wu. Then follows the jade book in question, composed of ten nephrite slabs containing an imperial document of 1648, in which the Emperor Fu-lin (Shun-chih) confers the posthumous honorary name Hing Tsu Chih Huang-ti on his ancestor in the sixth generation, Tu-tu-fu-man by name. It shows in which form the emperors used to bestow honorary titles on their ancestors. For the sake of reproduction, the ten slabs connected in one row had to be divided, so that the Chinese text arranged on four slabs on the right-hand side appears above and

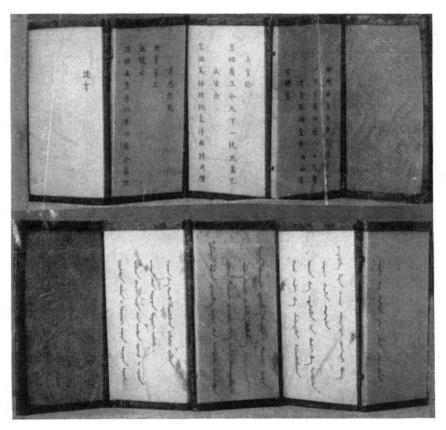

MANCHU AND CHINESE DOCUMENT OF THE EMPEROR SHUN-CHIH, CARVED IN JADE SLABS IN 1648.

the Manchu text which is on the left-hand side of the book appears below on our plate. On each of the two extreme slabs, a pair of dragons is engraved. The original, apparently taken from the imperial palace in 1900, was reported to me in 1905 to be in some private collection in Japan; I do not know what has become of it since.

The Manchu text of this unique document runs romanized as follows:

Ijishôn Dasan-i sunja-ci aniya suwayan singgeri, omšon biya-i ice-de šahôn coko. yüe, jakôn-de suwayan muduri inenggi.

soorin-be siraha hiyoošulara omolo Fu-lin, mafa fulin werihe wang-de hengkileme wesimbure, gisun. abka-i fejergi ba uhei toktobufi, amba doro-be mutebuhengge, mafai hôturi werihe[1] turgun. tondo ofi, doro kooli-be alhôdame, hiyoošulara, gônin-be akômbume, ts'a boo-bai-be gingguleme, jafafi amcame tukiyeme, Yendebuhe Mafa Tondo Hôwangdi fungnefi gung, erdemu-be tumen jalan-de tutabuha.

Translation. "In the fifth year of the period Ijishôn Dasan (*Shun-chih*, 1648), a year of the Yellow Rat, from the first day of the eleventh month, a day of the White Rooster, till the eighth day, a day of the Yellow Dragon.

"I, Fulin (*Shun-chih*), heir to the throne, thy grandson animated by piety, prostrating myself before the ancestor, the king who left me his blessing, announce as follows: The reason I brought under my sway the total empire and accomplished great deeds is due to the blessings bequeathed to me by the ancestor. Sincere, in accordance with law and usage, animated by piety, with all my heart, I confer upon thee, on this precious tablet, the posthumous honorary title 'the Sublime Ancestor, the Just Emperor' (*Yendebuhe Mafa Tondo Hôwangdi*, in Chinese: *Hing Tsu Chih Huang-ti*), whereby thy merits and virtues may be handed down to the ten thousand generations."

Also the autographs of famous calligraphists were formerly sometimes reproduced on jade slabs. A specimen of the writing of the famous Wang Hien-chih or Wang Ta-ling (344–388 A. D.) of the Tsin period was thus preserved. The jade tablet itself is lost, but a rubbing from it preserved in the collection of Mr. Shên in Wu-kiang (Su-chou fu, Kiangsu) is reproduced in the Chinese Journals *Shên chou kuo kuang tsi*, No. 2, Plate V, and *Kuo suei hio pao*, Vol. 4, No. 1.

[1] Written in the text erroneously *derihe*.

V. JADE IN RELIGIOUS WORSHIP — THE JADE IMAGES OF THE COSMIC DEITIES

In Chapter XVIII of the *Chou li*, dealing with the functions of the Master of Religious Ceremonies (*Ta tsung po*), it is said: "He makes of jade the six objects to do homage to Heaven, to Earth, and to the Four Points of the Compass. With the round tablet *pi* of bluish (or greenish) color, he does homage to Heaven. With the yellow jade tube *ts'ung*, he does homage to Earth. With the green tablet *kuei*, he renders homage to the region of the East. With the red tablet *chang*, he renders homage to the region of the South. With the white tablet in the shape of a tiger (*hu*), he renders homage to the region of the West. With the black jade piece of semicircular shape (*huang*), he renders homage to the region of the North. The color of the victims and of the pieces of silk for these various spirits corresponds to that of the jade tablet." (Biot, Vol. I, pp. 434–435.)[1]

The *Chou li*, further, gives us information regarding the pieces of jade to be placed in the coffin of a deceased member of the imperial house. The Steward of the Treasury (*t'ien fu*) was in charge of these treasures. "He fastens silken cords through the apertures with which these six pieces are perforated.[2] These are the *kuei*, the half-*kuei* or *chang*, the circular disk *pi*, the jade tube *ts'ung*, the tablet in shape of a tiger *hu*, and the tablet in shape of a half-circle *huang*. He removes the circular disk *pi* from the tube *ts'ung*. These objects are thus arranged to be deposited with the corpse in the coffin." The commentary adds the following valuable remark to this passage: "When the body is placed in the coffin, the *kuei* is to the left, the half-*kuei* is at the head. The tablet in the shape of a tiger is to the right. The tablet in shape of a half-circle is at the feet. The circular disk is under the back. The jade tube *ts'ung* is on the abdomen. In this way, one figures a representation of the *fang-ming* or brilliant cube which serves as emblem in the sacrifices. The circular disk *pi* and the octagonal

---

[1] I believe that this passage also explains the term *Leu tsung*, the Six Venerable ones to whom Shun sacrifices in the *Shu king* (Ch. *Shun-tien*, 6), a term which has been a great puzzle to all commentators (see their different opinions in Chavannes, Se-ma Ts'ien, Vol. I, p. 61). Nobody, however, has thought of this series of Heaven, Earth, and the four Quarters, which, in my opinion, would be an explanation to make reasonable sense.

[2] The commentary is quite right in alluding to the perforations in the two lower ends of these pieces which in fact, as we shall see, occur in nearly all burial jades.

120

tablet *ts'ung* are, by their separation, symbolical of Heaven and Earth.'' (BIOT, Vol. I, p. 490.)

It will be recognized that there is a correlation between the jade objects used in nature worship and those buried in the grave. Heaven, Earth, and the Four Quarters were six cosmic powers or deities, and the jade carvings serving their worship were nothing but the real images of these deities under which they were worshipped, as we shall see in detail when discussing the jade pieces in question. The idea upheld hitherto that the ancient Chinese possessed no religious images is erroneous: they had an image of the Deity Heaven, of the Deity Earth and of the Four Deities representing the Four Quarters and identified at the same time with the four seasons. We must, of course, not suppose that all religious images must be anthropomorphic and represent the figure of some human or animal being. Anthropomorphic conceptions are lacking in the oldest notions of Chinese religion, and therefore, there are no anthropomorphic images. The ancient Chinese had an abstract metaphysical mind constantly occupied with the phenomena of heaven and deeply engaged in speculations of astronomy and mathematics. Their religion is essentially astronomical and cosmological, and everything is reduced by them to measurable quantities expressed by numbers and to a fixed numerical system. They did not conceive of their cosmic gods as human beings, but as forces of nature with a well defined precinct of power, and they constructed their images on the ground of geometric qualities supposed to be immanent to the great natural phenomena. The shapes of these images were found by way of geometric construction, a jade disk round and perforated representing Heaven, a hollow tube surrounded by a cube Earth, a semicircular disk the North, etc. The West forms the only exception, being worshipped under the image of a tiger, the first and oldest example in China of a personal image of a deity. The geometric abstract aspect of the divine images is in perfect harmony with the whole geometric culture of the Chou period established on the interrelations of celestial and terrestrial phenomena formulated by numerical categories and holding sway over the entire life and thought of the nation in all matters pertaining to government, administration, religion and art.[1] The grave means only a change of abode, and if the corpse is surrounded by the images of the six cosmical gods, this signi-

[1] I beg to refer the reader to the epoch-making researches of LÉOPOLD DE SAUSSURE which have appeared in various instalments in the *T'oung Pao* for 1909, 1910, and 1911 under the title, Les origines de l'astronomie chinoise. DE SAUSSURE has with brilliant acumen proved the ancient and indigenous origin of Chinese astronomy and elucidated the fundamental ideas composing the culture of the Chou dynasty. He deserves congratulations for his work which marks a new era in sinological research, and an English translation of this should be given to the world.

fies the continuation of his after-life existence in partnership with the gods of his former life. Man himself is only part of this cosmos and the product of cosmical effects; so he remains also in the grave under the ruling influence of these cosmic powers as on earth.

It is therefore impossible to separate, in a consideration of this subject, the jade images in their relation to the cult from their relation to the grave; their relations to life and death are mutually connected and must be examined together. We shall treat first of the images of Earth as the most significant of all, then of those of Heaven and the Cardinal Points.

I. JADE IMAGES AND SYMBOLS OF THE DEITY EARTH

The three specimens illustrated on Plate XV, Figs. 1–3, though differing in dimensions and proportions, belong to the same type. They are based on the same geometrical construction and may be defined as tubes or cylinders to which four salient triangular prisms are attached in such a way as to form a rectangular wall around. The Chinese express this much simpler by saying that they are square or angular outside and round inside. The cylinder overlaps the quadrangular part on both ends and appears there as a projecting rim or lip. The piece in Fig. 1 consists of a much decomposed grayish-green jade, in color much resembling a very light seladon glaze. The entire surface inside and outside is full of fissures and cracks filled with hardened lime and loess, and corrosion has altered the stone into a very soft material. It is 5.8–5.9 cm high; the width of the sides varies between 7 and 7.3 cm; the circle of the cylinder is imperfect, the diameter varying from 5.6 to 6 cm. To judge from its appearance and, as will be seen, also on the ground of historical evidence, this and the next specimen are justly ascribed to the Chou period (B. C. 1122–255). The latter (Fig. 2, Plate XV) is very much veined like agate or marble, white with reddish-yellow stripes and spots,[1] but has the smooth polish of ancient jade. It is only 4 cm high, 6.7 cm wide with a diameter of 5.5 cm.

The third specimen of this type (Fig. 3, Plate XV) is much smaller than the two others, 4.1 cm high, 3.6 cm wide, and 2.8 cm in diameter at the opening; the projecting lips are wider here than in the two others. Another peculiar feature is formed by two crescent-shaped incisions inside, just about 1 cm below the rim, the one opposite the other, in appearance like thumb-nail impressions, the one cut in, the other cut

---

[1] This is doubtless the same color as described by the Chinese under the name "yellow" in connection with the jade objects called *ts'ung*, as will be seen farther on.

EXPLANATION OF PL. XV.

Figs. 1-3. Jade Images of Earth.
Figs. 4-5. Marble Symbols serving in the Sacrifices to Earth.

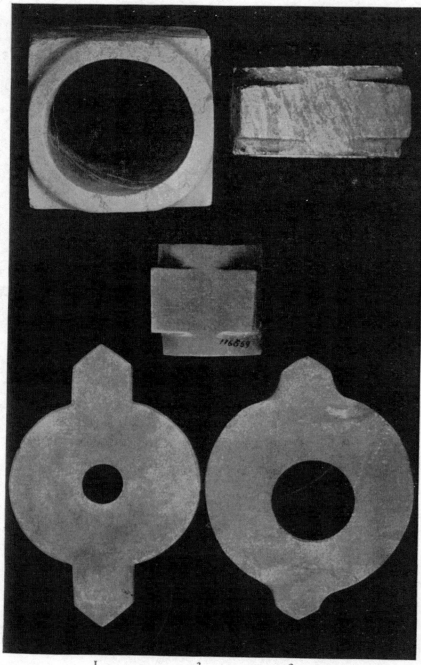

JADE IMAGES AND SYMBOLS OF THE DEITY EARTH.

out, thus appearing as positive and negative, or in Chinese as *yang* and *yin* (male and female). The jade of this piece, in a good state of preservation, shows only a few small spots possibly caused by chemical action underground, and is of a uniform light-green leaf-color. This is a kind of jade much employed in the Han period, and I have full confidence in the report given me that this piece has been unearthed from a grave of that period. It is a miniature model made at that time from

the larger specimens of the Chou period as represented in Figs. 1 and 2, which goes to show that much of the original symbolism attached to them was then lost. In Si-ngan fu where I obtained these three objects, they are called *kang t'ou, i. e.,* wheel-naves of a chariot, and the people there are unanimous in the opinion that they have been used as mortuary objects in connection with the dead, having been placed on their breast. We shall recognize from a study of the ancient texts that this information is correct;

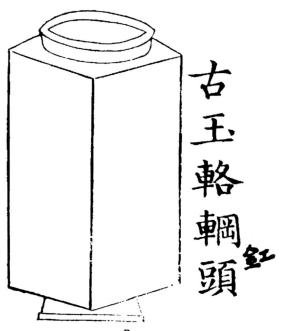

FIG. 41.
Alleged Jade Wheel-Nave (from *Ku yü t'u p'u*).

but aside from this purpose, these objects served also a most important religious function in worship and symbolized the deity of Earth.

A type corresponding to our specimen on Plate XV, Fig. 1, is in the Bishop collection (Vol. II, p. 103) and has been defined by Dr. BUSHELL as "part of a chariot-wheel nave," an explanation furnished to him by the *Ku yü t'u p'u*, which, however, as will be seen, is erroneous. Dr. BUSHELL describes the object as being of nephrite, brown and black mottled with russet spots and patches, the seams and fractures being of a dark dead-oak-leaf color, the whole covered with a russet patina. He continues by saying that it is a thick massive object of square section, with four oblong sides externally, but with the corners truncated and reduced so as to leave a thick round lip projected at each

右輖頭長二寸方一寸三分玉色甘黃璃斑

右
輖頭
虹

勻布周身朴素無文臣謹按車經云輖頭損

隅頭之飾王公之車即古之金根車也有方

有圓有六方八方之式已下諸輖頭皆古玉

輅損頭之飾漢魏巳下則無之矣

end; the interior is uniformly hollowed out into a cylindrical cavity, into which the end of the axle would be run.

Let us first examine the material of the *Ku yü t'u p'u* and see on what ground the claims of this book are based. The six specimens of this type there described are here reproduced in Figs. 41–46. They are all headed "Wheel-hubs of the ancient jade chariot." The latter (*yü lu*) was one of the five imperial carriages. The *Ku yü t'u p'u* has been written rather carelessly, and faulty characters in it are not infrequent. Thus, the character *kang* here used in the word *kang-t'ou* "wheel-hub" is unauthorized and not registered in K'ANG-HI's Dictionary; it ought to be *wang* (GILES, first ed., No. 12517) or *kang* (GILES, second ed., No. 5892). The description accompanying the first piece (see text opposite) reads as follows: "The wheel-hub here figured is two inches long and 1.3 inches wide. The color of the jade is yellow with red spots evenly distributed. All over it is plain [1] and unadorned. After careful investigation I find that the Book on Chariots (*Ch'ê king*) says: 'Wheel-hubs adorned on the sides with jade [2] pieces were the privilege of the sovereign's and the princes' carriages which are identical with the ancient gilded carriages.' There are styles of square ones, round ones, hexagonal and octagonal ones. The wheel-hubs figured below are all ornaments of the nave of the ancient jade carriage. Later than the Han and Wei periods there are none of this kind (*i.e.* they are all prior to the Han dynasty)."

That this argumentation is weak, is self-evident. Nothing seems to be known about the Book on Chariots here quoted which is not an ancient recognized text, but probably a production of the Sung period. In the *Chou li* and *Li ki*, on which we are bound to rely for the facts of ancient culture, there is no such statement to be found (see below). Then while in that quotation the jade ornaments are referred to the gilded carriage (*kin lu*), the author or authors of the *Ku yü t'u p'u* ascribe them, nevertheless, to the jade carriage (*yü lu*); these carriages, however, were two distinct types, and the jade ornaments could have belonged only to the one or to the other, as everything of this sort was conscientiously and minutely regulated and nothing left to arbitrary choice. The suspicion arises that the mere designation "jade carriage" has allured the author to run after this will-o'-the-wisp.

The jade in Fig. 42 is also stated to be yellow with red spots, and the rectangular sides are decorated with bands consisting of hexagons (in Chinese: "six-cornered balls"), a design called "refined and lovable,"

---

[1] The character *p'o* (GILES No. 9416) stands here for *p'u* (No. 9509).

[2] The word *yü* "corner, angle" does not make any sense; I presume that it is mistaken for *yü* "jade." The whole quotation would not be to the point, if no reference to the jade ornaments of the carriages were made.

this object being attributed to the Chou period; likewise Fig. 43 which is said to be of a pale-blue jade with red and carnation spots. The jade in Fig. 44 is lustrous white and slightly red; the piece is hexagonal in shape and decorated with a pattern called the iris design (*yü lan wên*). Similar designs may still be seen in the ornamental arrangement of lattice-work in Chinese windows. The specimen in Fig. 45 is described in the meagre text joined to it as hexagonal, though from the drawing it rather gives the impression of being octagonal; it is undecorated and of lustrous-white red-spotted jade. The object in Fig. 46, apparently, does not belong at all to this class, as it is rounded, and not square, outside, but it is here listed also as a wheel-nave described as of green red-spotted jade with an upper and lower band filled with "the pattern of sleeping silkworm cocoons" (*wo tsan chih wên*).[1] So far the *Ku yü t'u p'u*.

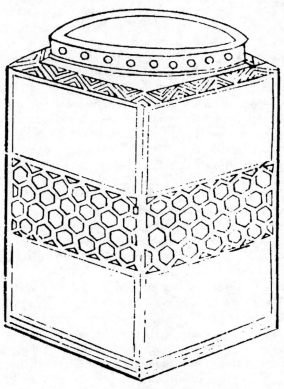

FIG. 42.
Alleged Jade Wheel-Nave (from *Ku yü t'u p'u*).

The *Li ki* (ed. COUVREUR, Vol. II, p. 82) says of the big carriage (*ta lu*) only that it was the prerogative of the Son of Heaven. There were five kinds of carriages in use at the imperial court, the first of which was called jade carriage (*yü lu*) from being adorned with jade (COUVREUR, Shu king, p. 353; BIOT, Vol. II, p. 482). The text of the *Chou li* does not make any statement in regard to the character of these jade ornaments, and the opinion of the commentary that "the extreme ends of the principal parts of the chariot were provided with jade" is by no means clear (BIOT, *Ibid.*, p. 122);

[1] See Chinese Pottery of the Han Dynasty, p. 5.

1
2
BRONZE FITTINGS OF CHARIOT WHEEL-NAVES

and in the chapter where the manufacture of chariots and of the wheel and nave in particular is discussed in detail, no mention is made of jade. Aside from this, it is highly improbable for technical reasons that jade should have been employed in forming the wheel-nave to allow the axis to pass through, as the material is too hard and not flexible enough. Then, these pieces occur in such different lengths and shapes that also for this reason this mode of employment must be doubted.

Fortunately, we possess a number of bronze wheel-naves coming down from the Chou period, and, if the jade pieces in question had ever served this purpose, we should expect to find some kind of agreement in the shape of these two types differing only in their material. This, however, is not the case. Nobody will fail to recognize in the specimens of bronze (Plate XVI) a wheel-nave, while this interpretation is not

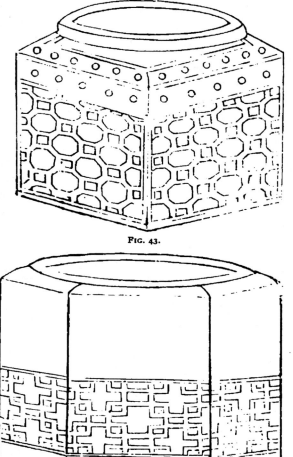

FIG. 43.

FIG. 44.
Alleged Jade Wheel-Naves (from *Ku yü t'u p'u*).

plausible for the objects in jade. The so-called bronze wheel-hub will be better designated as forming the metal mounting of the wooden hub.

It consists of three parts, a hollow tube open at one end and closed at the other to admit the end of the axle, a perforated piece curved like a bow or saddle and a peg surmounted by an animal's (apparently tiger's)

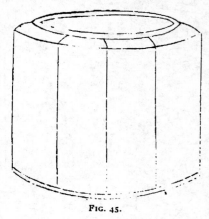

head with wing-shaped ears, which is stuck through the saddle and tube and affords a firm hold to the wooden part inserted into it. It will be recognized at a glance that this contrivance may serve as an excellent protection to the nave, that it is efficient and decorative at the same time. The saddle-like piece forms simultaneously the body of the animal, the feet of which are conspicuously moulded in front and leave two loop-holes for the passage of a cord or strap passing through an

FIG. 45.

FIG. 46.
Alleged Jade Wheel-Naves (from *Ku yü t'u p'u*).

opening in the animal's head. The tube has been cast in two halves, and two bands of scroll-work are laid out on the lower side. On the disk in front two frogs and two toads are brought out in flat relief; the two latter in circular shape running parallel with the edge-line of the circle, and the two frogs joining their heads in the centre. No doubt, this arrangement was intentional and suggestive of a turning motion of these creatures when the wheel moved.   This one seems to be an exceptionally good specimen, for the four others in our collection are not as elaborate and have not the saddle.   Two of these have a plain disk, one has it chased with scroll-work which is brought out also on the mantle of the cylinder.   In length, they are 8, 8.5, 11.5 and 18 cm.   The wings surmounting the tiger-head are presumably connected with the idea of the winged wheel.   All these peculiar features show that there is no connection between this type and those pieces of jade which are simply so designated from a very slight outward resemblance.   What neither the *Ku yü t'u p'u* nor Dr. Bushell have tried to explain, is just this very peculiar form of these jade pieces which are hollowed out into a tube, but quadrangular outside; the bronze wheel-nave mountings are naturally round or cylindrical.   It was by no means an easy task to carve a piece of jade into this singular shape to which a particular significance must have been due, but which has no meaning and no sense in the purpose alleged.   The Chinese, surely, would not have wasted so much labor without aim and *raison d'être*.

The so-called jade chariot was reserved for the emperor and used by him only on occasions when he offered a sacrifice.   Suppose that it was the wheel-naves which were adorned with jade (a supposition not warranted by any ancient text), and that the objects figured in the *Ku yü t'u p'u* are to be identified with them, how is it that this work can figure six specimens of this type, that Wu Ta-ch'êng can even produce thirty-six, that one is in the Bishop collection, and three in my own, making a total of forty-six?   It is not likely that such a number has survived from the time of the Chou when this object must have been of greatest rarity and was made but individually for a much restricted imperial use.   The specimens in question have all been found in graves, but certainly not in imperial graves.   How should it have occurred that these objects of an alleged imperial prerogative came to be dispersed among the graves of the people who could not have been entitled to them?   And what was their meaning and purpose in the graves?   To this question, the speculative theory of the *Ku yü t'u p'u* does not give any answer.

This whole theory is absurd and simply based on a misunderstanding which, as we shall see, probably goes back to a definition of the dictionary *Shuo wên* which remarks that these objects *resemble* wheel-naves which certainly does not mean that they are such.   A friend of mine pursuing ethnological studies in Peking once showed me a knife which he had just picked up in a hardware store, and passed it as the instrument by means of which the Chinese cut fish into pieces.   On inquiring the basis of this information, he said that it was called a fish-knife (*yü tao*), and I was able to explain to him that it derives this name merely from its peculiar form, being curved like a fish, and that, for the rest, it answers all purposes of a common pocket-knife. The authors of the *Ku yü t'u p'u* must have been in a similar position, having heard these objects popularly called "wheel-naves" and then concluding from this name that they really were.

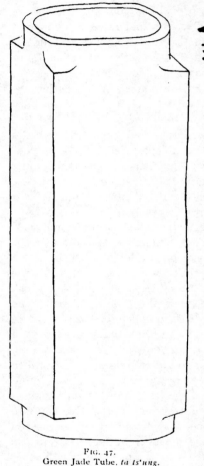

大琮

青玉滿身黑文水銀浸

When I ascertained that this identification was entirely unfounded, I began, by reading the *Chou li*, to arrive at the conclusion that these objects must be regarded as the ancient insignia called *ts'ung* (GILES No. 12026), and then discovered that Wu Ta-ch'êng had reached the same result. Therefore, I now give the word to this scholar and pass his material and notes in review.

The two specimens illustrated in Figs. 47–48 are derived from Wu Ta-ch'êng's book and correctly identified by him with the *ta ts'ung* (GILES No. 12026) mentioned in the *Chou li* (BIOT, Vol. II, p. 527). Both are hollow cylinders, round inside and square in cut outside, with two short projecting round necks at both ends; the former is plain, without any ornamentation,

FIG. 47.
Green Jade Tube, *ta ts'ung.*

of a dark-colored or dark-green jade with black veins all over, and as he says, saturated with mercury; the other piece of a uniformly black jade is decorated along the four corners with nine separate rectangular fields in relief carvings. The ornamentation in each field is the same; two bands consisting of five lines each, two knobs below, and a smaller band filled with spirals and groups of five strokes alternately horizontal and vertical. These prominent corner ornaments are called with a special name *tsu*.[1] The protruding ring-shaped necks receive the name *shê*,[2] a word which is used also in the jade tablets called *chang* to denote the triangular point at their upper end. WU TA-CH'ÊNG recalls the fact that in the *Chou li* twelve inches are ascribed to the *ta ts'ung*, with four inches in diameter to the outer rings having a thickness of one inch, and that it is the emblem of the empress, under the name *nei chên tsung* "the venerable object of the power of the inner (*i. e.* women's) apartments," corresponding to the *chên kuei* of the emperor. The identification of these specimens with this *ta ts'ung* is based on the finding that their measurements agree with the data of the *Chou li*.

Then follow eight yellow *ts'ung*, here illustrated in Figs. 49–56, identical with those stated in the *Chou li* as having been used for the worship of Earth.[3] It is easy to understand from these specimens what the commentaries mean by calling them octagonal or rather eight-cornered (*pa fang*) whereby the earth is symbolized, a notion which has given rise to such grotesque drawings in later days.[4] Of importance is the definition

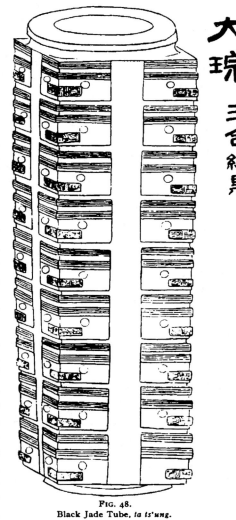

FIG. 48.
Black Jade Tube, *ta ts'ung.*

[1] GILES No. 11590 *tsang*, in this case to be read *tsu* (see farther below).

[2] GILES No. 9793; translated by BIOT (Vol. II, p. 527) arrow or projection.

[3] Compare BIOT, Vol. I, p. 434 (*i huang ts'ung li ti*).

[4] GINGELL has figured one of these on p. 38 which is drawn as the figure of an eight-pointed star (!). The translation of the word *ts'ung* by "octagonal tablet," as was the fashion up to now, will, of course, have to be dropped; it is neither octagonal nor a tablet, according to our way of thinking and speaking. But then there are people ready to say that the Chinese language is not ambiguous. It is rather

given in the *Po hu t'ung*, the work of PAN KU who died in 92 A. D.:
"What is round in the interior and provided with teeth outside,

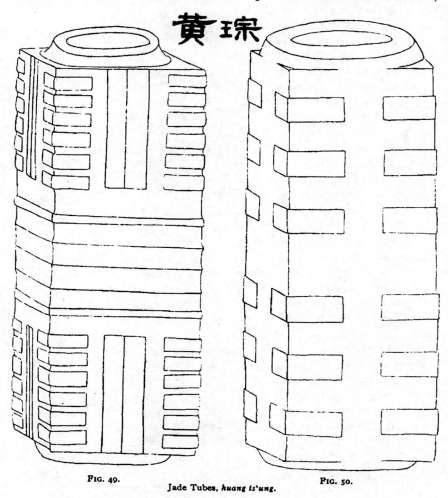

FIG. 49.
FIG. 50.
Jade Tubes, *huang ts'ung.*

is called a *ts'ung.*" Now the six raised rectangles which appear
in the upper and lower part at each corner in the first yellow *ts'ung*

amusing to observe that even a scholar of the mould of Wu Ta-ch'êng did not understand the meaning of the expression *pa fang* "eight sides" referring to the really
existing eight corners or eight triangular sides around the openings of this object
which, according to the commentary of the *Chou li*, symbolize the earth.   Alluding
to this statement, our author remarks that all *ts'ung* existing at present have only
four sides, and that it is therefore not appropriate to speak of eight sides; and in
another passage, he regrets that he has not yet seen the octagonal ones.

(Fig. 49) and the six raised rectangles in the corners of the second ts'ung (Fig. 50) are regarded as "teeth" and compared with the teeth of a saw. We said before that they are called by a special designation *tsu*, a character composed of the radical *horse* and a phonetic complement *tsie;* but CHÊNG K'ANG-CH'ÊNG explains that it must be read *tsu* written with radical 120 for silk (GILES No. 11828), and this word *tsu* denotes a silk band or cord, and because these objects were wrapped up with bands, that name was applied to them. WU TA-CH'ÊNG says, and this seems quite plausible, that the bands were tied around the deep-lying portions between the projecting rectangles, so that it would follow that the word *tsu* refers to the spaces between these, and not to the rectangles themselves. Judging from the character of these specimens and the traditions regarding the jade type *ts'ung*, there is no doubt of their identification. Wu Ta-ch'êng furnishes us also a clue as to their name "chariot wheel-nave," by making the following intro-

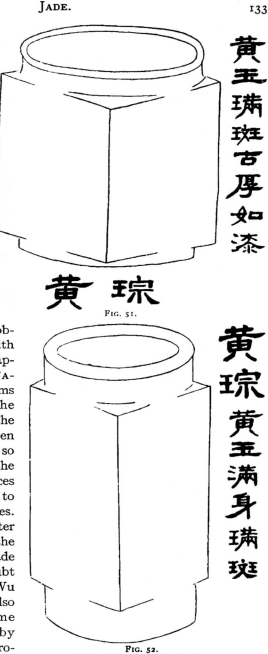

黄 琮

FIG. 51.

FIG. 52.
Yellow Jade Tubes, *huang ts'ung.*

ductory remark: "The large ones among the ancient jades going under the name wheel-naves (*kang t'ou*, GILES No. 5892) among the present generation are all identical with the *ts'ung*. The *Shuo wên* says: The *ts'ung* is an auspicious jade, eight inches big, *resembling* a wheel-nave." The case is therefore very plain. The object is likened to another similar one; it is called after this *simile* and finally taken for the real thing with which it is only compared.

Six further specimens of *huang ts'ung*, varying in length, all of yellow jade with red spots, plainly polished and unornamented, are figured by Wu Ta-ch'êng and here reproduced in Figs. 51–56. These were all doubtless connected with the worship of Earth. The specimen illustrated in Fig. 57 is not numbered in this series, but only tentatively defined as a yellow *ts'ung*, because it is also made of yellow jade interspersed with red spots on all sides; but it is, as our author himself remarks, different from that type in that it shows the shape of a ring in the exterior as well as in the interior,

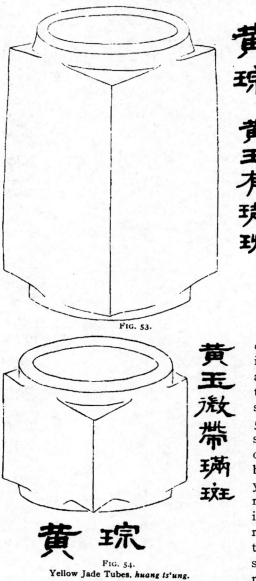

FIG. 53.

FIG. 54.
Yellow Jade Tubes. *huang ts'ung.*

and he compares it, not very happily, with the opening of the bronze goblet *ku*. Then follows a group of twelve *ts'ung* (Figs. 58–69) defined as *tsu ts'ung* without further explanation.

Five of these are of white, four of dark-green, and only three of yellow color. Some are ornamented with "teeth" as the emblems used in the worship of Earth, others bear geometrical band-ornaments in relief; others again are ring-shaped and have four undercut rectangles in the

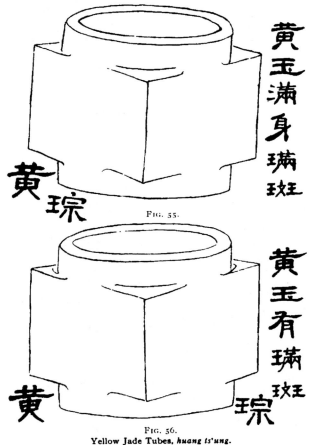

FIG. 55.

FIG. 56.
Yellow Jade Tubes, *huang ts'ung.*

style of medallions on the outward side. The *Chou li* (BIOT, Vol. II, pp. 527, 528) mentions two kinds of *tsu ts'ung*, one five inches long for the empress, and another seven inches long for the emperor, both used as weight-stones. A silk cord was drawn through the perforation and attached to the scale-yard. When, *e.g.*, silk cocoons were offered to the empress, she availed herself of this object in weighing the silk.[1] It is

[1] It is noteworthy that the empress weighs silk by means of a weight-stone representing the image of Earth. Silk is a gift of Earth, and the rearing of silkworms and spinning of silk is ascribed by tradition to a woman's initiative. In the modern imperial worship, silk is offered in the sacrifice to Earth and buried in the ground, a usage restricted to the spirit of Earth (EDKINS, Religion in China, p. 31).

possible that the one or other of these specimens has been utilized for such a purpose.

Another series of eight objects, called only *ts'ung*, is here added, four of dark-green, two of white, and two of green and white jade. As they are plain and do not exhibit any forms differing from the previous yellow *ts'ung*, they are not reproduced in this paper. From their colors we may infer that they have not been employed in the

FIG. 57.
Yellow Jade Ring, *huang ts'ung*.

worship of Earth, but must have served another end; but this question is not discussed by our author.

In the *Chou li* (BIOT, Vol. II, p. 528), another kind of *ts'ung* is mentioned under the name "festooned *ts'ung* (*tuan ts'ung*)," eight inches long, which are offered by the feudal princes to the spouse of that prince to whom they pay a visit. These apply to princes of the first rank, while those of the second and third ranks offer pieces six inches, and those of the fourth and fifth ranks pieces four inches in length. It is very possible that types of this kind occur among the *ts'ung* mentioned, and this gradation according to rank finds its echo in the varying sizes of the actual specimens, but it would be in vain to attempt here special identifications, as all detailed descriptions of the ancient pieces are lacking.

We have so far recognized various uses of the jade object called *ts'ung*. Foremost, it is the symbol of Earth as a deity, it is symbolic of the shape of the earth which is round in its interior and square on the outside;[1] that is, in other words, it represents the real image of the deity Earth. In this case, the jade selected was always of a yellowish tinge

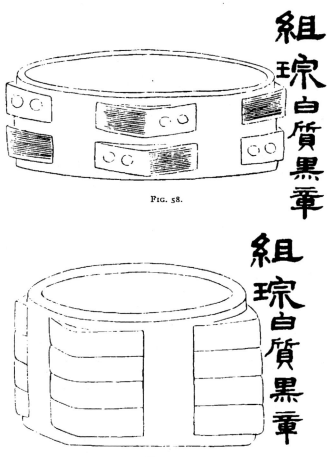

FIG. 58.

FIG. 59.
Tubes, *tsu ts'ung*, of White Jade with Black Veins.

or at· least striped or spotted yellowish. A jade object of the same type, but black in color or dark-green with black veins, was the sovereign emblem of the empress; a similar type was used by her as a

---

[1]Liu Ngan, who died in B. C. 122, expressed the general view held by ancient philosophy with the words: "Heaven is round, and Earth is square; the principle of Heaven is roundness, and that of Earth squareness" (DE GROOT, The Religious System of China, Vol. VI, p. 1264).

scale-weight, and if ever employed by the emperor, it seems to be a secondary development by transfer from the female sphere, and there is no instance of his ever having made actual use of it in person.[1]   If offered as a token of respect by feudal princes, it was only presented by them to a princely consort. This emblem, therefore, has always referred to female power.

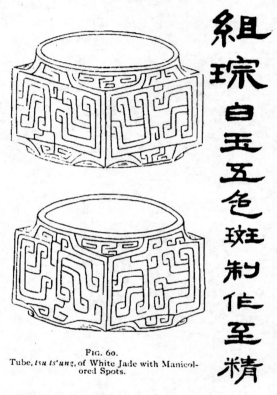

But its use is not exhausted therewith. It also entered into relations with the grave and the dead. We read in the *Chou li* that the superintendent of the jade tablets at the imperial court was charged with six objects of jade to be placed on the corpse in the coffin (BIOT, Vol. I, p. 490). One of these was the object *ts'ung* which, as the commentary explains, was placed on the abdomen of the corpse. The circular ring-shaped piece of jade *pi* was placed under the back and was to symbolize Heaven, while the jade *ts'ung*, also in this case,

FIG. 60.
Tube, *tsu ts'ung*, of White Jade with Manicolored Spots.

symbolized Earth. The remaining four jade pieces were emblematic of the Four Quarters. The dead person was, accordingly, confided to and protected by the great powers of the universe. The same powers and influences which had controlled all his thoughts and actions during life-time, to which he looked up with a feeling of awe and reverence, held sway over him also in the grave. It must be emphasized that the images of Heaven and Earth and the Four Quarters, if arranged in the coffin, were not intended as personal amulets to protect or preserve the body, as for this end a good number of other objects were available, but that the idea was implied that man took with him his gods into the life hereafter, that he meant to live the other life in the same

[1] But also the emperor partakes of the nature of Earth (see below).

manner as this one, surrounded by the same gods and worshipping them as before.

This fact will account also for the comparatively large number of these specimens which have survived, and which have all been dug up from graves. If my informants at Si-ngan fu asserted that they have been found on the breast of the corpse, this is not a fundamental error

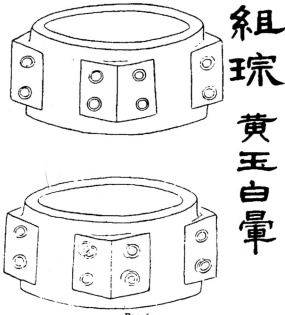

Fig. 61.
Tube, *tsu ts'ung*, of Yellow Jade with White Veins.

as the decayed condition of the skeletons and disturbance of the graves by earth-slips and other natural causes render it difficult to clearly recognize the original position of the objects.

The *Chou li* mentions in two passages (Biot, Vol. I, p. 487, and Vol. II, p. 528) another object of jade serving in the sacrifices offered to Earth and for the joined sacrifice offered to the Four Venerable ones, *i.e.* the spirits of the Mountains and Rivers. This object is called *liang kuei yu ti, i.e.* two jade tablets *kuei* having a central foundation; the latter is a perforated disk from which an appendage resembling in shape the tablet *kuei* emerges at the upper and at the lower end, the whole being cut out of one piece of jade. As the Chinese illustrators have preserved a fairly correct drawing of this plain object, I am in a position to identify with it two specimens in our collection which I was fortunate enough to acquire.

These two objects are illustrated on Plate XV, Figs. 4–5.   The material of both is called jade by the Chinese, though, judging from

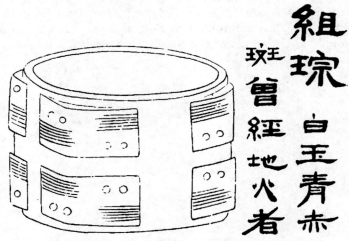

FIG. 62.
Tube, *tsu ts'ung*, White Jade with Green and Red Spots.

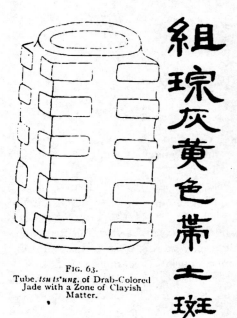

FIG. 63.
Tube, *tsu ts'ung*, of Drab-Colored Jade with a Zone of Clayish Matter.

their appearance, it rather seems to be a species of marble.[1]   The one (Fig. 4) has a lustrous white color and is smoothly polished; it is full of earth incrustations and corrosions which have partially altered the surface into a yellow brown.   From the point of the upper *kuei* to that of the lower *kuei* it measures 19 cm in length; the diameter of the circular portion varies from 10.5 to 12 cm; it is therefore not a perfect circle, being larger in width than in length.   The central perforation, however, forms a perfect circle with a diameter of 2.5 cm, the hole having been drilled from both sides.   The thickness averages 2 cm, but in the lower *kuei*

[1] This is also the judgment of Dr. O. C. Farrington, Curator of the Department of Geology, Field Museum.

it only reaches 1.7 cm. The long sides of the *kuei* are 2.5 cm long, the two ridges 2.2 cm. All edges of the circular portion as well as of the two *kuei* have been chamfered, except the ridge-poles of these

FIG. 64.
Tube, *tsu ts'ung*, Green Jade with Russet Spots and ''a Zone permeated by Mercury.''

which are pointed. Altogether, this piece has been worked with great care, and the dimensions have been ascertained and regulated from conscientious measuring, and with appliance of rule and compass; a line

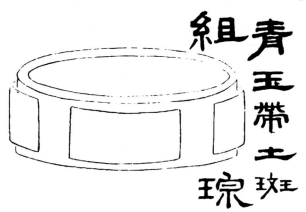

FIG. 65.
Tube, *tsu ts'ung*, of Green Jade with Zone of Clayish Matter.

connecting the two points of the *kuei* will pass exactly through the centre of the circular perforation.

The same cannot be said of the other piece (Fig. 5, same Plate) which looks somewhat grotesque in its clumsy irregularity. The connecting-line of the points of the *kuei* passes here the extreme right portion of the inner circle, leaving there a small segment; nor do the two *kuei* lie in one vertical zone. It will be seen that the perforation (with a diameter of 5.5 cm) is not in the centre of the disk with a diameter varying between 13 and 13.5 cm (1.1–1.5 cm thick). The

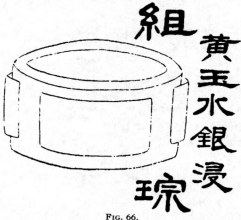

FIG. 66.
Tube, *tsu ts'ung*, of Yellow Jade, "permeated by Mercury."

kuei are very short and of irregular shapes, the sides not being straight, as in the preceding piece, but curved. The surface is much decomposed, owing to chemical action underground, and has, for the greater part, been changed into a dark yellow-brown, while in some spots, particularly on the edge, a light apple-green of the original (?) color is still visible.

These two objects must be distinguished from those previously described; the latter are real images of Earth. These, however, are merely symbols accompanying the sacrifice, and not objects of worship. Their position in the cult is similar to that of the jade tablets which were sent

FIG. 67.
Tube, *tsu ts'ung*, of Green Jade with Black Zone interspersed with White Specks.

along with any official presents. It is a symbol of rank, the degree of rank being expressed by the two attached *kuei*. This is evidenced by the fact that in the ordinary sacrifices to Heaven and in the extraordinary sacrifices offered to the Supreme Ruler Shang-ti a similar jade disk was employed, but set with four *kuei* (BIOT, Vol. I, p. 486), arranged crosswise. A disk, but not perforated, with but one *kuei* at the upper end, served in the sacrifices to the Sun, the Moon, the Planets and Constellations (*Ibid.*, p. 488), and a perforated disk to which a *chang, i. e.* half of a *kuei*, was attached, served for the sacrifices offered to the Mountains and Rivers. The latter could be used also in ceremonious offering of

articles of food to strangers who had come on a visit. This is sufficient proof for the mere representative character of these symbols.

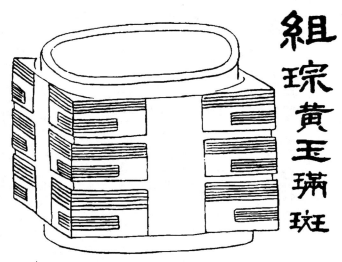

FIG. 68.
Tube, *tsu ts'ung.* of Yellow Jade with Russet Spots.

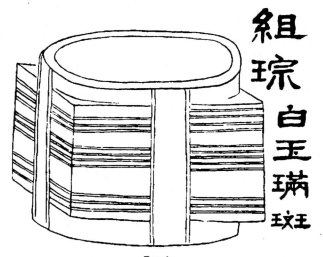

FIG. 69.
Tube, *tsu ts'ung,* of White Jade with Russet Spots.

Being in possession of the one relating to Earth, I could easily arrange
for having reconstructions made of the others which have been executed

in plaster in the Field Museum and are here reproduced on Plate XVII.[1]

The deity of Earth occupies a prominent place in the religion of the ancient Chinese and yields in importance only to Shang-ti, "the Emperor Above," the ruler of Heaven. It is simply called *t'u* and sometimes *ti* "Earth" and frequently characterized by the attribute *hou*, meaning a sovereign. The sex is not expressed by this word; a sharp distinction between male and female deities does not occur in the oldest religious concept of China where anthropomorphic notions were but weakly developed. Primarily, Earth was neither a distinctively female nor a distinctively male deity, but rather sexless; nevertheless, it falls under the category of *yin*, the negative, dark, female principle, as already indicated in the Book of Mutations (*Yi king*) where the notion of *yin* is defined as the action of Earth (*yin ti tao*). It is certainly doubtful whether the word *yin* conveyed in the beginning a clear sex-notion which may be regarded as a philosophical abstraction of later times; but there is no doubt that the combination *yin yang* signified the combined action of Heaven and Earth in the production and transformation of beings, or the creative power of these two great forces. In the sacrifices to the deity Earth, all paraphernalia are derived from the sphere of *yin*, and the jade image of the deity symbolizing its shape doubtless partakes also of the character of the female element.

It has been emphasized that only under the Han dynasty (B. C. 206–221 A. D.) the word *hou* adopted the meaning of "princess, empress," and that first in a hymn of that period the deity Earth is conceived of as female and is designated definitely as "the fertile mother."[2] Only as late as under the Emperor Wu (B. C. 140–87), the cult of the Empress Earth is instituted at *Fen-yin* where she was adored under the statue of a woman.[3] But this action does not prove that the deity has been considered a male previously; it only means the termination of a long-continued development, the final official sanction manifestly expressed by this imperial approval of a general popular feeling presumably cherished for centuries. An important deity occupying to a large extent the minds of the people does not change in a day from a

---

[1] The *Ku yü t'u p'u* (Ch. 8–10) gives a wonderfully rich selection of these four symbols in all varieties decorated with all sorts of impossible ornaments. It would be a waste of time to discuss these bold forgeries of the Sung period. No ancient text nor any of the older commentaries contain a word about these pieces having ever been decorated. In the symbols of Earth, the *faux pas* has been committed of drawing instead of the circular perforated disk a solid quadrangle; of course, also the Sung artists knew that the earth was square, — *mais du sublime au ridicule il n'y a pas un pas.*

[2] W. GRUBE, Religion und Kultus der Chinesen, p. 35.

[3] CHAVANNES, Le T'ai Chan, p. 524.

EXPLANATION OF PL. XVII.

Fig. 1.   Emblem serving in Sacrifices to Heaven.
Fig- 2.   Emblem in Sacrifices to Sun, Moon, Planets and Constellations.
Fig. 3.   Emblem in Sacrifices to the Mountains and Rivers.

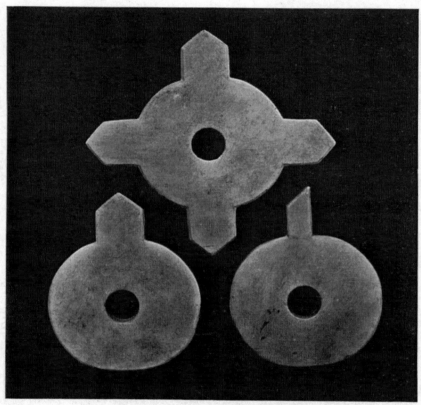

2　　　　1　　　　3
RECONSTRUCTIONS IN PLASTER OF SACRIFICIAL EMBLEMS.

male into a female, nor can it be all of a sudden proclaimed as a female, unless it was imbued with such latent qualities in popular imagination which gradually led to this end with forcible logic. We are therefore bound to assume that female elements and characteristics must have been slumbering in the conception of this deity ages before the time of the Han.

A sharp distinction must be made, at the outset, between the deity of Earth (*t'u*) and the Spirit of the Soil (*shê*). The latter has been made the subject of a very remarkable and fruitful study by Prof. CHAVANNES.[1] The Spirit of the Soil is a decidedly male god of partially anthropomorphic concept; but it is a god restricted in power as to space and to time, it is a god of territorial groups, of social communities occupying a more or less limited area of the soil. There is a complex system of an official hierarchy of a plurality of gods of the soil graduated according to rank and power. The individual families harbor their god of the soil, the territorial communities have their own, the territorial officials have their own, and the feudal lords and the emperor have their own. The imperial god of the soil is, so to speak, the particular property of the dynasty, and his power vanishes with the extinction of the latter. The new rising dynasty chooses a new god of the soil of its own by erecting to him a new altar, and neutralizes the action of his predecessor by building an enclosure around his altar. The gods of the soil, accordingly, are individual gods of a local and temporary existence connected with the coeval living owners of the soil, living and dying with them.

The deity of Earth, however, is infinite in space and time. It comprises the totality of the entire known earth, the limits of which were unknown; it is permanent and eternal like Heaven, and the second great cosmic power of nature acting in harmony with Heaven towards the welfare of the whole creation. It is an almighty great abstract deity like Heaven and the object of veneration and worship on the part of the people, and in particular of the emperor, through all generations. It is *the* telluric deity, whereas the Spirit of the Soil merely shares the function of a terrestrial tutelary genius.

CHAVANNES (*l. c.*, p. 524) has adopted the point of view that the sacrifice to Earth does not go back to times of a great antiquity, and that only since the time of the Han Emperor *Wu* the dualistic cult of Heaven and Earth has assumed a prodigious importance. I am inclined to think that such a view cannot be upheld.

The dual concept of Heaven and Earth as the deified omnipotent powers of nature seems to me to have formed an essential constituent

[1] Le Dieu du Sol dans la Chine antique, in his book *Le T'ai Chan*, pp. 437 *et seq.*

of the most ancient religious notions of the Chinese.   As early as in
the *Shu king* (*Chou shu, T'ai shih shang*, I, 3; ed. COUVREUR, p. 172)
we read in the beautiful speech of *Wu Wang:* "Heaven and Earth are
like *father* and *mother* of all beings, and of all beings, it is man alone who
is endowed with reason.   Who excels in intelligence and sagacity, is
made the supreme sovereign; the supreme sovereign is the *father* and
the *mother* of the nation."[1]   In this text, Heaven and Earth are characterized as living creative forces comparable to the propagating
faculty of a father and mother.   The Chinese have certainly not
yet spoken directly of the "Mother Earth" in those early days, but
what is apparent is the fact that the idea of Earth being or acting like
a mother was subconsciously latent in their minds, and that the emperor partakes not only of the nature of Heaven but also of Earth.[2]

And then the passage in the *Li ki* (IX, I, 21; ed. COUVREUR, Vol.
I, p. 587; LEGGE, Vol. I, p. 425): "The sacrifices to Earth were made
to honor the beneficent actions of Earth (*ti*); for Earth (*ti*) carries all
beings, while Heaven holds the constellations suspended.   We derive
wealth from Earth,[3] we derive the regulation of our labors from Heaven.
For this reason, we honor Heaven and love Earth, and we therefore
teach the people to return them thanks."   The relations of the people,
i. e. the farmers, to these two factors upon which they depended for
their existence could not have been better expressed; they honored
Heaven and loved Earth, as they honored their father and loved their
mother, and therewith the farmer's emotional religion was bound up.
It was a wide-spread deep-rooted national sentiment, it was a subject
of instruction, the key-note of the lessons given to the people.

In "the Doctrine of the Mean" (*Chung yung*), the word *ti* occurs
constantly as the correlative of *t'ien*, the phrase *T'ien Ti* "Heaven and
Earth" being now the component parts, and now the great powers,
of the universe, as a dualization of nature, producing, transforming,
completing.[4]   It has been said that Chinese religion does not know of

[1] Compare *Yi king* (*Shuo kua chuan*, 10): "*Kien* is Heaven and is therefore
called the father; *kun* is Earth and is therefore called the mother."

[2] The philosopher WANG CH'UNG (first century A. D.) says likewise: "The emperor
treats Heaven like his father and Earth like his mother.   In accordance with human
customs, he practises filial piety, which accounts for the sacrifices to Heaven and
Earth."   (FORKE, Lun Hêng, Part I, p. 517.)   This symbolism penetrated the whole
life of the sovereign.   The rectangular wooden body on which his chariot rested
represented the Earth, and the circular shape of the umbrella planted on the chariot
represented Heaven; the wheels with their thirty spokes symbolized sun and moon,
and the twenty-eight partitions of the umbrella the stars (BIOT, Vol. II, p. 488).

[3] *I. e.* it is the giver of all earthly goods produced by the soil.   Hence, in a passage
of the *Lun yü* (LEGGE, The Chinese Classics, Vol. I, p. 168), the word *t'u* "earth"
assumes the meaning of comfort, worldliness, of which only the average small-
minded man thinks, whereas the superior man aspires for good qualities.

[4] LEGGE, The Chinese Classics, Vol. I, pp. 460, 461.

a creation or creator; this saying should be thus formulated, that there is no creation myth preserved to us in the Chinese traditions. These have been handed down by practical philosophers or scholars who hardly took an interest in the religious notions and legends of the masses. We therefore have merely a one-sided and biased version of their religion abridged and curtailed after an eclectic method stamping out everything that did not fit the Confucian system. A number of passages in the *Yi king* will allow the inference that there has been an ancient idea of Heaven and Earth having created the universe by their combined action. Even the most ancient texts express themselves in an abstract style of dignified philosophic speech which does not reflect the people's language. It will not be a heresy to imagine that in popular thought this process was simply conceived of as a parallel to the human act of generation, as so universally found among many primitive peoples, and this presumption will adequately account for the dropping into high literature of such comparisons of Heaven and Earth with father and mother.[1]

It is difficult to see why Chavannes denies that a regular cult was devoted to Earth in the same way as to Heaven. King Süan in the *Shi king* (LEGGE, Vol. II, p. 529) presents his offerings to the Powers above and below, and then buries them. The *Li ki* (*Wang chih* III, 6; ed. COUVREUR, Vol. I, p. 289) says expressly that the Son of Heaven sacrifices to Heaven and Earth (*t'ien ti*). The *Chou li* (BIOT, Vol. I, p. 487 and Vol. II, p. 528) mentions the perforated circular jade piece with a *kuei* attached at the upper and lower ends, described above, which serves in the sacrifices to Earth. The archæological finds exhibiting a jade image of Earth and a jade symbol used in the sacrifices in its honor point in the same direction and afford a still weightier evidence. One of the crimes of the last tyrant of the Shang dynasty was that he sacrificed neither to Heaven nor to Earth, nor to the souls of his ancestors (*Shu king*, ed. COUVREUR, p. 181). It was and is only the emperor who possessed the privilege of sacrificing to Heaven and Earth, and hence it is clear that Earth ranked with Heaven on the same level, that they were correlate to each other, and that the cult and the sacrifice devoted to the deity of Earth was of just as great importance as that to Heaven.

Also at the present time, the emperor stands in the same relation to Earth as to Heaven. In prayer to both, he styles himself a "subject,"

---

[1] This point of view has been rejected by PLATH (Die Religion und der Cultus der alten Chinesen, Part I, p. 37) on the ground of objections the validity of which I fail to see. His interpretation that the term "father and mother" merely refers to parental care, and that therefore the sovereign is called father and mother of the people, is the mere outcome of a rational subjectivism which is not borne out by the wording and thoughts of the Chinese texts.

but only to these, and not to any other spirit.[1] The spirits of the
great mountains and rivers are treated in the ritual by the emperor as
ministers subject to him.

In the *Chung yung* (Legge, Chinese Classics, Vol. I, p. 404) Confucius says: "By the ceremonies of the sacrifices to Heaven and Earth
(*kiao shê*) they served God (*Shang-ti*), and by the ceremonies of the
ancestral temple they sacrificed to their ancestors. He who understands the ceremonies of the sacrifices to Heaven and Earth, and the
meaning of the several sacrifices to ancestors, would find the government of a kingdom as easy as to look into his palm!" As Legge understands, the service of one being, God, was the object of both these
ceremonies. Would it mean that in the view of Confucius at least,
Earth was also subject to the Supreme Ruler?[2] The difficulty arising
from the Chinese texts is to decide whether such notions spring up
from popular religion or are the outcome of individual philosophical
speculation.

Though the god or gods of the Soil and the deity of Earth are two
distinct types moving on opposite lines of thought, there are nevertheless mutual points of contact in the cult rendered to them and ideas
fusing from the one into the other, for, after all, the god of the soil
invariably roots in the ground which is part of the earth. Of chief
interest to us with reference to the present subject is the image under
which the god of the Soil was revered. The material of which it was
made was common stone in distinction from the nobler substance
of jade reserved for Earth. This shows the wide gulf separating the
two in general estimation. Jade is the product of earth, but at the
same time the essence of Heaven perfected under supernatural influences. Stone is simply a species of earth and the most solid object
found within the domain of things created by the soil; it was therefore
selected as the material to figure the spirit of the Soil.[3] Though there
is no doubt that these images go back to a great antiquity, there is
no description given of them earlier than in 705 A. D. It was then
proposed to make them five feet long (which is the number corresponding
to Earth) and two feet wide (two corresponding to the female principle

[1] J. Edkins, Religion in China, p. 31.

[2] The passage has been one of great controversy among the Chinese commentators.
To overcome the difficulty, it has been proposed by some that the word *Hou-t'u* "the
sovereign Earth" has been suppressed for the sake of brevity after *Shang-ti*, and
they accordingly translate, "By the sacrifices *kiao*, they did homage to the Ruler
Above, by the sacrifices *shê*, to Earth," a view adopted by Couvreur (Les quatre
livres, p. 44).—If J. Ross (The Original Religion of China, p. 65) infers from this
sentence that God is thus removed to a greater distance from man, and approached
through the visible media of Heaven and Earth, this is too rational an explanation
and no longer in agreement with Chinese thought.

[3] Chavannes, Le T'ai Chan, p. 477.

yin); the upper part should be rounded off to symbolize the birth of the beings; the basis was to be made square to symbolize Earth; half of its body should be sunk into the ground that it may root in the earth, in such a way that the parts underground and above are of equal size. Under the Sung (960–1279 A. D.) this regulation was still in force, and we hear that the stone was shaped like a bell.[1] The image of the god of the Soil was accordingly based on a geometrical construction like that of the deity of Earth; in a twofold manner, it partakes of the nature of Earth, in the symbolism of the figures five and two, and in the square form of the basis derived from the idea of the square shape of the earth. It deviates from the image of Earth in the crude material chosen, in its much larger size, in being solid (not hollowed out), and in its stationary character by which it is fixed in a definite place, half under the soil and half above it, to indicate its relations to the land and to the inhabitants of it.

In the ceremonies called *fêng* and *shan* performed on the summit and at the foot of the T'ai-shan, the sacrifice *fêng* was addressed to Heaven, and the sacrifice *shan* to Earth. These sacrifices were performed for the first time in B. C. 110. The altar on which the sacrifice to Heaven was performed consisted of a circular terrace fifty feet in diameter and nine feet in height. The ceremony *shan* in honor of Earth took place on an "octagonal" altar, a phenomenon analogous to the "octagonal" image of Earth. The offerings were buried in the soil in a pit dug for this purpose to reach the subterranean deity, while those for Heaven were burned on a scaffold to be carried above by the smoke.[2]

The word "octagonal" must not be understood in the sense that a horizontal cut made through the altar was a figure with eight sides; it was a rectangular block of earth, a solid with two square and four rectangular faces; the Chinese expression merely arises from the eight corners or angles. This condition of affairs we may infer from the description of this altar which was gilded on the upper face and painted on each of its four sides with the color appropriate to that particular direction, *i. e.* green for the east, red for the south, white for the west, and black for the north.[3] And an altar of the same construction was dedicated, at the epoch of the Han, to the great god of the Soil who had his quarters in the imperial palace; with the only difference that the earth on the top of the altar was yellow.[4] In all these cases, the altar derives its shape from the object of worship and indeed coincides in appearance with the image of the deity itself.

---

[1] Compare CHAVANNES, Le T'ai Chan, p. 478. [2] CHAVANNES, *l. c.*, p. 21.
[3] CHAVANNES, *l. c.*, p. 195. [4] *Ibid.*, p. 451.

The same ideas still pervade modern China. In the temple of Agriculture (*Sien Nung t'an*) in Peking, north of the ground where the emperor ploughs in the spring, there is a square terrace, five feet high, and fifty feet on each side, from which the ceremonies of ploughing are watched; there are further two rectangular altars there devoted to the spirits of Heaven and Earth, respectively. In the Tai Miao, the ancestral temple of the imperial family, there is still the altar to the spirits of land and grain, fifty-two feet square and four feet high, built of white marble; the terrace is laid with earth of five colors, distributed in the above mentioned way among the cardinal points, yellow being in the centre. The inner wall is 764 feet long, and is built with bricks glazed in different colors on each of its four sides, according to position.

It is of great interest to note the manner in which the image of Earth has been reconstructed in the imperial temple of Earth (*Ti t'an*) of the present dynasty. In the *Huang ch'ao li k'i t'u shih*[1] (Ch. 1, p. 22 b) it is stated that the image in use there is the *huang ts'ung* with reference to the passage in the *Chou li* quoted at full length with the commentatorial annotations that yellow represents the color of Earth, and that the *ts'ung* is square. Consequently, in the K'ien-lung epoch, all stress was laid on these two features,—yellow and square, and on this basis, an ideal reconstruction was attempted by mere intuition. The result is shown in Fig. 70, reproducing the wood-cut of the *huang ts'ung* in the imperial Code of Rituals. It is, as described in the text, quadrangular, some four inches and somewhat more in diameter, seven-tenths of an inch thick in the centre and two-tenths of an inch thick along the edges, the upper edge being convex, and the lower side having the shape of a segment (the figure is certainly misdrawn in order to show the appearance of this lower side). "The ornaments," it is said, "are like mountain-formations, also they serve in symbolizing Earth." The unilateral arrangement of this pattern is curious. Sentiment may have prevailed that the bare quadrangular yellow jade piece was, after all, insufficient to be a worthy representative of the deity, and may have suggested the addition of this hill ornament.

On the imperial Altar of the Tutelary Deities of the Soil and the Harvest (*Shê Tsi t'an*) of the present dynasty, the two are worshipped under jade images consisting of a quadrangular solid foundation to the upper and lower end of which a *kuei* is attached (*liang kuei yu ti*, see

---

[1] A finely illustrated handbook describing the objects of the cult and the state paraphernalia of the reigning dynasty, drawn up by order of the Emperor K'ien-lung in 1759, and revised in 1766 (see A. WYLIE, Notes on Chinese Literature, p. 72). A copy of this rare and important work was procured by me for the John Crerar Library of Chicago (No. 589); another copy is preserved in my collection in the American Museum, New York.

p. 139). In the *Huang ch'ao li k'i t'u shih* (Ch. 1, p. 35 b) where the two are sketched, it is stated that the *Chou li* does not mention any jade devices devoted to those two gods, and that this usage goes back to the ritual established in the K'ai-yüan period (713–741 A. D.) of the T'ang dynasty. The regulations of the Emperor K'ien-lung provide for the Great God of the Soil (*T'ai Shê*) an image of white jade with yellow stripes "to symbolize the virtues (forces) of Earth" (*t'u tê*),

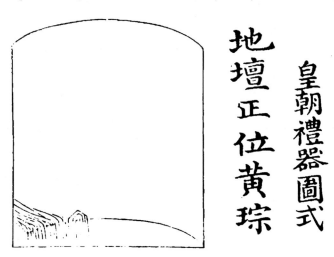

FIG. 70.
Jade Image of Earth in the Temple of Earth, Peking (from *Huang ch'ao li k'i t'u shih*).

and for the Great God of the Harvest (*T'ai Tsi*) an image of green jade, "because green implies the idea of budding." Both pieces are much the same in shape, except that the foundation of the latter is square, while that in the image of the God of the Soil is shaped like the image of Earth in the Ti-t'an (Fig. 70) without the hill-ornaments. The *kuei* attached are the same as in the ancient Chou specimens (Plate XV, Figs. 4 and 5). The sides of the central squares measure three inches in length and three-tenths inch in thickness, while the points of the *kuei* are only two-tenths of an inch thick and their sides somewhat thicker. The symbolism expressed by the formation of these two images is clear: the square nucleus is emblematic of Earth; the pointed *kuei* is an emblem of male potency and fertility and therefore connected also with the worship of the Spring (see below).

The *ta ts'ung* which was the jade emblematic of the power of the empress at the time of the Chou period and the *huang ts'ung* which was the image of the deity Earth, give us occasion to grasp the meaning of an interesting group of pottery which has apparently derived its peculiar shape from those jade objects. A specimen of this group, I had already figured in "Chinese Pottery of the Han Dynasty" on Plate XIII (p. 60) when I was under the impression that this cylindrical vessel was related to the type of granary urn. Having now made a new most comprehensive collection of Chou and Han pottery for the Field Museum which contains a good many types heretofore unknown, I can here introduce on Plate XVIII a gray unglazed quadrangular jar with loose cover (28.2 cm high and 13.5 cm wide) which bears two large characters painted white[1] in ancient script, reading *ku tou* "grain vessel." I cannot discuss in this connection, as we do not deal here with the subject of pottery, why this jar is classed among the *tou* or tazza. But for two other reasons, this inscription is of great significance to us, as it reveals that the object of this urn was to be filled with grain to serve as food for the inmate of the grave. Grain is the gift of the deity Earth, and the Spirit of Harvest (*Hou tsi*) was the natural outgrowth of the Spirit of the Soil. And this may account for the reason why an attempt is made to imitate in this piece of pottery the jade image of Earth. We see here the same short straight neck in the centre of the upper surface leaving room for four corners as in the jade image. The four walls are sloping inward at their lower ends so that the bottom is narrow.

At the time of the Sung dynasty, we find a class of vessels displaying a still more striking resemblance to the jade type. Four of these pieces from our collection are selected here for illustration (Plates XIX and XX). In the first two pieces, the coincidence is perfect, for here we have the eight corner-pieces, rectangles moulded in relief and a "tooth" in each of them. There is not only a straight neck over the opening, but also one of identical shape set on the bottom, so that in outward appearance these two pieces exactly agree with the jade *ts'ung*. They are, of course, square in cut, though the corners inside have been chamfered, probably with some intention to reach a closer similarity with the originals. The one in Fig. 1, Plate XIX (28 cm × 13.5 cm), is covered with a finely crackled buff-colored heavy glaze; the other in Fig. 2 (24.5 × 10 cm) is glazed gray with a slight yellowish tinge; both have presumably been made in the kilns of Ju-chou.[2]

The piece of Sung pottery in Fig. 1, Plate XX (28.5 cm × 11 cm), is remarkable in that the yellowish tinge is more intense in the glaze, and that thus its relation to Earth is more forcibly brought out. On

[1] There is a goodly number of unglazed Han pottery with ornaments and inscriptions painted on in white, red and black paints.

[2] BUSHELL, Description of Chinese Pottery, p. 40 (London, 1910).

HAN POTTERY QUADRANGULAR JAR, WITH INSCRIPTION ''GRAIN VESSEL,'' ITS
SHAPE IMITATING THE IMAGE OF EARTH.

1

2

SUNG GLAZED POTTERY VASES SHAPED INTO THE IMAGE OF EARTH.

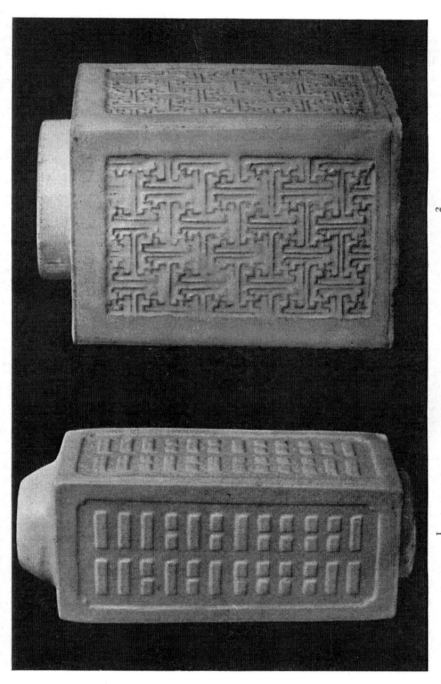

1

2

SUNG GLAZED POTTERY VASES SHAPED INTO THE IMAGE OF EARTH.

each of the four sides, the *pa-kua*, the eight mystical trigrams for divination, are brought out on panels sunk into the surface. It is conceivable that the ancient "teeth" could be supplanted by the more popular figures of the trigrams. But the original character is also here preserved in the rim on which the bottom stands. This one is wanting in the fourth piece on Plate XX, Fig. 2 (17 cm × 11.5 cm), which has a plain square bottom and its four sides decorated with a continual svastika pattern in relief; for the rest, its shape is related to the preceding pieces.

It is not known for which purpose jars of this type were employed. There is, however, in the *Fang-shih mo p'u* (Ch. 3, p. 4 b; published by Fang Yü-lu in 1588) the figure of a vase supposed to be of jade and belonging to the same type as described. On the surface of it, the three characters *shih* (Giles No. 9958) *ts'ao p'ing*, "Vase for the plant *shih*" are inscribed. This is a species of *Achillea*, the stalks of which were used in ancient times for the purpose of divination, mentioned in the *Shi king*, *Chou li* and *Li ki* (Bretschneider, Botanicon Sinicum, Part II, p. 244). No explanation of this piece is given. Wu Ta-ch'êng figures a small quadrangular jade vase of this type built up in four sections which he calls *fang li*, or by a popular name (*su ming*) *shih ts'ao li*.[1] It therefore seems that a popular belief appears to exist which connects these jars with the practice of divination; it is perhaps presumed that the stalks of the plant were preserved in them. As far as I know, there is no record to this effect in the ancient texts. At all events, this point of view deserves attention, in case the material will increase in the future. The possible employment for divination might account also for the application of the *pa kua* on the above mentioned jar. Wu has unfortunately not examined his specimen and devotes no comment to it; I do not feel authorized to say how old it may be, but can say only that it is old, as Wu would not have included it in his collection otherwise. It thus presents the missing link between the Chou jade images of Earth and the Han and Sung pottery pieces, in that it is a jade vessel of the "Earth" type. The legend on our Han jar, "grain vessel," is of principal importance, for it is a contemporaneous interpretation and proves the typological unity of this whole series.

[1] This word is written with a character (not in Giles) composed of the classifier *yü* and *li* "strength." It is explained in K'ang-hi after the *Shuo wên* as "a stone ranking next to jade."

2.   JADE IMAGES AND SYMBOLS OF THE DEITY HEAVEN

There are three kinds of annular jade objects, called *pi* (GILES No. 8958), *yüan* (No. 13757) and *huan* (No. 5043).   The former is a disk with a round perforation in the centre, the two latter are rings. The difference between the three is explained in the dictionary *Êrh ya:* "If the flesh (*i. e.* the jade substance) is double as wide as the perforation (*hao*), it is called *pi;* if the perforation is double as wide as the jade substance, it is the ring *yüan;* if the perforation and the jade substance are equally wide, it is a ring of the kind *huan.*"   This is a good point for guiding collectors in defining their specimens, although, as measurements on actual specimens show, these definitions are by no means exact, but to be taken *cum grano salis.*   The Chinese, also, determine these three groups from the general impression which they receive from the relative proportions of the dimensions of the ring and the perforation.

The definition that the jade disk *pi* symbolizes Heaven, is first given by CHÊNG K'ANG-CH'ÊNG of the second century A. D. in his commentary to the *Chou li* (BIOT, Vol. I, p. 434).   The *Shuo wên* states only that *pi* is an auspicious jade and a round implement; of dictionaries, the *Yü pien* by KU YE-WANG (523 A. D.) is the first to register the entry: "an auspicious jade to symbolize Heaven."   All Chinese archæologists of later times have adopted this explanation.   A singular position is taken by PAN KU in his work *Po hu t'ung* who says: "The *pi* is round on the outside which symbolizes Heaven, and square inside which symbolizes Earth."   Thus, he is quoted in K'ANG-HI's Dictionary, but this is only a clause culled from a longer exposition. If we turn to the edition of the *Po hu t'ung* in the *Han Wei ts'ung shu*, we find in Ch. 3, p. 17 as follows: "The *pi* which is used in soliciting the services of talented men is that kind of *pi* which is square in the centre and round in the exterior.   It symbolizes Earth, for the action of Earth is peaceful by producing all objects of wealth, and hence the *pi* is fitted to enlist talent.   The square centre is the square of the female power (*yin tê fang*); the round exterior is the female principle *yin* attached to the male principle *yang*.   The female power is flourishing in the interior; hence its shape is found in the interior, and its seat is in the centre.   Therefore, this disk implies the shape of Heaven and Earth and is employed in accordance with this.   The interior square symbolizes Earth, the exterior circle symbolizes Heaven."   Also LI SHIH-CHÊN, the author of the *Pên ts'ao kang mu*, speaks of such circular *pi* with square central perforations; but, as far as I know, jade disks of such description have not yet turned up. If they existed,— and there is no reason to doubt the correctness of Pan Ku's statement,— they were certainly not identical with the jade disks *pi* having a round perforation and representing only the image of Heaven. The clue to their meaning is given by Pan Ku with the words that the services of talented men were enlisted with them; in other words, they were tokens of reward, or plainly money, and we thus come to the surprising result that they coincided in shape with the well-known Chinese copper coins (*cash*). For this reason, we have to take up this matter again in speaking of the subject of jade money.

The sovereigns of the Chou and Han dynasties were accustomed to make contributions to the funeral of deserving princes and states-men by presenting them with coffins, grave-clothes, jades and other valuables for burial. The bestowing of the jade disk *pi* on the minister Huo Kuang, together with pearls and clothing adorned with jade for his grave, is mentioned in the *Ts'ien Han shu* (DE GROOT, The Religious System of China, Vol. II, p. 410); but not "emeralds," as there trans-lated, as these were unknown to the Chinese in the Han period. In the Old World, the emerald occurs only in Egypt and in the Urals, and its occurrence in India or Burma is not well authenticated (MAX BAUER, Precious Stones, London, 1904, p. 317). The Chinese made its ac-quaintance only in recent times from India; in the "Imperial Dictionary of Four Languages," it is called *tsie-mu-lu* (Manchu *niowarimbu wehe* "greenish stone") corresponding to Tibetan *mar-gad* and Mongol *markat*, both the latter derived from Sanskrit *marakata*, which itself is a loan word from Greek *smaragdos;* to the same group belongs the Persian *zumurrud*, to which the Chinese word seems to be directly traceable.

As jade was a valuable gift, it was also a valuable offering to the gods. We saw that it was offered to the great deities of Heaven and Earth. But on important occasions it was given to other gods too, especially to the gods of the rivers, if their assistance was invoked for the success of an enterprise. When, under the reign of P'ing Kung (B. C. 557–532), the two states Tsin and Ts'i were at war, the troops of Tsin were obliged to traverse the Yellow River. Sün-yen, holding two pieces of precious jade attached to a red cord, invoked the Spirit of the River in the following words: "Huan, the king of Ts'i, full of confidence in his fortresses and defiles, proud of the multitude of his people and soldiers, has rejected our amity and destroyed the treaties of peace concluded with us; he annoys and tyrannizes the state of Lu. Hence, the last of your servants, our prince Piao is going to lead the army of the vassals to punish him for his insolence. I, Yen, his minister, am in charge of assisting his plan. If we shall succeed, you, sublime Spirit, will be honored by our success. I, Yen, shall not cross your river again. Deign, venerable Spirit, to decide our fate!" With these words, he dropped the jades into the water and traversed the river (TSCHEPE, Histoire du royaume de Tsin, p. 258). The God of the Yellow River was invoked also in making a solemn oath (TSCHEPE, Histoire du royaume de Ts'in, pp. 42, 68). Jade pieces were joined also to offerings rendered to the gods. In the year B. C. 109, the emperor visited the spot where the Yellow River had caused a breach in the bank and caused a white horse with a jade ring to be thrown into the river to appease the God of the River (CHAVANNES, Se-ma Ts'ien, Vol. III, p. 533).

Early references to the use of the disk *pi* in sacrifices are given in the *Shu king* and *Shi king* (LEGGE, The Chinese Classics, Vol. II, p. 529).

There are two principal types of the disks *pi*, plain ones the impression of which mainly lies in their beautiful colors and in their size, and smaller ones with elaborate decorations.

The former were chiefly used for three purposes; for the worship of Heaven, to be offered to the Son of Heaven by the feudal princes on a visit to the court (BIOT, Vol. II, p. 524), and for burial (p. 120), the symbolism being the same in all cases, as this disk was the emblem or image of round Heaven. There was perhaps a diversity in the dimensions of the three, the measurement of the imperial disk being given as nine inches; for that used in the cult of Heaven, no measurement is given in the *Chou li*. In all probability, there was also a difference in the colors, as the actual specimens show us. In the collection of WU, there are two specimens called by him *ta pi* "large disks," which he identifies with those presented by the vassal princes; of the one, he does not state the color; the other is of green jade. He further has a *ts'ang pi* of green (*ts'ing*) jade used in the worship of Heaven, and several other large *pi*, one of yellow and one of white jade, both mixed with russet spots.[1] These, I suppose, served for interment, being placed under the back of the corpse.

In our collection, there are three large ancient disks *pi* of nearly the same size which may be considered as images of the deity Heaven. They are extraordinary in dimensions and workmanship. While the one in Plate XXI was, in all likelihood, actually connected with the worship of Heaven because of its greenish color, the two others may have served only burial purposes.

The disk shown in Plate XXI has a diameter of 22 cm, the perfora-

---

[1] There would be no sense in reproducing these figures, as they consist of two concentric circles only.

JADE DISK REPRESENTING IMAGE OF THE DEITY HEAVEN.

JADE DISKS SYMBOLIC OF THE DEITY HEAVEN.

tion being 5.8 cm in diameter, and is 4–9 mm thick. The wall of the perforation is slanting, as may be recognized in the illustration. The coloring of this jade is very curious; a rhomboid central section is colored dark-green while four segments stand out in a light-brown with greenish tinge. A finger-shaped depression will be observed in the lower part, and there is another on the opposite face; presumably, the object was held during the ceremony by placing the fingers in these grooves.

The stone of which the disk in Plate XXII, Fig. 1, is carved resembles in appearance the marble of *Ta-li fu* (in Yün-nan Province) but is indeed jade. The two faces are different in coloring. The one not shown has a white background veined with parallel black streaks in the portion on the right-hand side, a black segment in the upper left corner, and is covered throughout with a veining of russet lines and with patches of the same color. The other face has a yellowish-white background with black patches of curious outlines. In the wall of the perforation there is also a black ring not completely closed. The diameter of this piece measures 20 cm, that of the perforation 5.5 cm, its thckness 1.2 cm.

This disk (in Plate XXII, Fig. 2) is distinguishedialso by its curious coloring. A black background is covered all over with yellow and greenish patches and veined with red serpentine lines. The diameter of the disk is 21.5 cm, that of the perforation 5.5 cm (which is the same measurement as in the preceding specimen). In general it is 1.2 cm thick, but in three places flattened out along the edge and there only 0.5 cm thick.

It appears from the *Huang ch'ao li k'i t'u shih*[1] (Ch. 1, p. 1) that the jade disk *ts'ang pi* is still employed as the image of the deity Heaven in the Temple of Heaven (*T'ien t'an*) in Peking. According to the regulations of the Emperor K'ien-lung, its diameter is fixed at $6\frac{1}{10}$ inches, the perforation having a diameter of $\frac{4}{10}$ inch; it should be some $\frac{7}{10}$ inch and more thick. It is not stated what kind of jade is used for the purpose, except that it is called blue (*ts'ang*). The illustration shows a disk covered with a mass of irregular cloud-shaped veins which are apparently in the stone, *i. e.* a veined stone is chosen for this image.

Because of their convenient form, these disks *pi* were utilized for various other purposes, among others, as we read in the *Chou li*, for the investiture of the feudal lords of the fourth and fifth ranks (p. 86). There are so many varieties of these *pi* mentioned in the ancient texts that it is difficult, if not impossible, to assign a clear definition to each and every given specimen, and even Wu has resigned here to identifica-

[1] See above p. 150.

tions. The two types with "grain" and "rush" patterns mentioned
above, however, stand out clearly. In Fig. 71, a *ku pi* "disk with grain
design" of green jade with red spots is illustrated after Wu who con-
fesses that he does not know its meaning. The four monsters laid

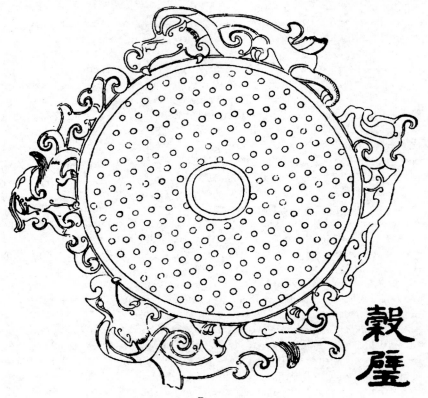

FIG. 71.
Jade Disk with "Grain" Pattern, *ku pi*.

around the edge in open work are all different. The one on the left is
winged and bird-headed; the two above and below, with projecting
fangs, have each a wing on the back. There are also plain *ku pi* in
Wu's collection, with geometrical designs occupying the one, and with
the "grain" pattern taking up the other face.

Figure 72 is a disk styled *p'u pi*, of green jade with russet patches,
with a pattern of "plaited rushes," as Wu explains. Indeed, this
pattern is nothing more than a basketry design bordered by the basket
rim, and presents the interesting case of the transfer of a textile pattern
to a jade carving. This is then, doubtless, what the *Chou li* understands

by the design of "rushes" on the jade token of the feudal princes of the fifth rank. This case is very instructive from another viewpoint in showing how apt the Chinese themselves are in misunderstanding the brief style of their ancient texts; and not only that, but also in basing artistic designs on such misunderstandings. They jumped at the conclusion that the pattern merely styled "rushes" was a living real plant design, and delineated four live rushes on the drawings of these

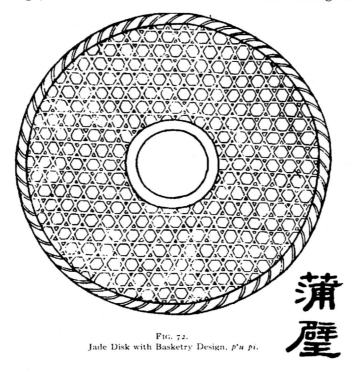

FIG. 72.
Jade Disk with Basketry Design, *p'u pi*.

jade emblems (Fig. 19). But the "rush pattern" was so named because it was patterned after a basket of plaited rush-work. WU illustrates another *p'u pi* displaying two hydras on the upper face and the basketry design on the lower face.

Why was this design applied to the jade image of Heaven? A kind of rush called *shên p'u* is mentioned in the *Chou li* among the vegetable dishes presented to the Son of Heaven, being one of the five salted preparations; this rush was plucked when it began to sprout in the water (BRETSCHNEIDER, Botanicon Sinicum, Part II, No. 375). The *Li ki* (ed. COUVREUR, Vol. II, p. 449) compares a prosperous administration with the rapid growth of rushes and reeds, and this may account for the employment of the ornament on the disks bestowed on the feudal lords of the fifth rank.   On the other hand, rush baskets were used in ancient times to hold offerings for sacrifices[1] (*e. g.* CHAVANNES, Se-ma Ts'ien, Vol. III, p. 617), so that the rush-basket design on the image of Heaven may have well illustrated a sacrificial offering to the Deity.

FIG. 73a.
Jade Disk, *pi*, Upper Face.

Figure 73, derived from WU, belongs to those *pi* which cannot be properly determined now as to their mode of use.   WU explains the one face (*a*) as a dragon pattern, the other (*b*) as a tiger design.   The genuine ancient traditions regarding the significance of these ornaments are lost; the minds of the modern Chinese are turned towards another sphere of ornaments of different style and a different psychical basis. WU's explanation is no more than a modern reflection.   In the upper portion of *a* we observe a conventionalized bird's head, and for the

[1] Several types of bamboo baskets are among the objects still used in the Confucian cult and in the imperial worship of the cosmic powers.

rest a conglomeration of geometric designs of spirals, small circles, triangles and lozenges, a composition of distorted elements which may have suggested an animal's body to a primitive mind. Compare a similar style in Fig. 83 (p. 178) and Fig. 92 (p. 187).

While the design on the upper face (Fig. 73 *a*) is characterized by

Fig. 73*b*.
Jade Disk, *pi*, Lower Face.

asymmetry and a studied irregularity, that on the lower face (*b*) is striking for the very rigidity and symmetry of geometrical arrangement. The circular zone is divided into eight equal compartments[1] filled with exactly the same designs. Two and two of these belong together as may be ascertained from the presence of two eyes, two nostrils and two curved bands apparently forming a mouth. But what these four heads were to be in the minds of the ancient designers

[1] From a comparison with the diagram given in Mayers's Chinese Reader's Manual, p. 346, it appears that the arrangement is identical with that of the nine (counting the center as one) divisions of the celestial sphere (*kiu t'ien*), and this may be intentional on this disk representing Heaven.

escapes our knowledge, and it would be preposterous to speculate as to their meaning. They differ from the tiger-heads represented on the tiger jade-tablets (Figs. 81–84).

The design on the jade disk in Fig. 74 which is of intense interest is interpreted by WU as "a nine-dragons-pattern with three heads *en face* and six heads in profile." But these six heads are manifestly bird-heads, and the arrangement is such that there are three pairs of bird-heads, and that each of these pairs alternates with a dragon-head.

FIG. 74.
Jade Disk, *pi*, decorated with
Dragon and Bird Heads.

Each dragon-head has a bird-couple opposite to it, and as, besides, the bird-couple has the same space allotted to it as a dragon-head, the principle of arrangement is one by six (not by nine), a number corresponding to Heaven represented by the disk. From a viewpoint of mere decorative art, this design is a fillet or interlaced band terminating in birds' and animals' heads. In the decorative art of the Amur tribes many traditions of very early Chinese ornamentation have survived to the present day, and when engaged in a research of these peoples in 1898–99, I had occasion to study their ornamental designs in connection with the verbal explanations received from their makers. In "The Decorative Art of the Amur Tribes," pp. 8–16, I discussed at length the great importance of interlacement bands in their ornamentation, and pointed out that in many cases they bear out a symbolic realism. Thus, *e. g.*, on the sacred spoons of the Gilyak used at their bear-festivals, carvings of bears appear together with fillet-ornaments forming loops; in one, the image of the bear is bound around its body with two ropes crossing each other over its back, referring to the first of the ceremonies of the bear-festival, when the bear is taken from its cage, tied with

FIG. 75.
Jade Disk. *ku pi*, with Band-Ornaments.

ropes, and led to the scene of festivities. In this case, the band-ornament may be considered as the continuation of the ropes with which the carved standing bear is bound, and this may be the underlying reason for the employment of this ornament on spoons specially designed for the banquet of the great bear-festival. Moreover, those spoons carved with figures of bears are decorated with svastikas on the bowls and representations of the sun alluding to solar worship together with the solar character of the bear. In other objects, knotted and looped band-ornaments refer to the use to which these objects are put; *e. g.* a double-knotted band on a girdle-ornament of antler serving to fasten the girdle implies a reference to the knots in which the ends of the girdle are tied, and the symbolism of elaborate knotted bands on an awl of elk-bone used to loosen knots speaks for itself. Among the Tungusian tribe of the Gold on the Amur, I found the representation of a dragon surrounded on all sides by interlaced band-ornaments (*l. c.*, p. 40 and Plate XIII, Fig. 1). I wrote at that time: "The band-ornament is so placed around the monster as to suggest that the animal might be bound with ropes; it is very likely the embodiment of the rain-dragon soaring in the clouds, but hampered by its fetters in pouring out its blessings on the thirsty land." I venture to apply this idea to the design in Fig. 74, which is justified in view of the fact that Chinese and Tungusian art rest on a common historical basis and have influenced each other to a large extent. On this disk *pi* representing the image of Heaven, an atmospherical phenomenon seems to be represented. The birds in connection with the dragons symbolize clouds, as we see from the Han bas-reliefs on stone (LAUFER, Chinese Grave-Sculptures of the Han Period, p. 29). The dragons are fettered by bands, that is to say, they do not send rain, they are in a state of repose. It is the picture of a sky slightly clouded, but serene, over which silver-bright bird-shaped cirri are hovering, encircling dragon-heads in majestic tranquillity.

In this context also the jade disk in Fig. 75 is interesting. It is decorated with four interlaced bands on a background of cross-hatchings. It is a *ku pi*, the grain-pattern *ku* being displayed on the lower face.

We now examine the decorated *pi* in our collection.

The disk of white jade represented in Fig. 1 of Plate XXIII shows two dragons of the hydra type (*ch'ih*) facing each other, carved in high relief and undercut so that they freely stand out from the surface which is smoothly polished. The lower face is decorated with knobs of the same description as in Fig. 2 identical with the "grain" pattern *ku*, except that they are not laid in concentric circles as there, but run in seventeen vertical rows. The diameter of the disk is 9 cm, that of the perforation 1.5 cm; it is 7 mm thick, 1.3 cm, the relief included. The white spots appearing in the illustration are substances of white clay embedded under the surface of the jade. This piece may have been a badge of rank for the feudal princes of the fourth rank.

The disk in Fig. 2 of the same Plate (5 cm in diameter, 7 mm thick) is of a peculiar drab-colored jade (light in weight) with a stratum of black in the upper right portion. The ornamentation on the one face consists of three concentric circles of raised dots (9+15+20), on the obverse of two circular rows of double spirals, six in the outer and four in the inner row. In all probability, it served also as a badge of rank.

1
3
5

6

2
4
7

DECORATED JADE DISKS OF THE TYPE *Pi.*

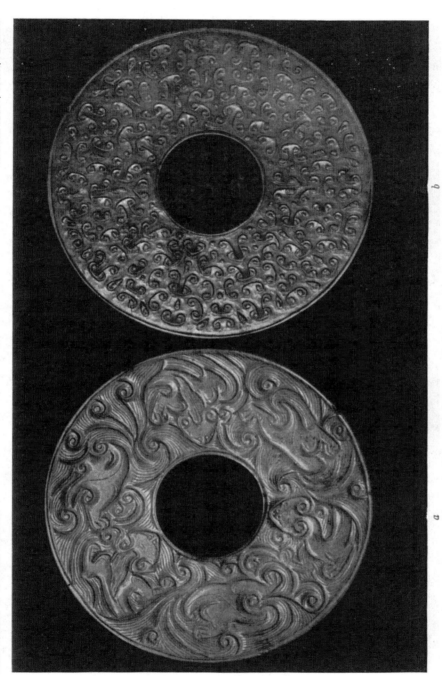

DECORATED JADE DISK *pi*, UPPER AND LOWER FACES.

On the disk in Fig. 3 (7 mm thick) a running *ch'ih*, a hydra of lizard shape in flat relief, is brought out; on the obverse four pairs of double spirals. The diameter is 6 cm, that of the perforation 2 cm, so that it equals the width of the jade ring and the definition of the *pi* does not fit exactly this case.[1] But judging from the style of ornamentation, it belongs to this class. The color of jade is apple-green in various shades clouded with numerous russet specks on both faces and the edge.

The disk in Fig. 4 (8 cm in diameter and 4 mm thick) is of a gray-white lustrous jade with a yellow-brown zone in the centre (called *tai* "girdle" by the Chinese) cut out of the stone in this way intentionally. The ornaments are here all engraved and perfectly identical on both faces. The two designs above and below the circle are similar to those on the dance hatchets of the Han period (Figs. 2 and 3) and even seem to imitate the outline of such a hatchet.

The flat disk (Fig. 5 of Plate XXIII, diameter 4.5 cm) of irregular thickness (1–2 mm) is covered with angular meander ornaments scratched in on both sides. It is a light-green transparent jade with a black yellow-mottled stratum in the left lower portion.

The rectangular piece of green jade in Fig. 7 (4.7 cm long, 3.4 cm wide, 3 mm thick) appears to me a fragment, though the four lateral sides are polished. The decoration consists of rows of single spirals known under the name "sleeping silkworm cocoons" which, as indicated by the circular line below on the left, were arranged in a circular band; hence it is probable that this piece originally formed part of a disk. The lower face is unornamented.

Another fragment which is easier to recognize as belonging to this class is shown on Plate XXV, Fig. 3, which, if wholly preserved, would doubtless be a specimen of great beauty; it is made of a sea-green jade ground into a very thin slab not thicker than 1 mm and even only ½ mm at the outer edge. On the lower face, the same geometrical ornaments as on the upper face are carefully and deeply engraved, and it seems as if the artist, when the pattern on one face was finished, held the thin transparent plaque against the light to make a tracing of these ornaments on the opposite face. When found, all the engraved lines of this piece were filled with vermilion.

The disk, the two faces of which are illustrated on Plate XXIV, *a* and *b*, is a brilliant carving of light-gray jade (11.2 cm in diameter; diameter of perforation 3.5 cm; 5 mm thick). The one side (*a*) is decorated with geometric ornaments in undercut flat reliefs, the single components consisting of the double spiral single or grouped by three, raised dots and pointed wedges acting as space-fillers between.

---

[1] Pieces of these proportions are classed also by Wu among the *pi*.

On the other face (*b*) three dragons of hydra type with heads of geometric cast, the eyes being indicated by spirals, are displayed in the midst of sea-waves, the water covering their waists. As these jade disks originally represent the image of Heaven, there can be no doubt that it was by the idea of the dragon's association with Heaven that the application of dragons to these disks was suggested.

In the introductory notes to this chapter, we mentioned the two types of jade rings *yüan* and *huan*. In grouping them here with the disks, we follow the usage of Chinese archæologists. There is little known about the rings *yüan*. Wu Ta-ch'êng gives two definitions. The one derived from the *Êrh ya shih wên* quoting the *Ts'ang hie pien* (a work of the Han period, see Bretschneider, Botanicon Sinicum, Part I, p. 200) says that *yüan* is the name of a jade girdle-ornament. The other taken from the commentary to the *Shuo wên* by Tuan explains that in enlisting a man's service a jade tablet *kuei* is used, in summoning a man, the jade ring *yüan*. Wu remarks that numerous *pi* of ancient nephrite have survived to the present time, but only a few rings *yüan*. There are two in his collection, and four in our own.

In their make-up, these four jade rings *yüan* represented on Plate XXV (Figs. 1–4) show identical features and only differ in the colors of the jade; they are all unornamented. That in Fig. 1 (10.2 cm in diameter and 3–5 mm thick) is of a grayish-green transparent jade with blue tinges, and brownish specks along the edge. The circle is irregular.

The ring in Fig. 2 (11 cm in diameter and 4 mm thick) is of a white–mottled soap-green jade with black streaks and spots scattered here and there.

The ring illustrated in Fig. 3 (9 cm in diameter and 4 mm thick) is of a color similar to that in Fig. 1, but lighter in shade, interspersed also with russet and white specks.

In the ring shown in Fig. 4 (7.6 cm in diameter) the whole surface is decomposed, the polish, except a small apple-green portion, has disappeared; and one side is completely weathered out, and shrunk to 7 mm compared with the original thickness of 11 mm.

The jade rings *huan* belong typologically to the same class, but ideologically to another group. They were worn as ornaments suspended from the girdle, and for this reason we shall deal with their symbolism in the chapter on Girdle-Ornaments. We here review merely the material in our collection.

The large ring in Plate XXVI measures 15.3 cm in diameter, and 6 mm in thickness, the perforation being 5.8 cm in diameter. It is of a sea-green transparent jade sprinkled with white clouds in the lower

EXPLANATION OF PL. XXV.

Figs. 1, 2, 4, 5. Jade Rings of the Type *yüan*.
Fig. 3. Fragment of Jade Disk *pi*.
Fig. 6. Unique Jadeite Ring with Projecting Ridge.
Fig. 7. Agate Ring.

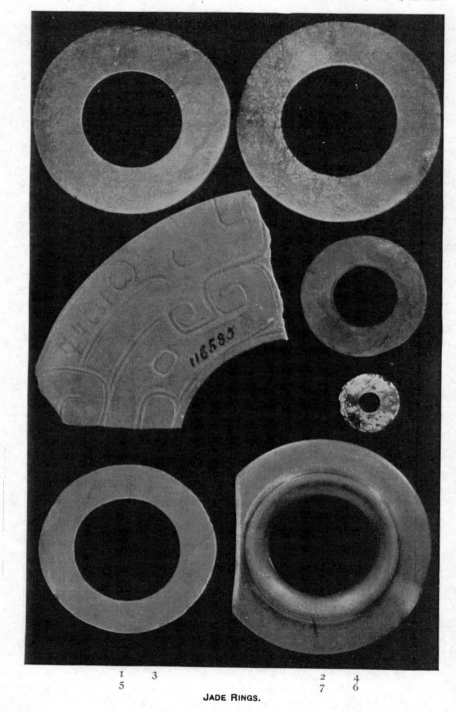

JADE RINGS.

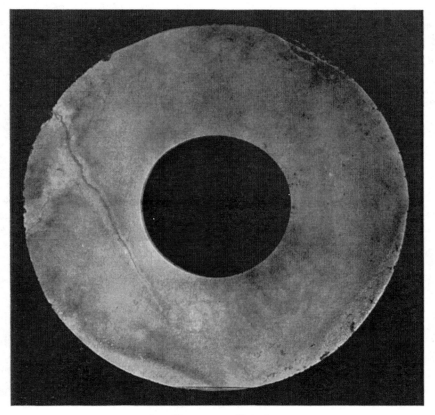

JADE RING OF THE TYPE *huan.*

right portion. The jade still appears enclosed in its matrix, a mass of brown rock. At the lower end in our illustration an incision will be noticed which is just large enough to admit the thumb-nail, and the ring was doubtless grasped or carried in this way. On the lower face, saw-marks are visible.

It is hardly credible that a piece of this dimension and weight should have been worn in the girdle. I am inclined to think that it served a religious purpose. The *Ts'ien Han shu* relates that a jade ring (*yü huan*) was placed in front of the soul-tablet in the ancestral temple of the Han Emperor Kao-tsu (B. C.206-195).

The ring shown on Plate XXV in Fig. 6 (diameter 6 cm) is a unique specimen, none like it being illustrated in any Chinese book. There is a projecting ridge over the perforation on both faces, and on one side, a segment is cut off from the circle in an almost straight line; this edge is rounded, well polished and exhibits a milk-white color, while otherwise this jade has a deep-brown-red agate-like hue. Dr. O. C. Farrington, curator of geology, Field Museum, defines it as jadeite, the specific gravity being 3.3.

Figure 7 on Plate XXV represents a small flat ring of red agate (2.8 cm in diameter) covered all over under the surface with white stripes. Mr. A. W. Slocom of the Department of Geology has examined this specimen and arrived at the conclusion that these are clayish substances already embedded in the stone at the time of carving, and that it is impossible to adopt the view of the Chinese that they are due to the effect of burial or underground action.

Hitherto we have studied the tablets *kuei* and the disks *pi*. There is also a type forming a combination of these two and therefore designated as *kuei pi*. The *Chou li* (BIOT, Vol. II, p. 524) attributes to it a length of five inches and says that it serves in sacrificing to the Sun, the Moon, and the Stars. A good idea of the shape of this symbol will be conveyed by our specimen in Fig. 6 of Plate XXIII, though it is only a miniature edition of the Han period.

It has the annular disk in common with the other pieces on the same plate, but has in addition the so-called tablet *kuei* attached to it, the whole being carved out of the same stone. The upper projecting prong is the handle of the *kuei*, and its point is directed downward. It will be seen from the illustration that it is carved in two sloping planes joining in the centre, while it is flat on the lower face. On the circular part, seven small circles are engraved, connected by lines; this is the usual design representing the constellation of the Great Bear (*Ursa major*) and points to the use of the object in star-worship. On the back are roughly incised two rampant creatures with dragon-like bodies, but with bird-heads;[1] between them a flaming ball or jewel as is often seen between two dragons or two lions; it is very likely an emblem of the sun.

Measured over the *kuei*, it is 8 cm in length and 1.5 cm in width; the disk is 3.8 cm in diameter and 6 mm thick. The ground color of the jade is white in which brown shades of an agate-like appearance are sprinkled, from a light iron-rust or blood color to a dark-brown with black specks.

In the Bishop collection No. 318 (Vol. II, p. 104; the plate is in Vol. I) there is a very curious *kuei pi* teeming with elaborate decorative designs. Dr. BUSHELL defines it as a "sacrificial tablet, previous to the Han dynasty." The central portion, perforated in all other specimens, is here solid and countersunk. The *pi* is decorated with the raised knobs of the "grain" pattern which Bushell calls "mammillary protuberances" by confounding them with the so-called nipples on ancient bronze bells and mirrors, quite a different matter (see HIRTH, Chinese Metallic Mirrors, pp. 250, 257). The upper part of the *kuei* is surrounded by two dragons carved in open work, the lower part filled with wave patterns; on the other face appear the three-legged crow representing the sun and the hare pounding drugs with a pestle in a mortar, — designs springing up only in the Han period. There are three stars on the top and mountains below, — completing the evidence that an attempt has been made at a reconstruction of the Chou tradition, — sun, moon and stars, with the addition of sea and mountains. It is certainly an ancient piece, judging from the character and color of its jade and from its technique, but it cannot be anterior to the Han period.

Another *kuei pi* in the Bishop collection No. 325 (Vol. II, p. 107; plate in Vol. I), also made "previous to Han dynasty" is decidedly much later than the Han period. It is covered with an inscription of four characters in an antique style (but of such a style as proves nothing in favor of antiquity) read by Bushell *ts'ien ku shang hia* and translated by him "The thousand ages of the above and below, that is, of heaven and earth." As I before pointed out, inscriptions on ancient jade pieces are always open to suspicion. But more than this, I am suspicious of the design of five bats as symbols of the five kinds of happiness, arranged on this *kuei*, — which, to my knowledge, occurs neither in the Chou nor in the Han period. From a consideration of the jade court-girdles of the T'ang dynasty, it will appear that the bat as an emblem of happiness on objects of jade occurs not earlier than in the T'ang period.

[1] Compare Fig. 74, p. 162.

3. JADE IMAGES OF THE NORTH, EAST AND SOUTH

We saw from the *Chou li* that the quarter of the North is worshipped under a jade symbol called *huang*, and that the same object is buried with the dead, being placed at the feet of the corpse. The commentators to the *Chou li* explain that the shape of the *huang* was half of the perforated disk *pi*, and that it symbolized the winter and the storage of provisions[1] when vegetation has ceased on earth and only half of Heaven is visible. The latter expression is explained by the saying that the constellations are the ornament of Heaven, the plants the ornament of earth, that in the winter when the plants have withered away, only the constellations remain in Heaven, and that hence the saying arose that only half of Heaven is visible. From the *Li ki* (ed. COUVREUR, Vol. I, p. 545) we learn that in the same way as the tiger-shaped jades *hu*, also the jade *huang* was presented by the emperor to a prince jointly with a wine-cup.

The *ta huang*, *i. e.* the large *huang*, a special kind of this jade ornament in the shape of a semicircle and coming down from the Hia dynasty,[2] was in the ancestral temple of Chou-kung in Lu, and otherwise an object due only to the emperor (*Li ki, Ming t'ang wei*, 24; ed. COUVREUR, Vol. I, p. 738). WU TA-CH'ÊNG has identified one specimen in his collection (Fig. 76) with this large *huang* mentioned in the *Li ki*, though he does not give any special reasons for so doing. He points out the great rarity of this type and emphasizes that it is not identical with the top-piece of the ancient girdle-pendant, designated by the same word *huang*, of which we shall deal in a subsequent chapter.

The *huang* in Fig. 77, derived also from WU, is of white jade with red speckles. "In its shape," remarks the author, "it differs from the large *huang;* it has two perforations in one end and one in the other, but I do not know how it was used; in view of its dimensions, it cannot certainly have been a girdle ornament." I believe that pieces like this one were employed for burial purposes as indicated by the *Chou li.*

The carving in Fig. 78 of white jade with "a yellow mist" is in the shape of a fish engraved alike on both faces. There is an oval perforation in the middle, and there is one in the head of the fish indicating its mouth and another in the tail. In its make-up, says WU, it is of antique elegance, and there is no doubt that it is a jade of the Chou time. To explain the design of the fish in this *huang,* WU refers to the

---

[1] Compare the Hymn to the Winter in the *Ts'ien Han shu* (CHAVANNES, Se-ma Ts'ien, Vol. III, p. 616).

[2] This is an addition of the commentary. WU cites also the *Ch'un Ts'iu* to the effect that the Duke of Lu was in possession of the *ta huang* of the Hia dynasty.

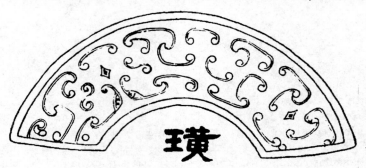

FIG. 76.
Jade Image *huang* of the North.

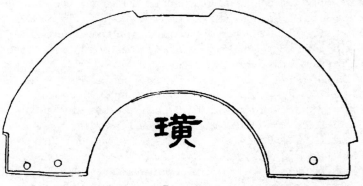

FIG. 77.
Jade Image *huang* of the North.

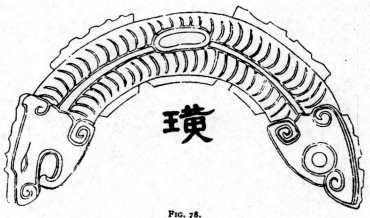

FIG. 78.
Jade *huang* in the Shape of a Fish.

FIG. 79.
Jade of the Type *huang*.

FIG. 80.
Jade of the Type *huang* (from *Ku yü t'u p'u*).

legend of *Lü Shang* or *T'ai-kung wang* catching a carp in whose belly a jade carving *huang* was found with the inscription: "The family Ki will be elevated to the throne, and Lü will aid it." Wên-wang then met him and recognized in him the minister whom his ancestor had selected as the only man capable to make the house of Chou prosper.[1] Here we have an example, — and there are several others,[2]— of jade used in a prophecy. It is possible that this legend may have induced an artist to produce a *huang* in the shape of a fish, but it does not appeal to us as a general explanation, for in our collection there are two burial jades of similar shape in fish-form (Plate XXXVIII, Figs. 4 and 7)

[1] Compare PÉTILLON, Allusions littéraires, p. 249. CHAVANNES, Se-ma Ts'ien, Vol. IV, p. 35. This legend is not narrated in Se-ma Ts'ien, but in the Bamboo Annals (*Chu shu ki nien*).

[2] CHAVANNES, Se-ma Ts'ien, Vol. II, p. 183.

used as amulets for the protection of the corpse, and we have two large jade carvings of fishes originating from graves (Plate XLI, Figs. 1 and 2) so that we can but imagine that the fish must have had a particular relation to the dead.

Figure 79 is derived also from Wu, a *huang* of white jade in which "mercury is absorbed;" but the author says nothing to elucidate this specimen.

The *Ku yü t'u p'u* offers a variety of these *huang* in Ch. 24, the first of which is here reproduced in Fig. 80. It would be all right as a specimen, if the alleged inscription on the *huang* of Lü Shang, above referred to, were not carved on the back of this piece in a neat style of ancient characters, so that in all likelihood the whole affair is a daring forgery of the Sung period. And it is positively asserted in the text that this object is identical in fact with the legendary tablet of Lü Shang discovered in the carp's belly! All the other *huang* are of the same shape with a variety of patterns and all nicely inscribed; it is not worth while speaking of them.

The *Chou li* says (p. 120) that homage is rendered to the region of the East with "the green tablet *kuei*." We remember that Wu Ta-ch'êng has figured a specimen under this name (Fig. 28). But the commentaries to the *Chou li* insist upon this jade emblem being pointed like a lance-head, and remark that it symbolizes the spring (correspond-ing to the East) and the beginning growth of creation (*wu ch'u shêng*). It was, in consequence, a symbol of fertility indicating the awakening of nature in the spring, and therefore also green in color. And in its origin, as we pointed out also for other reasons, it was no doubt a phallic emblem.

With the red tablet *chang*, worship is paid to the region of the South corresponding to the summer to which the red color refers. We gained some idea of the appearance of this symbol from Figs. 34 and 35. As far as I know, no specimen which would allow of identification with this particular symbol of worship has survived. The commentators interpret that the *chang*, being half of the tablet *kuei*, symbolizes the dying of half of the creatures during the summer. It was, accordingly, found by way of geometric construction and symbolism. In the Ritual Code of the Present Dynasty (*Huang ch'ao li k'i t'u shih*, Ch. 1, p. 14 b) it is expressly remarked that no unanimity regarding the shape of this *chang* has been reached, and that it is therefore no longer used in the worship of the sun. In 1748 it was stipulated that the disk *pi* of red color with a diameter of $4\frac{6}{10}$ inches, a perforation of $\frac{4}{10}$ inch in diameter, and $\frac{5}{10}$ inch in thickness should be employed in the imperial Temple of the Sun (*Ji t'an*) for the worship of the Sun. In other words, this means that the ancient image of the Sun of the Chou period had sunk into oblivion, and that it was then reconstructed on the basis of the still known jade image of the Deity Heaven, with the difference that it was on a smaller scale (4.6 : 6.1) and of red (not blue) color. The illustration in the above work shows a jade disk of the same construction as that of Heaven, proportionately smaller, with the same peculiar cloud-like veining in the stone (p. 157).

4. JADE IMAGES OF THE DEITY OF THE WEST

Under the name *hu* (GILES No. 4922) the Chinese archæologists present us a number of ancient jade carvings which differ widely in shape and design. There can be no doubt about the meaning of the word: the character is composed of the symbols for "jade" and "tiger," and the word, accordingly, signifies a jade carving in the shape of a tiger, or a jade on which the figure of a tiger is carved. It is but rarely mentioned in the ancient texts, twice in the *Chou li*, once in the *Li ki*, once in the *Tso chuan*, and first defined in the dictionary *Shuo wên* (100 A. D.) "as an auspicious jade being the design of a tiger, used to mobilize an army." This, however, was a custom springing up under the Han dynasty, connected with the ancient bronze tallies called *fu;* but the shapes and designs of the latter differ from the jade carvings *hu*, and I have no doubt that they represent a type of objects developed independently from the *hu*.[1] During the Chou period, the jade tablet *chang* provided with a "tooth" was employed to levy and move soldiers, the teeth being the emblems of war. From the *Li ki* (ed. COUVREUR, Vol. I, p. 545) and from the *Tso chuan* (LEGGE, Chinese Classics, Vol. V, pp. 739, 741)[2] we learn only that these carvings were given as presents; when an emperor offered a prince a jade in the shape of a tiger or the semicircular jade *huang*, he presented them together with a wine-cup. In these cases we see that these carvings were precious objects. To understand their real meaning, we have to turn to the *Chou li*. There we are informed that the tiger-jade of white color is used in the worship of the western quarter (see above p. 120). The commentator CHÊNG adds that it has the shape of a tiger, and that the tiger in his ferocity symbolizes the severity of the autumn. Again, we learn from the *Chou li* (see p. 120) that, together with five other objects of jade, the tiger-jade was buried in the grave at the right side of the corpse, *i. e.* facing west in the grave. Also there, the figure of a tiger was the emblem or image of the West. It is striking that WU TA-CH'ÊNG does not call any attention to these passages of the *Chou li* and does not investigate this subject; he is content with quoting the definition of the *Shuo wên*, which

---

[1] LÜ TA-LIN, the author of the *K'ao ku t'u* (Ch. 8, p. 3) quoting this definition of the *Shuo wên* remarks that such a statement is not to be found in canonical literature (*king*), and that it is not known on what source Sü-shên, the author of that dictionary, bases his statement. The Han, adds LÜ TA-LIN, availed themselves of bronze tallies representing tigers (*t'ung hu fu*) to mobilize troops, and it may be that his statement is derived therefrom.

[2] LEGGE translates "a piece of jade with two tigers cut upon it," and GILES follows him. All surviving specimens, however, show only one tiger, even when the carving is executed on both faces, which suggests but one animal to the Chinese mind; also from a purely grammatical viewpoint, *shuang hu* can mean only "two tiger-jades."

does not relate in fact to these objects, and the passage from the *Tso chuan*.

The tiger is in this case a celestial deity symbolizing the cardinal palace of the West and at the same time the autumn, as the Green Dragon corresponds to the spring (East), the Red Bird to the summer (South) and the Tortoise to the winter (North). According to L. DE SAUSSURE (*T'oung Pao*, 1909, p. 264) the tiger as symbol of the autumn is Orion. The same author explains the association of this animal with the autumn by referring to the legend according to which Huang-ti tamed tigers for purposes of war so that the name of the tiger remained associated with that of the warrior, and finally by the fact that the tiger descends in the fall from the mountains to invade the human habitations. This explanation is quite correct. SCHLEGEL (Uranographie chinoise, p. 572) has aptly pointed out that, according to the Chinese naturalists, the tigers pair towards the autumnal equinox, and that the tigress brings forth the young ones towards the end of April, so that they will appear about the month of May, the first month of the summer. I can confirm this observation from a personal experience: while once strolling around in the province of Hupeh in the latter part of May, I met a party of sturdy Chinese hunters who had just caught alive a couple of graceful tiger-cubs in the high mountains and offered it to me for sale. They were two jovial little creatures tame like cats, and certainly only a few weeks old. During the summer, they remain in their mountain resorts, — and therefore the tiger is a solar animal (*yang wu*),[1] the lord of the mountains (*shan kün*), the chief of all the quadrupeds (*pai shou chih ch'ang*), — until they have gained sufficient strength to enter on their début in the autumn. For this reason, the image of the tiger is a sign of the zodiac to mark the commencement of the summer, and the beginning of the fall is signalized by the awakening of the young tiger to its full strength. We now understand why in the Han period the tiger symbol was used as a token of command over the army.

Figure 81 derived from the work of WU TA-CH'ÊNG represents a tablet of white jade perforated in the apex showing a band below the aperture on which the figure of a recumbent tiger is engraved in flat relief. The reverse displays the same design. I am under the impression that this object presents the oldest form of these tiger-jades used in the cult. The tablet was doubtless suspended freely from a cord passing through the aperture so that the design of the animal could be viewed from both sides.

---

[1] He is seven feet in length, because seven is the number appertaining to *yang*, the male principle, and for the same reason his gestation endures for seven months (MAYERS, The Chinese Reader's Manual, p. 60).

Figure 82 illustrates a carving of white jade filled all over with clay matter and dug up in Shensi (*Kuan chung ch'u t'u*). It is in the shape of a knob displaying a symmetrical tiger's face of geometrical mould on the front, and a pair of feet carved out on the back. It will, of course, be difficult to say in each particular case to what use a given piece was turned, whether it was intended for the cult or as a gift or

球

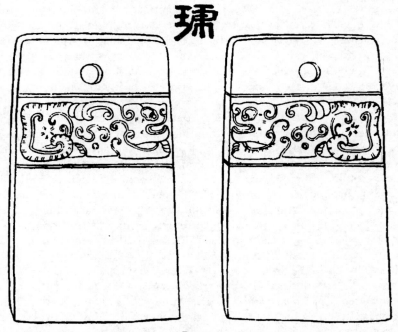

FIG. 81.
Jade Tablet, Upper and Lower Faces, with Design of Tiger, used in the Worship of the West.

as an ornament with a specific meaning which may escape our knowledge. It is not out of the question that in certain localities of China where there was an abundance of tigers the animal was actually worshipped under an image like these, and that these images were used, as rude drawings in modern times, to ward off evil demons (compare GRUBE, Religion und Kultus der Chinesen, pp. 123, 177).

The four strokes engraved twice on the forehead of this tiger-face (Fig. 82) are noteworthy, as they were interpreted later as the written character *wang* "king" which is always found on the modern conventional tiger-heads as they occur on shields of plaited rattan, on soldiers' buttons or on amulets in the form of paintings or woodcuts.

The carving in Fig. 83, also of white jade, and identical on both sides, shows a strongly conventionalized figure of the tiger with an arrangement of spiral ornaments on the body such as is met with also on other jade pieces connected with the symbolism of the quarters.

In Fig. 84 a jade carving of a reclining tiger is reproduced after WU, which I believe was the type buried on the right side of the corpse, as described in the *Chou li;* for the lower face of this specimen is flat and is provided with two oval cavities intercommunicating below the surface for the passage of a thread, and exactly the same method is employed with other jade burial pieces of this description in our collection, indicating that they were fastened to the grave-clothes. For this reason, I believe to be justified in this identification.

The *Ku yü t'u p'u* (Chs. 22 and 23) has deluged us with no less than twelve of these jade carvings of tigers. This very number is apt to cause suspicion, and our suspicion must increase, as all these specimens have their definitions carved into their lower faces and even some the character *hu* on the front. This was certainly not customary in the Chou period. The specimens of WU demonstrate that there must have existed a large variety of these tiger-jades. The

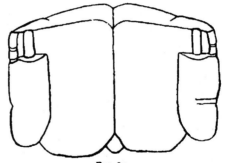

FIG. 82.
Jade Carving of Tiger's Head, Upper and Lower Faces.

Ku yü t'u p'u, however, has only one type different from those of WU and conjugated through the paradigma of the usual decorative scheme. This shows that these twelve specimens are merely artificial and imaginary reconstructions, devoid of archæological value. In Fig. 85 one may be given as example which is sufficient for all. This monster does not bear any resemblance to the traditional representations of the tiger, but is evidently a fish-monster with dorsal fins. This supposition is confirmed by two other figures of *hu* in the same book which are actually covered with fish-scales styled "whale-pattern" (*king wên*). These alleged tiger-jades are therefore not tigers at all and have nothing to do with the ancient *hu* of the Chou period; they are simply ornamental fantasies and inventions of the Sung epoch. The present specimen in Fig. 85 is said in the accompanying text to be of red jade, a statement

FIG. 83.
Jade Carving of Tiger.

plainly contradicting the tradition of the *Chou li* that they were made of white jade, as the specimens of Wu are in fact.[1] The body of the monster is filled with meanders, the "thunder pattern" (*lei wên*), and the explanation is that the tiger has a voice like thunder and was therefore adorned in ancient times with the pattern of thunder, a statement not warranted by any ancient text.

[1]The white color is necessary, as it constitutes the symbolism of "the white tiger" (*pai hu*) who presides over the western quadrant of the celestial sphere.

The explanation for this invention of the *Ku yü t'u p'u* is not far to seek. If we turn to the *Po ku t'u* (Ch. 26), we find there four engravings of bronze antiquities (Figs. 86–89) styled *k'ing* "resonant stones." It appears from this name that these objects are imitations in bronze of the original jade sonorous instruments, of which we shall have to speak later on. Here, the shapes and designs of these instruments interest us in connection with the pattern of the *Ku yü t'u p'u*.

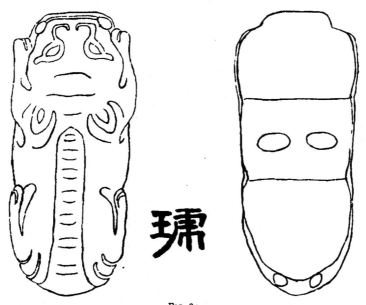

FIG. 84.
Jade Carving of Tiger, Upper and Lower Faces, buried to the West of the Corpse.

Here we observe in fact a fish monster with dorsal fins ending in a bird's head in Fig. 87, and the bodies are filled with compositions of meanders and spirals. Here, they are in their proper place with a significant function, as these instruments were struck with a wooden mallet to produce sounds "like thunder." Now the object in Fig. 88 is called *Chou hu k'ing,* *i. e.* a resonant bronze plaque in imitation of a tiger-shaped jade of the Chou dynasty. The *Po ku t'u* says that this *k'ing* is made in the shape of a *hu* and therefore regarded as and called a *hu,* that anciently these *hu* were employed in the worship of the Western Quarter and derived their shape from that of a tiger, and that also in the interior the figure of a tiger is outlined to characterize the nature of this object. We certainly do not know in how far the drawing of the *Po ku t'u* is correct, and if so, whether this bronze type represents a really faithful reproduction of an ancient tiger-jade, or which is more probable, a much more elaborate design. It can hardly be imagined

FIG. 85.
Jade Carving of Tiger, Reconstruction (from *Ku yü t'u p'u*).

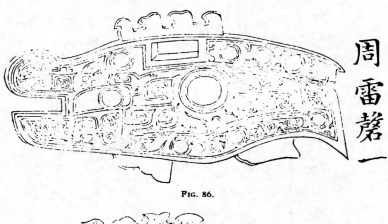

FIG. 86.

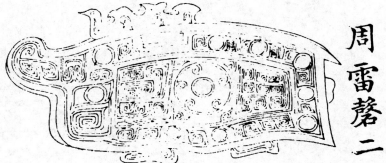

FIG. 87.
Bronze Sonorous Instruments of the Chou Period (from *Po ku t'u*, edition of 1603)

that this complicated structure adorned the jade specimens of the Chou, especially if we compare it with the simpler types of WU. The style of drawing in Fig. 88 is, in my opinion, traceable to ancient Siberian art (with a distant relationship to the famous fish of Vettersfelde), but this question does not concern us here. We mean to establish

FIG. 88.
Bronze Sonorous Instrument of the Chou Period in Shape of Tiger (from *Po ku t'u*, edition of 1603).

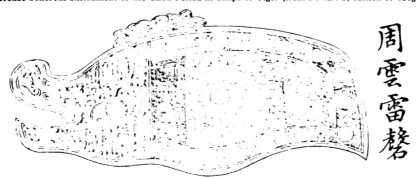

FIG. 89.
Bronze Sonorous Instrument of the Chou Period (from Same Source).

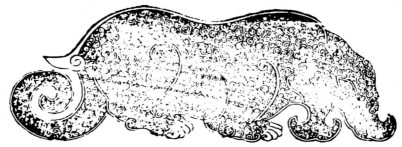

FIG. 90.
Jade Carving of Tapir (alleged Tiger, from *K'ao ku t'u*).

only the case that specimens of this kind have furnished the models to the authors of the *Ku yü t'u p'u* or to the artisans responsible for the collection there described, in framing their type of tiger-jade.

Figure 90 represents a jade carving derived from the *K'ao ku t'u* (Ch. 8, p. 2) and there styled *hu* "jade tiger." But it will be readily seen that this animal is by no means a tiger, but is provided with the head of a tapir.

Under the regulations of the reigning dynasty (*Huang ch'ao li k'i t'u shih*, Ch. 1, p. 51 b), the image of the white tiger has been abolished in the imperial temple of the Moon, because there is no consensus of opinion concerning its real shape. It has been replaced by a disk of white jade (*pai pi*), $3\frac{6}{10}$ inches in diameter with a quadrangular perforation of some $\frac{2}{10}$ inch and more, and some $\frac{3}{10}$ inch and more thick. Judging from the illustration, the stone is pure white, unornamented and unveined.

The tiger was regarded as possessing the power of chasing away demons, as stated in the *Fung su t'ung* (Ch. 8), a work by YING SHAO of the end of the second century A. D. But there are earlier testimonials on record of tigers watching the grave, as the following story will show; and the burial of tiger-heads cast in bronze as practised in the Chou period, as we shall see, was very likely connected with a similar notion.

Ho Lü, the king of Wu (B. C. 513–494) was buried in a triple coffin made of copper. In front of his tomb, a water-course six feet deep was dug; in the coffin, ducks and geese of gold, pearls, and his three precious swords were placed. The tomb was surmounted by the stone carving of a tiger and hence called Tiger's Hillock. The tiger was there to protect the grave, as we see from the legend told concerning the Emperor Ts'in Shih Huang who once passed that place and desired to take the three swords of Ho Lü; but a live tiger then crouched over the grave to guard it; the Emperor seized a sword to kill it; he missed it and struck the stone, the mark still being visible. Then the tiger fled, but the Emperor despite his boring a hole into the grave did not find the swords. According to another tradition, this grave had already been opened before and dishonored by the people of Yüeh on one of their invasions into the kingdom of Wu.[1]

In the collection of Mr. Sumitomo in Osaka, Japan, there is an ancient Chinese bronze of the type of the kettle *yu* shaped into the figure of a tigress suckling a human child.[2] This unique and extraordinary work is doubtless intended to illustrate an ancient legend of the country of Ch'u. Jo Ngao, prince of Ch'u (B. C. 789–763), was married to a princess of Yün who bore to him a son, Tou Po-pi. On his father's death, the boy followed his mother into her native country

[1] A. TSCHEPE, Histoire du royaume de Ou, pp. 99–100 (Shanghai, 1896).
[2] Published in The *Kokka*, No. 163, 1903, Plate II.

Yün, where he was brought up. In his youth, he had an intrigue with a princess of the court there, and the fruit of this clandestine union was a child who subsequently was to be a famous minister in the state of Ch'u. The grandmother ordered the infant to be carried away and deserted on a marsh, but a tigress came to suckle the child. One day when the prince of Yün was out hunting, he discovered this circumstance, and when he returned home terror-stricken, his wife unveiled to him the affair. Touched by this marvellous incident, they sent messengers after the child and had it cared for. The people of Ch'u, who spoke a language differing from Chinese, called suckling *nou*, and a tiger they called *yü-t'u;* hence the boy was named *Nou Yü-t'u* "Suckled by a Tigress."[1] He subsequently became minister of Ch'u. The time to which this tradition is ascribed is the end of the eighth century B. C., and it seems very likely that the bronze referred to presents an allusion to this event and was cast, in commemoration of it, soon afterwards while the story was still fresh in the minds and imagination of the contemporaries. The work exhibits the brilliant technical faculty of bronze-casting of that period, and I should go still further to say that it must have originated from the hands of an artist of Ch'u who created it for the glorification of his country; fertile poetic imagination distinguished the people of Ch'u from the Chinese, as shown by their songs preserved in the *Shi king* and by the famous elegies of K'ü Yüan. If we look upon this production as an artwork of Ch'u, we readily appreciate the fact that this piece is unique and was saved from the doubtful honor of being copied or imitated in later times; the subject was not apt to appeal to the Chinese. In speaking of it in this connection, it was my intention to point out the early deification of the tiger; also in this case, he is a guardian-spirit watching over the life of a child as he drives away the enemies from a grave. As a *deus protector*, his image appears also on Chinese bronzes at an early date. I secured for our collection a bronze *tui* of the Chou period surmounted by the full figure of a tiger; a beautiful bronze ewer excavated near the city of Ho-nan fu in the summer 1910, the spout of which is formed by a finely modeled tiger-head with open jaws spurting forth the water when poured out; and a colossal bronze vessel made in the Court-atelier of the Sung emperors with a cover worked into an imposing tiger-head pointing to an ancient model of

---

[1] Compare LEGGE, Chinese Classics, Vol. V, p. 297, and TSCHEPE, Histoire du royaume de Tch'ou, p. 34. The word *nou* seems to be related to Tibetan *nu* in *nu-ma* "breast" and *nu-ba* "to suck;" in *yü-t'u, yü* is a prefix, and *t'u* possibly points to Tibetan *s-tag* (pronounced *ta*) "tiger." Nevertheless, this very expression, *i. e.* the position of the words, shows that the language of Ch'u did not belong to the Tibeto-Burman, but to the Shan group; for a Tibetan language could make only *yü-t'u nou*, but never in the reverse order.

the Chou period. There is, further, in our collection a unique bronze two-edged sword of the age of the Chou, on the handle of which the fine figure of a tiger is engraved, while the blade is decorated with parallel stripes in black lacquer, in all probability denoting the stripes of the tiger's fur. There can be no doubt of the intention of the symbolism here brought out; it seems to convey the belief that this sword was to be imbued with the strength, spirit and prowess of the tiger, and that his double picture essentially increased its power. We have seen that tiger-heads are represented on the bronze pegs of the Chou wheel-naves (Plate XVI) where they apparently act as defenders of the chariot; they are likewise familiar to us on the bronze vases of the Chou and the Han, and particularly on the Han mortuary pottery vases as relief-heads on the sides. We also remember the running tigers designed in hunting-scenes on the relief-bands of those vases. They occur again as full figures in the bronze tallies used in the military administration of the Han,[1] and in many other examples. This manifold utilization of the tiger furnishes evidence for the ancient belief of the Chinese in his divine powers and attributes, which culminated in the erection of his image for the worship of that celestial and terrestrial region ruled by his spirit, the West. It may be well to bear this in mind in view of the great importance that the tiger has later assumed in Taoism which is composed of popular notions going back in their foundation to a large extent into times of earliest antiquity.

Another anthropomorphic creation of the ancient Chinese is the monster *t'ao-t'ieh* which frequently occurs in conventionalized designs on early bronzes. Dr. BUSHELL (in BISHOP, Vol. II, p. 106) made the following suggestive remark on this subject: "The gluttonous ogre with a fang projecting on each side, no doubt figures the all-devouring storm-god of the Chinese, with a background of clouds. The scrolls indicate thunder (*lei*). The ancient hieroglyph representing this was composed originally of a cross with the 4 ends terminating in spiral curves. The Chinese believe that bronze vessels were first cast by their old sovereign Yü the Great, and moulded by him with the forms of the storm-gods of the hills and the waters, so that the people might recognize their dreaded features and avoid them." It is matter for regret that Dr. Bushell did not develop his view by giving the material on which he based it. We hear indeed of a Master of the Wind and a Master of the Rain in the third century B. C. and in Se-ma Ts'ien,[2] and Bushell's opinion would furnish a sensible explanation for the

[1] CHAVANNES, Se-ma Ts'ien, Vol. II, p. 466.
[2] CHAVANNES, Se-ma Ts'ien, Vol. III, p. 444.

fact that the *t'ao-t'ieh* always appears on a background of meanders emblematic of atmospheric phenomena. But it seems doubtful if this view of the matter covers the whole ground of what the Chinese have to say in regard to this monster. Prof. Hirth (The Ancient History of China, pp. 84–87) has devoted a very ingenious discussion to this subject which should be taken as the starting-point for further investigations. It is very possible, as Hirth thinks, that the Tibetan mastiff lent its features to the iconographic types of this monster, since the mastiff was also deified in the shape of pottery figures to watch the grave in the Han period, and we now know, also at the time of the T'ang dynasty. I do not mean to have said anything conclusive here, but merely wish in this connection to point to this monster as an anthropomorphic conception, a subject which is deserving of a thorough investigation. It is time to insist on viewing this monster in the light of a deity, and not merely to regard it as a purely decorative emblem of vases.

5. Jade Images of the Dragon

The dragon, in intimate connection with the growth of vegetation, appears as a deity symbolic of fertile rain, of rain-sending clouds, of thunder and lightning; it is therefore invoked in times of drought with prayers for rain. A carving of jade, cut out in the shape of a dragon is (in this case) placed on the altar.[1] This object is called *lung* (Giles No. 7491),[2] the character being composed of the classifier *yü* (jade) and the complement *lung* (dragon). It seems that such offerings of

Fig. 91.
Jade Carving of Dragon used in Prayers for Rain.

dragon-figures carved from jade come into existence as late as the Han period; they are not made mention of in the *Chou li* or in the *Li ki*.

Wu Ta-ch'êng has figured three pieces of this type here reproduced in Figs. 91–93. The first, made of white jade with black stripes, is shaped in a half-circle, the centre of which is occupied by a trapezoidal escutcheon-like medallion overlapping the ring; it is filled with designs known under the name "sleeping silkworm cocoons" doubled up, in five rows containing 7, 6, 5, 4 and 3 of them. From this centre-piece two dragon-heads spring forth, their faces turned towards each other;

---

[1] The source for this statement is the dictionary *Shuo wên*: "The *lung* (dragon-shaped jade) is the jade for prayer on behalf of a drought; it has the form of a dragon, and a jade designed with this motive has also the dragon's voice." Hence the phrase *ling lung* denotes the tinkling of jade. In earliest times, prayers for rain were addressed to Shên-nung, the father of husbandry (Legge, *Shi king*, Vol. II, p. 378).

[2] The definition given of this word in Prof. Giles's Dictionary is perfectly correct, while that given by Couvreur (Dictionnaire de la langue chinoise classique, p. 588) "tablet bearing the figure of a dragon and serving to demand the cessation of rain" is erroneous.

the necks are decorated with scales, and above them, fish-tails become
visible, so that the dragons must be conceived of as coiled. It will
be noticed that there are diversities in the delineation of the two heads;
the one on the left has a fish-pattern over the eye and fish-scales on
the upper jaw below the nose, besides, two spirals in the lower jaw and
another over the hind-head. These elaborate details are wanting in

FIG. 92.
Jade Carving of Dragon.

the head on the opposite side; perhaps the two are to be looked upon
as male and female dragons.

In Fig. 92 derived from WU's book which is made of a white jade
with reddish spots, the two dragons almost form a ring, the heads
being separated by a narrow space only as in the half-rings *küeh*. The
geometric treatment of the dragons is remarkable, everything being
dissolved into bands and spirals, and but for the eyes which are plainly
marked, it would be hard to guess the figure of the dragon in this design. The carving in Fig. 93 is stated to consist of white jade over
which a yellow "mist" is spread, interspersed with russet spots; here
the dragon forms a complete ring, and a *Ju-i* sceptre (see Plate LXVIII)
is inserted between head and tail.

WU does not give a period for these three specimens. They are of
a highly elaborate artistic character and doubtless productions of the

age of the Han. This agrees with the fact that these objects were adopted into the cult at that time. The dragon when invoked for rain as the embodiment of the fertilizing power of water thus became a veritable deity. For full details on these prayer ceremonies for rain see G. SCHLEGEL "Uranographie chinoise," pp. 453–459 and DE GROOT, "Les fêtes annuelles," p. 361. If we look upon the dragon

FIG. 93.
Jade Carving of Dragon.

as a deity, we shall at once arrive at a better understanding of the various conceptions of the dragon in religion and art: the manifold types and variations of dragons met with in ancient Chinese art are representatives of different forces of nature, or are, in other words, different deities. At this point, the investigation must set in; we cannot expect to understand Chinese art properly, without being cognizant of all the religious conceptions leading up to its creations.

In the Chinese Journal *Shên chou kuo kuang tsi* (published in Shanghai by the *Shên chou kuo kuang shê*), No. 4, there is the reproduction of an inscribed tablet of jade offered in 928 A. D. to the dragon of the Great Lake (*T'ai hu*) by Ts'ien Liu, king of Wu; this tablet had been found by a fisherman in the K'ien-lung period (1736–1795). Other known tablets of this kind are one of bronze dated 738 A. D., one of silver originating from Ts'ien Liu, and perhaps another jade tablet likewise due to the latter. Similar offerings of the Sung and Yüan periods are known only from literary and epigraphical records, but no actual specimen of those times has survived (compare PELLIOT, *Bulletin de l'École française d'Extrême-Orient*, Vol. IX, 1909, p. 576). It therefore seems that the jade image of the dragon remained restricted to the Han period and was substituted at later ages by prayers inscribed on jade or metal tablets. A survival of the ancient custom may be seen in the large paper or papier-maché figures of dragons carried around in the streets by festival processions in times of drought to insure the benefit of rain.

VI. JADE COINS AND SEALS

In ancient China, jade took also the place of valuable money,[1] and was occasionally also turned into coinage. We remember the statement of Pan Ku that jade disks were given away as a stimulus to scholars and statesmen, if their services were demanded by a particular state (p. 154). KUAN-TSE, the minister of Huan, duke of Ts'i (B. C. 693–642), speaks in his book on political economy of the trade then existing between the different parts of China and the outside countries, mentioning jade, gold and pearls as objects of barter. The former kings, he says, because these things came from afar and were obtained with difficulty, made use of them according to the respective value of each, pearls and jade being estimated highest, gold placed in the second class, knife money and spade-shaped coins ranging in the lowest class.[2]

A jade coin is illustrated on Plate XXVII, Fig. 1. It is a combination of the round *cash* type and the so-called knife on which the back is marked by two parallel incised lines; the blade is broad and blunt, is running in a curve, and pointed below. It is noteworthy that the hole is round, and not square as in the copper coins.

The four characters on the obverse read *ta ts'üan wu shih*, "Great money, fifty."

On the reverse, the symbols of sun and moon are engraved, the sun in the shape of a circle above the hole, and the moon as a crescent below it.

The coin is cut out of a pure-white jade, 7 cm long, 3–4 mm thick; the diameter of the disk being 2.5–2.7 cm. This coin originated from Wang Mang, the Usurper (9–13 A. D.). The same legend as above is found on a common copper coin with square hole and without the knife-handle, on the reverse of which the figures of the sun, moon and the dipper are shown; in another type it is *Ursa major*, a two-edged sword and a tortoise with a snake (see *Kin-shih so, kin so*, Vol. 4). This jade coin is of the same style as the copper coins of the usurper Wang Mang. The identity of the types will be recognized from Figs. 2 and 3 of the same Plate. On the circular portion in Fig. 2 the two characters *i tao* "one knife" are inlaid with gold. On the handle or knife-part, three characters are cast in high relief, reading *p'ing wu*

[1] For analogous examples see H. SCHURTZ, Grundriss einer Entstehungsgeschichte des Geldes, p. 107. Weimar, 1898.

[2] Compare BUSHELL in BISHOP, Vol. I, p. 23, and J. EDKINS in *Nature*, 1884, p. 516.

190

Explanation of Pl. XXVII.

Fig. 1. Jade Knife-Coin of Wang Mang.
Figs. 2-3. Copper Knife-Coins of Wang Mang.
Figs. 4-5. Jade Seals.

1
4
2
3
5

COINS AND SEALS.

ts'ien, "weighing five-thousand." The reverse is blank. The coin in Fig. 3 bears on the obverse the two characters *ch'ih tao*, "contract knife" in slight relief, and on the knife the two characters *wu pai* "five hundred," while the reverse is also blank.[1]

It will be noticed that a raised rim borders the entire coin and also the square hole.

While there is in general no marked difference between the jade and bronze types of this coin, an essential variation remains in the circular and square perforations. It may be appropriate to recall here the theory of L. C. HOPKINS[2] according to which the *cash* is a mere reproduction in metal of the emblematic perforated jade disk *pi*. This view finds a certain support in the statement made in K'ang-hi's Dictionary that the first metal coins were shaped like these *pi;* but the form of the perforation is passed over with silence, and as, in view of Chinese geometric symbolism, a strong contrast must be supposed to exist between a round and a square hole, this important point is left unexplained by the theory of Hopkins. The combination of the *cash* with the knife underlying our specimen certainly was only a personal whim of the fantastic Wang Mang; we may cut out the knife, and then we have a jade *cash* with a round hole, and such seems to have really existed in times before Wang Mang. It seems possible and plausible that this coin may have sprung from the jade disk *pi* with which it agrees except in dimension. A square perforation has not yet been found in a *pi*, but the *Po hu t'ung* mentions *pi* with a square inside, without saying, however, that this square was a perforation (see above p. 154). The square would indicate Earth in Chinese sentiment, but the Chinese have left to us no explanation as to why the holes of their coins have always been made square.

In the numismatical work *Kin ting ts'ien lu* (Ch. 1, p. 1 b) a square metal coin rounded off at the corners with a round hole is engraved (Fig. 94) with the statement that it was issued by the legendary Emperor Shên-nung, the round hole being looked upon as a special characteristic of this type. While we need not accept the association of this coin with Shên-nung or any period of a similar antiquity, I

---

[1] The *Wang Mang* coins are still highly appreciated by Chinese collectors and bring good prices according to the scarcity of the single types. The following price-list has been quoted to me in Si-ngan fu as the present standard of valuation:

Wang Mang	1000 *cash* = 200 cash (about $0.12).
"	" 200, 300, 400, 600 *cash* = 4 Taels each (about $2.40).
"	" 500 and 5000 *cash* = 6 Taels each (about $3.60).
"	" 700, 800, 900 *cash* = 15 Taels each (about $9.00).
"	" 40 *cash* = 50 Taels each (about $30.00).

A piece of the last type is in the collection of H. E. Tuan Fang.

[2] The Origin and Earlier History of the Chinese Coinage (*Journal Royal Asiatic Society*, 1895, p. 330).

have no doubt that this example presents a very ancient coin, perhaps of the Chou period, being an older type than the circular coins with square hole.

It is curious that also the first copper coin of Japan is said to be a piece quite plain, circular in shape and having a round (rather ellipsoidal) hole in the centre. "Illustrations of this coin are found in nearly every old book treating on coins," remarks VAN DE POLDER,[1] "and it is always stated to have been struck in the time of Mombu Tennō (697–709 A. D.); but it is impossible to find out either the exact date or its size and weight."

FIG. 94.
Ancient Metal Coin with Round Perforation (from *Kin ting ts'ien lu*).

Aside from the jade coin of Wang Mang, there is another report relative to the reign of the Emperor Wu (265–285 A. D.) of the Tsin dynasty, who received from the country of Yin-k'in in the northern part of Turkistan a thousand strings of jade coins shaped like rings, each ring weighing ten ounces and bearing on the obverse the inscription: *T'ien shou yung ki*, "May you live as long as Heaven and eternally prosper!"

The *Ku yü t'u p'u* exhibits a large series of jade coins and medals in which little or no confidence can be placed.

Two jade seals are shown on Plate XXVII, Figs. 4 and 5. The one in Fig. 4 carved from grayish-green jade filled on all sides with some clayish substance is a private seal of the Han period bearing on the lower face the name of the owner Ngan Yi (GILES, Nos. 44 and 5397). The upper part of the seal is perforated, so that it could be suspended from the girdle. The other seal in Fig. 5, though carved from jade of the Han period (a plant-green brown and red-mottled jade), is a recent work, and presumably not older than the *K'ien-lung* period (1736–1795). The serpent-like coiled dragon carved in high relief on the top is in the style of the Han period; there is no name engraved on the lower face. The usual material for seals in China is copper or bronze, and, as there is an extensive number of such metal seals in our collection, comprising the periods of the Ts'in, Han, San kuo, T'ang, Sung and Yüan, including also a series of ancient clay seals, there will be occasion to revert to this subject.[2] The shapes and designs of the

[1] Abridged History of the Copper Coins of Japan (*Transactions of the Asiatic Society of Japan*, Vol. XIX, p. 427, 1891).

[2] An exposition of the imperial and official seals and their functions will be found in P. HOANG, Mélanges sur l'administration, pp. 57 *et seq.*, Shanghai, 1902.

jade seals do not differ from those of the metal ones; in fact, they are derived from them so that they must be treated in connection with this subject. Wu Ta-ch'êng has figured also a number of interesting jade seals of the Mongol period engraved with Tibetan and Mongol characters, the latter of the so-called "square" form (*P'ags-pa*) which I shall deal with in another place.

VII. PERSONAL ORNAMENTS OF JADE

This subject is vast and complex, but I have attempted to treat it as comprehensively as possible, as far as the present state of our knowledge permits. Stress is naturally laid on the ornaments used in the earliest times of Chinese antiquity; without arriving at an adequate understanding of these, we cannot hope, either, to appreciate those of the present age. Personal ornaments always exercised a deep influence upon social life, being full of hidden emblematic significance and speaking a language of their own understood by the donor and the wearer. Their shapes and their designs are living realities. In their execution, the Chinese genius shows at its best and rivals that of the greatest lapidaries of all ages. Neither the cut gems of Greece nor the much praised Netsuke of Japan come up to the ideal standard of these humble carvers, their refinement of taste, their vigor and elegance of design, their zest for linear beauty, and their almost superhuman mastery of the tough stone material with their crude implements. With all our progress in technical matters, we often stand in bewilderment before these gems, puzzled as to how they did it. A rich source of instruction may open up here also for our art-designers and craftsmen who are desirous of forsaking the old ruts and of receiving a new stimulus. The ethnological and art-historical importance of this material cannot be overvalued: it is a sort of object-lesson for the study of decorative forms and designs.

1. THE GIRDLE-PENDANT

Jade pieces were worn as girdle-ornaments in most ancient times, and the wearers rejoiced in their tinkling and clattering while walking, and imagined they heard real musical tones produced by these. "The gentlemen in times of antiquity," says the *Li ki* (*Yü tsao* III, 6) "were sure to wear jades suspended from the girdle. Those on the right side emitted the notes *Chih* and *Kio* (the fourth and third notes of the Chinese gammut), and those on the left gave the notes *Kung* and *Yü* (the first and fifth notes of the gammut)." This was, so to speak, the accompaniment to music actually played; for certain tunes were performed in connection with the emperor's walks, and in all his movements, the pieces of jade sounded their tinklings. So also the gentleman, when in his carriage, heard the harmonious sounds of its bells; and when walking, those of his pendent jades; and in this

194

way evil and depraved thoughts found no entrance into his mind. Jade here appears as the embodiment of excellent qualities and instilling virtue into the heart of its owner.[1] Its musical qualities caused a joyful disposition, which may have led to the observance of laying the pendants aside during the time of mourning, while otherwise they were obligatory (*Yü tsao* III, 8).

Special regulations were mapped out for the heir-apparent. "When in the presence of his father, he was not allowed to let his girdle-jades hang down freely nor to have them sound. Therefore, to prevent this, he tied them together on his left side, while he wore on his right side the usual instruments destined for the girdle. In his private apartments, he allowed his girdle-ornaments to hang down as usually, but at court, he tied them up. When an official purified himself by fasting, he took the girdle-ornaments, but tied them up." (*Yü tsao* III, 7.)[2] The tinkling of the jades as an occasion of joy had to give way to the respect for the emperor, and during the religious act of abstinence. In all other cases, it was a kind of moral obligation to wear the girdle-jades, for "a gentleman, without special reason, never took the jade off from his body, regarding it as emblematic of virtuous qualities" (*Ibid.*, III, 8).

In a song of the *Shi king* (Legge, Vol. I, p. 102), a royal lady of the house of Wei, married in another state, expresses a longing for her former home and the scenes of her youth where she could freely ramble with her companions, in elegant dress and happy chats. And she remembers in her sorrow the girdle gems of her friends, how they moved to their measured steps! In another song (Legge, Vol. I, p. 198) in praise of a king, his subjects enjoy hearing the jades at his girdle emit their tinkling, suggestive to them of wishing him long life. The joy over the success of a victorious general is described by the tinkling of the bells of his horses and the sounding of the gems of his girdle-pendants (Legge, Vol. II, p. 286).

It is not quite clear whether the jade pieces used for girdle-ornaments were strung on silk threads or sewed on to silk bands as appliqué work. In the colors of the silk,[3] and in the kinds and qualities of jade

---

[1] In a song of the *Shi king* (Legge, Vol. I, p. 136) the beauty of a lady is praised which is heightened by the fine gems of her girdle-pendant appearing as she moves in her carriage. And the gems of her girdle-pendant tinkle. So far Legge. But I understand the text in the sense that both the man and the lady wear the tinkling girdle-pendants, and that the last verse *tê yin pu wang* does not merely mean "his virtuous fame is not to be forgotten." *Tê yin* is the voice of his virtue sounded by his tinkling jades which those cannot forget who had occasion to hear them.

[2] Legge's translation is here hardly correct, and I prefer to follow Couvreur.

[3] I omit the color-names of the silks, as they cannot be adequately translated. Those who will compare Legge's and Couvreur's translations, will notice their discrepancies in this regard.

there were in the Chou period five gradations according to rank. White jade considered the most precious was the privileged ornament of the emperor; jade green like the mountains was reserved for the princes of the first and second ranks (*kung hou*); water-blue jade was for the great prefects (*tai fu*); the heir-apparent had a special kind of jade called *yü* 'assigned to him; and a plain official had to be content with a stone inferior to jade called *juan min* (presumably prehnite) (*Yü tsao*, III, 9).

In a beautiful song of the *Shi king* (LEGGE, Vol. I, p. 134), suggestive of a pleasant picture of domestic life, a wife expresses her affection for her husband and encourages him to cultivate friendships with men of worth to whom she would offer jewels for the girdle out of regard for him.[1]   "When I know those whose acquaintance you wish, I will give them various girdle-ornaments.   When I know those with whom you are cordial, I will send to them various girdle-ornaments.   When I know those whom you love, I will repay their friendship (or thank them) by gifts of girdle-ornaments."   We see that presents of such ornaments were prompted by a feeling of amity and by a desire to keep up friendly relations.

It is an irreparable loss that the proper significance of many words designating either particular kinds of jade or ornaments of jade and occurring in the *Shi king* and other ancient texts is entirely unknown; the commentators are too easily satisfied in explaining them as a beautiful jade or an ornament.   If we had fuller definitions of them, we could make much more out of the symbolism which was probably associated with them.   Thus, *e. g.* in a little song of the *Shi king* (LEGGE, Vol. I, p. 203) Duke K'ang escorts his cousin of whom he was very fond, and gives him *k'iung-kuei* stones for his girdle as a parting gift; the translation "precious jasper" in this passage is a poor makeshift, for it is doubtless the question here of a specific ornament with a hidden meaning suitable to the occasion, which, however, is unfortunately lost.

In discussing the single jewels composing the girdle-pendant, we shall notice how deeply they are related to friendship and love by means of punning upon the words used to designate them (so-called phonetic rebus).   Here we speak of the significance of the girdle-ornaments in general.   They were an object of mutual attraction between the two sexes and naturally played a rôle in sexual imagination, as the desire to please the other sex is the keynote of all primitive orna-

---

[1] I do not understand the passage with LEGGE that "she would despoil herself of her feminine ornaments to testify her regard for them (*i. e.* her husband's friends)," which, even granted the greater freedom enjoyed by woman in ancient times, would never be congruous with Chinese customs.   The text implies only that she would gladly give any girdle appendages (not her own) as gifts to his friends, out of respect for, or to please, her husband.

ment. "O you with the girdle-gems strung on blue silken bands, long, long do I think of you; I cannot go to see you, but why do you not come to me?" laments a young lady mourning the indifference and the absence of her lover, in the *Shi king* (LEGGE, Vol. I, p. 144).

In the *Po hu t'ung* (Ch. 4, p. 4)[1] by PAN KU († 92 A. D.) it is on record: "The girdle-ornaments symbolized the respective occupations of people: the farmers wore girdle-ornaments in the shape of their plough-handles and shares; workmen those in the shape of axes and adzes; married women wore their needles and pins in the girdles to make known thereby that they were married women, but they suspended also jade objects from the girdle." This passage is interesting in showing that among the people girdle-ornaments indicated also their callings. We thus recognize altogether a fourfold symbolism associated with them during the Chou period: rhythmical movement and sonorous qualities enjoyed by the wearer and impressing his fellow-mates;[2] their character as trinkets of friendship and love; their indication of rank among the official class according to the material; and their character emblematic of the vocation among the people at large.

The general arrangement of this ancient girdle-pendant may be gathered from Fig. 95 derived from the *Ku yü t'u p'u* (Ch. 53). Although not correct in details, and nothing more than an attempt at reconstruction, yet it may give a fair idea of what the conception of the Chinese archæologists in regard to the appearance and disposition of the single ornaments is; for no complete sets have survived, only single components, and as we shall have to discuss a number of these, the wrong impressions gained from this afterthought of the Sung period may be corrected. We notice that this ancient girdle-pendant is a chatelaine consisting of seven separate articles, to each of which a special name is assigned. There is a top-piece or brooch called *hêng* serving as the support from which the six other ornamental pieces are suspended. There is a circular central plaque (*yü*) surrounded by two square ornaments (*kü*); below, an ornament in the shape of a segment (*ch'ung ya*) in the center, and two bow-shaped ornaments on either side (*huang*). For brevity's sake, and to avoid the Chinese names as much as possible, my nomenclature will simply be: head-piece, central piece, lower piece, central side-pieces, lower side-pieces.

In the ancient songs of the *Shi king* (*Wei fêng*, X; LEGGE, Vol. I, p. 107), there is a little ditty under the title "The Quince" as follows:

---

[1] Reprinted in the collection *Han Wei ts'ung shu*.

[2] At the sacrifices to Heaven and Earth during the Han dynasty, the dancers singing the hymns and accompanying them with a dance, wore also girdle-pendants of pearls and jade to chime in with the rhythm of the music (CHAVANNES, Se-ma Ts'ien, Vol. III, pp. 613, 617, 621).

"Who will give me a quince, I shall return to him a central side-ornament of fine jade for the girdle-pendant. It is not meant as an act of thanks, but I want to render our friendship ever-lasting.— Who

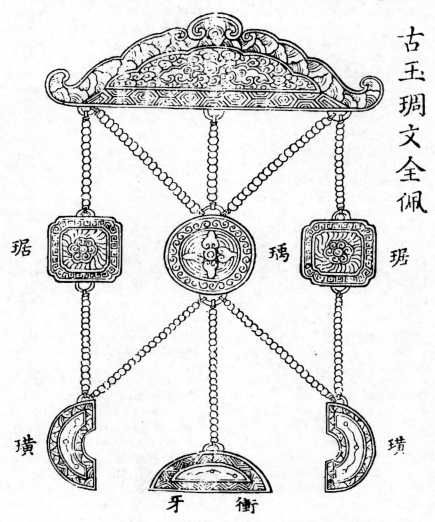

FIG. 95.
Complete View of the Ancient Jade Girdle-Pendant (from *Ku yü t'u p'u*).

will give me a peach, I shall return to him the red jade *yao*. . . [with the same refrain].— Who will give me a plum, I shall return to him the black jade ornament *kiu* . . . [with the same refrain]."

A. CONRADY[1] has happily recognized in the two words *kü* and *kiu* the foundation of a rebus: the former contains the word *kü* "to dwell, to remain," the latter the word *kiu* "long."[2] The verses from the *Shi king* prove that these stones were presented as tokens of a lasting friendship: thus, the gift of the jewel *kü* may have implied the wish: "May you remain my friend!" and that of the jewel *kiu:* "May we long be friends!" This symbolism of the gem *kiu* is especially noticeable in the pretty song "Hemp grows on the Hill" (LEGGE, Vol. I, p. 122): a maiden is longingly awaiting the presence of her lover who she imagines may be detained by another woman; she expects that he will soon appear and present her with *kiu* stones for her girdle. Here the desire for this very ornament is doubtless of symbolical meaning too: she is longing for the long-stay-with-me stone. And the gift of such a stone may have alluded also to lasting love. On the other hand, the jewel *kü* seems to have alluded to the happy union of two lovers possessing each other, as may be inferred from Song VII, 9 (LEGGE, Vol. I, p. 136) where the couple is driving in the carriage (note the word *t'ung* "together") and wearing the girdle-ornament *kü* as emblem of their communion.

We have no exact information in regard to the symbolism expressed by the arrangement of the seven jewels in the girdle-pendant of the Chou. But the names assigned to them allow of the conclusion that they related also to the nature-cult prevalent in that period. The name *hêng* for the top-piece appears as *yü hêng* "the regulator of jade" for one of the stars in the Great Bear, or for the three stars forming its tail (SCHLEGEL, Uranographie chinoise, p. 503), and the word *huang* for the lower side-piece is at the same time the designation for the semicircular jade symbol under which the quarter of the North was worshipped (p. 169). In short, the fundamental idea underlying this girdle-pendant seems to be associated with that cosmological formula of six terms, the basic dogma of the Chou culture pervading the official hierarchy, the astronomic and cosmic system, the sacrificial rites, the religious beliefs and the interment of the dead.[3]

We shall now submit the single parts of the girdle-pendant to a closer inspection. The head-piece is called *hêng* (GILES No. 3910), a word identical in sound with and supposed to be derived from *hêng*

---

[1] In the preface to *Stentz, Beiträge zur Volkskunde Süd-Schantung's*, p. 10, and *China* in *Pflugk-Harttung's Weltgeschichte*, p. 511.

[2] It should be remembered that the Chinese have a double enjoyment in these puns, by the ear and by the eye; the written character conveys to the mind as much of it as the sound.

[3] Compare L. DE SAUSSURE, Les origines de l'astronomie chinoise (*T'oung Pao*, 1910, pp. 257 *et seq.*).

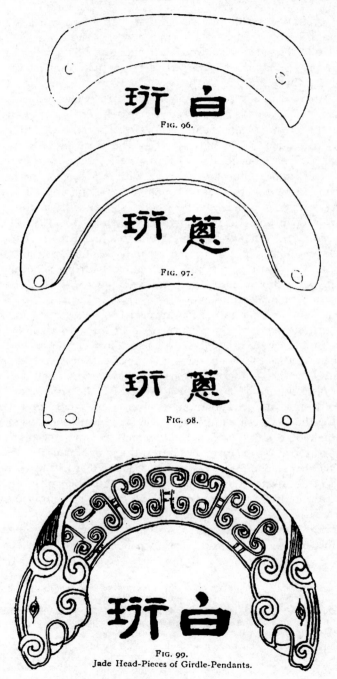

FIG. 96.

FIG. 97.

FIG. 98.

FIG. 99.
Jade Head-Pieces of Girdle-Pendants.

(No. 3912) "crosswise, horizontal" meaning also several objects placed in such a position as the yoke of a draught animal, horse or ox, the beam of a balance or steel-yard, and even the space between the eyebrows. The name for the ornament would therefore imply that it is placed transversely to serve as support to the six pendulous jewels. An ancient commentator remarks that it resembles in shape the resonant stone *k'ing*, but is smaller than this one. WU has succeeded in identifying several specimens with this head-piece. The first of these is shown in Fig. 96, made of white jade with russet spots; the second in Fig. 97, of green jade with black veins, defined as *ts'ung hêng* "onion head-piece" in allusion to a passage in the *Shi king*[1] where this phrase occurs. The word "onion" certainly refers to the peculiar green tinge of the jade, and BUSHELL (in BISHOP, Vol. I, p. 26) correctly explains that the simile relates to the color of the young sprouts, not to the bulb of the onion, the name corresponding to our "grass green."[2] It will be seen that these two pieces are perforated in the extreme ends for purposes of suspension. That in Fig. 98, likewise "onion-green," has two perforations in one end and one in the other. These three objects are plain, while that in Fig. 99 is decorated, the two hanging tips terminating in dragon-heads; it is of white jade spotted all over with red patches except the two dragon-heads which are pure-white.

In connection with the head-piece we deal also with the lower side-pieces of the girdle-pendant, because they much resemble the top-piece, as may be gathered from Fig. 100. In Fig. 101, an ornamented piece of this type in green jade is shown, displaying two dragons with bodies intertwined. A Chinese commentator remarks that, if the two pieces forming a pair are placed together, they make the perforated disk called *pi*, and maybe this ornament has originated in this way by cutting a disk into halves, from a purely technical viewpoint. It will be seen that it is somewhat difficult to distinguish between the head-piece and these lower side-pieces, particularly when they are undecorated. It seems that the only real difference between the two, in the latter case, lies in their dimensions, the head-piece being of larger size. A specimen in our collection may be identified with one of these types.

The objects united on Plate XXVIII are all burial jades of the Han period, not amulets, however, to preserve the body from decay, but personal ornaments of the dead, ornaments which a person had

---

[1] *Siao ya, T'ung kung*, IV, 2 (LEGGE, Vol. II, p. 286; ed. COUVREUR, p. 205).

[2] WU TA-CH'ÊNG explains this color by the two words *ts'ang tsui* "sky-blue and kingfisher-blue."

presumably worn during life-time and cherished, and which he therefore desired to have buried with him.

The piece in Fig. 1 (9 cm long, 2.9 cm wide and 3 mm thick) of this Plate is a grass-green jade full of earth incrustations, unornamented, in the shape of a crescent. The two perforations at the ends have been bored only from one side where they measure 5 mm in diameter, while it is only 3 mm on the lower side. The lower edge has been much affected and somewhat weathered out by chemical influences

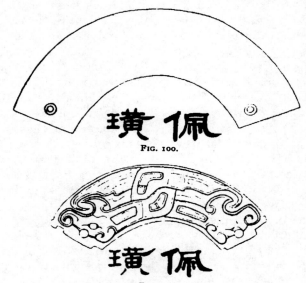

FIG. 100.

FIG. 101.
Jade Lower Side-Pieces of Girdle-Pendant.

which seem to have also darkened the original color in some places. In color, this specimen agrees with those designated by Wu as onion-green and in form with that in Fig. 100, explained by him as a lower side-piece of a girdle-pendant, and I feel therefore justified in applying the same identification to our specimen. The jade ornament published in the *Boston Museum of Fine Arts Bulletin* (Vol. VI, p. 55, 1908) must be reckoned, in my opinion, as belonging to the same class.

Figures 102 and 103 are head-pieces of girdle-pendants illustrated in the *Ku yü t'u p'u*, alleged to be Han, but in all likelihood not older than the T'ang period, especially the design in Fig. 103, an inverted lotus-leaf with turned-up edges.[1] But the *Ku yü t'u p'u* has well

---

[1] Pictorial influence is manifest in it and admitted in the Chinese text by the words: "It is refined and sublime like painting."

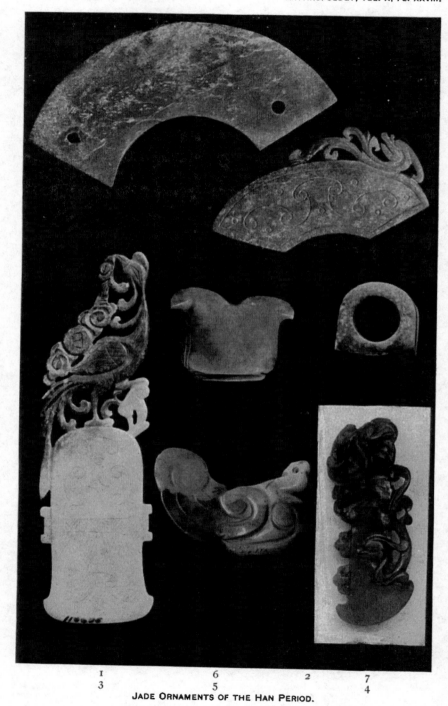

JADE ORNAMENTS OF THE HAN PERIOD.

deserved of us in transmitting another type of chatelaine in Fig. 104 with the remark that it is "a devil's work (*i. e.* very clever work) from Turkistan." The chains with their links carved from one piece of jade are remarkable; the same kind of work may be seen in one of our jade resonant stones (Plate LVII). Ten of these chains terminate in tiny jade bells. This is, of course, a special type of chatelaine being

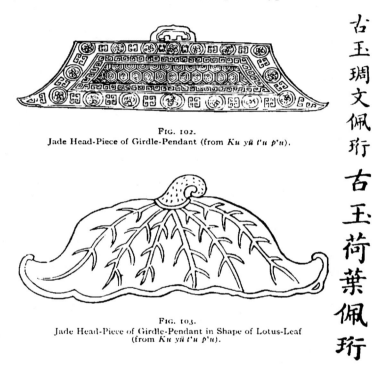

FIG. 102.
Jade Head-Piece of Girdle-Pendant (from *Ku yü t'u p'u*).

FIG. 103.
Jade Head-Piece of Girdle-Pendant in Shape of Lotus-Leaf
(from *Ku yü t'u p'u*).

in no historical connection with the ancient Chinese girdle-pendant. The editors of the *Ku yü t'u p'u* are, in this case, quite honest and confess that the age of this piece is not known. I believe we shall not err in assigning it to the T'ang period, since at that time there was lively intercourse between China and Turkistan, and the trade in jade from Khotan to China was at its height. Besides, this speci- men is of great historical value, as it seems to be the father of these modern silver chatelaines[1] current all over China, Tibet and Mongolia, and usually consisting of five pieces,— toothpick, tweezers, earspoon, small brush for oiling the hair and boar's tooth for parting the hair,

[1] Numerous varieties from China and Tibet are in our collection. See Fig. 107 in BUSHELL, Chinese Art, Vol. II.

scratching the head or loosening knots.[1]   If we realize the jade specimen in Fig. 104 with the chains made in silver and with these five implements attached to the ends of the five lower chains, we have the
whole affair as still in use.   And we may imagine that also at an early

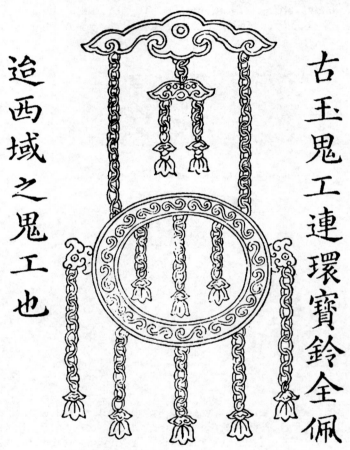

FIG. 104.
Jade Chatelaine from Turkistan (from *Ku yü t'u p'u*).

date they were commonly made of silver and but rarely of jade.   I
was hitherto under the impression that this silver chatelaine may be
Chinese in origin, but the evidence furnished above might lead one

[1] In the Chinese specimens, the two latter objects are replaced by a miniature
sword and halberd or some other kind of weapon. The tradition of the Turkistan
prototype is also preserved in tiny silver bells suspended from chains in many of
these chatelaines.

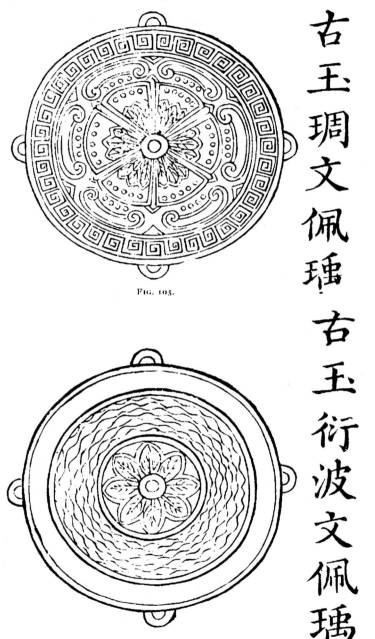

FIG. 105.

FIG. 106.
Jade Central Pieces of Girdle-Pendant (from *Ku yü t'u p'u*).

古玉珇文佩瑶. 古玉衍波文佩瑂

to suspect that it belongs to the Turkish-Tibetan culture-sphere.[1] Indeed, many Tibetan specimens of this kind have an originality of design not to be met with in the Chinese counterparts. At all events, it is certain that in ancient times the Chinese did not avail themselves of similar chatelaines which seem to spring up among them not earlier than in the T'ang period.

Of the central pieces of the girdle-pendant, only a few have survived. The *Ku yü t'u p'u* is able to reproduce only two (Figs. 105 and 106). The one in Fig. 105 with a meander pattern in the outer zone and a floral design in the centre explained as "banana-leaves fulfilling every wish" (*ju i tsiao yeh*) is made a work of the Han or Wei period; the other in Fig. 106 with a band of wave-ornaments (*yen po wên*) and a star-shaped leaf-pattern is alleged to be pre-Han.

Two of the central side-pieces (*kü*) used in the girdle-pendant may be viewed in Figs. 107 and 108, both ascribed to the Han time which is possible. The names given by the *Ku yü t'u p'u* for the ornaments are interesting. The eight groups of quadruple "square" spirals are styled *tieh shêng, i.e.* accumulated, superposed or repeated ornaments, the word *shêng* denoting an ornament in a woman's hair-dressing.[2] Fig. 108, as will be seen from the Chinese legend, is labeled *wei kio liao* "carved with the corners cut out" alluding to the four chamfered corners. The four plant designs in the interior are designated as "mallows;" the entire pattern is stated to be "of such elegance that there is no doubt that this is an object of the Han." This elegance of the art of the Han is frequently insisted on by these authors, and they are certainly right in their judgment.

Figure 109 represents a lower central jewel (*ch'ung ya*) used in the girdle-pendant, derived from the same work. It is segment-shaped as in Fig. 95, and decorated with designs of "sleeping silkworm-cocoons (*wo tsan*) and beads (*lien chu*)," a design "breathing the spirit of Han work." The Sung Catalogue gives, further, this ornament without any decoration.

We are not bound to assume that all the ancient girdle-pendants looked like the typical one in Fig. 95 which is nothing more than an attempted reconstruction of the Sung period. There is, in fact, no

---

[1] The type of these pendants is widely disseminated and occurs among the antiquities of Siberia and the Finno-Ugrians. A striking analogy to the ancient jade type of Turkistan is offered by silver types found in tombs near Kasan (see DE UJFALVY, Expédition scientifique en Russie, Vol. III, p. 151, and Vol. VI, Pl. XXIII).

[2] In modern Peking ornaments occurs the *fang shêng*, a geometrical figure consisting of two overlapping squares, employed as a rebus in the sense of "flourishing condition" (GRUBE, Zur Pekinger Volkskunde, p. 147). The meaning of *tieh shêng* in application to the above ornament may, accordingly, have been "duplicated or repeated abundance" or "may you thrive in all ways and directions as this meander!"

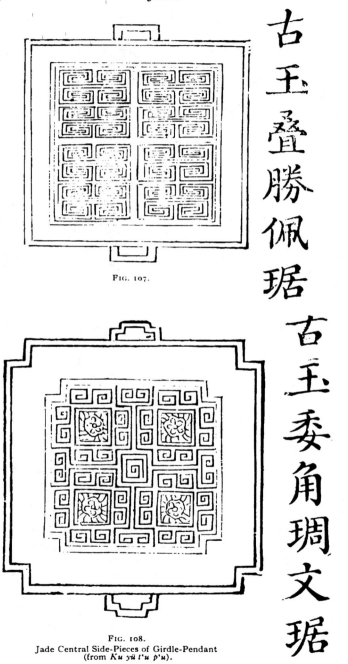

FIG. 107.

FIG. 108.
Jade Central Side-Pieces of Girdle-Pendant
(from *Ku yü t'u p'u*).

古玉疊勝佩琚

古玉委角琱文琚

ancient text by which the arrangement of the single ornaments in the given order would be backed up. We can only say that it *may* have been so, but that it *must not* have been that way. The connection of the single articles by means of bead laces is also doubtful. The verses quoted from the *Shi king* have shown us that there were other girdle-ornaments like the *yao* and *kiu* entering the girdle-pendants, but which are not known to the archæologists. Further, we know from literary records as well as from a large number of specimens which

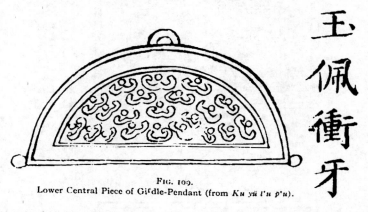

FIG. 109.
Lower Central Piece of Girdle-Pendant (from *Ku yü t'u p'u*).

have come down to us that there were other types of girdle-ornaments. Rings, half-rings and miniature imitations of implements are prominent in this group. I may first be allowed to review the material in our collection.

Figure 2 of Plate XXVIII is of a peculiar jade yellow in color and spotted brown and red; in shape similar to Fig. 1 on the same Plate, but not perforated. The same design of spirals engraved on the upper surface appears also on the lower face. To the upper edge, the figure of a hydra (*ch'ih*)[1] is attached in openwork carving, the whole being cut out of one stone. It is 7.2 cm long, 2.3 cm wide and 7 mm thick. This object belongs to the type of half-rings called *küeh* which played such an important rôle in the life of antiquity on account of a peculiar symbolism attached to it (see p. 210).

The two specimens reproduced in their original size in Figs. 3 and 4 of the same Plate are most remarkable for the brilliancy of their technique; they are both ornamental adzes worn as girdle-pendants. The piece in Fig. 3 is a flat carving showing exactly the same designs and colors on both sides. The axe itself is of a grayish-white color;

[1] Chinese explained to me this figure as a tiger, but I see no basis for this supposition.

the crane forming the handle gray-black; its beak and the cloud-ornaments on its left surrounding the sun-ball as indicated by the character
ji engraved in the circle are yellow; the fungus of immortality (*ling
chih*) on which the crane sets one of its feet, and the tip of its longest
tail-feather are again of the same gray-white color as the axehead.
The colors are by no means artificial, but natural in the stone, and it
is a source of astonishment how the artist planned the working out
of the rough stone into this harmonious arrangement of colors. The
agreement of its ornamentation with the large ornamental axes (p. 42)
will be noticed; the blade is brought out, but blunt.

This design apparently expresses a wish by means of punning; the
word *ling* in the name of the fungus standing for another *ling* meaning
"duration of life," *ji* "the sun" meaning also "day," so that we can
read the sentence *ji hao ling*, "May your days be long like those of
the crane!" The crane, as well known, is a symbol of longevity (compare GRUBE, Zur Pekinger Volkskunde, p. 95).

Figure 4 of this Plate XXVIII represents a little masterpiece of
carving. Around an axehead with gracefully sloping blade, as nucleus,
two dragons are carved out in open work, the one of the type of a
hydra resting its head in relief on the surface of the axe and winding
its body elegantly around the edge, while the handle is formed by a
coiled dragon in hollowed-out carving, leaning on the end of the axe
and continued on the lower side in high undercut relief. The jade
is of a dark grayish-green gradually assuming in about the middle a
tinge of delicate light-brown, passing into a deep chocolate brown in
the handle. The harmonious proportion of all parts, the clever utilization of the coloration in the stone and the fine execution of the carving
are equally worthy of admiration in these two pieces.

In Fig. 3, we recognized the crane in connection with the fungus
of immortality as a symbol of longevity. In Fig. 5 reproduced in
the original size we meet with a girdle-pendant carved in the shape
of a double fungus with spiral ornaments engraved in the surface;
the jade is gray in the handle and upper zone and light-brown in the
lower portion.

This fungus is a species of *Agaric* and considered a felicitous plant,
because it absorbs the vapors of the earth (see BRETSCHNEIDER, Botanicon Sinicum, Part II, p. 40 and Part III, p. 480).

In the *Li ki* (ed. COUVREUR, Vol. I, p. 643), it is mentioned as an
edible plant. As a marvellous plant foreboding good luck, it first
appeared under the Han dynasty in B. C. 109 when it sprouted in the
imperial palace Kan-ts'üan. The emperor issued an edict announcing
this phenomenon and proclaimed an amnesty in the empire except

for relapsing criminals. A hymn in honor of this divine plant was composed in the same year (CHAVANNES, Se-ma Ts'ien, Vol. III, p. 624; J. EDKINS, *Journal Peking Oriental Society*, Vol. II, p. 230). This event may have led at that time to the reception of this fungus as a motive of art in girdle-pendants. Subsequently, it became one of the magic emblems of Taoism and a symbol of long life. We shall meet them repeatedly in connection with this design of the fungus on later carvings of jade.

The jade agarics (*yü chih*) mentioned in the year 748 A. D. as "produced on the pillars of the Ta-t'ung Palace and shining through the hall with magic splendor" (HIRTH, Scraps from a Collector's Note Book, p. 78) were perhaps really carved from jade, although the word *ch'an* "to produce" used in this passage would seem to refer to a natural phenomenon and to favor the view expressed by HIRTH. There is indeed, if not a real, a fabulous kind of agaric called *yü chih* and supposed to grow on the sacred mountain Hua shan in Shensi. A special agaric is ascribed to each of the four sacred mountains with the addition of two others, making six kinds altogether (see BRETSCHNEIDER, *l. c.*, Part III, p. 418). An illustration and description of the *ling chih* is given by G. SCHLEGEL in *T'oung Pao*, Vol. VI, 1895, pp. 18–21.

The favorite girdle-ornaments were doubtless the ring and the half-ring. *Huan* (GILES No. 5043) "a jade ring" is written with the same phonetic and pronounced with the same sound as *huan* (No. 5047) "to return, to repay." It was accordingly the symbol by which the emperor summoned an exiled official to return, or the signal given for besieging a city (on account of the word *huan* Nos. 5040 and 41 "an enclosing wall").[1] In making such a ring over to a friend as a gift it doubtless meant an expression of thanks,[2] or implied also the philosophical symbolism underlying the ring,— all divine principles being supposed to run in a ring or circle without beginning or end (see GILES No. 4862). The opposite sense is connected with the incomplete or half-ring *küeh* (GILES No. 3222). This character is alternately used with the word *küeh* (No. 3219) meaning "to cut off, to slay; to pass sentence; to decide, to settle." CONRADY (*l. c.*, p. 9) has discovered the oldest authenticated use of this half-ring in the tragic case of Prince Shên-shêng who in B. C. 659 was sent by his father, to please a concubine, on a fatal war expedition and received from him a half-ring as girdle-pendant, signifying that he was cast off and should not return. The emperors availed themselves of this symbol in banishing

---

[1] See CONRADY, *l. c.*, p. 10.

[2] As proved by the story of Yang Pao and the gold bracelets of the Chou family (PÉTILLON, Allusions littéraires, p. 250).

a man to the frontier, and the exiled K'ü-yüan, the celebrated poet of the elegies Li-sao, according to CONRADY, wore such an ornament in his girdle.[1] But as *küeh* means also to decide, the wearer of this ring intended to indicate his ability to decide all sorts of intricate questions and problems; according to Chuang-tse the scholars used to wear such a ring in his time.

Pan Ku, the author of the *Pai hu t'ung* whom we quoted above in regard to the professional symbolism of girdle-pendants, makes the following remarks on the rings and half-rings:

"The objects which are to be worn suspended from the girdle make known one's intentions and display one's abilities. Hence he who cultivates moral conduct (*tao* "the way" in the sense of the Confucian school) without end, wears a ring. He who makes reason and virtue (*tao têh* in the sense of Lao-tse) the foundation of his conduct, wears the jewels *kun*.[2] He who is able to decide (*küeh*) questions of aversion and doubt, wears a half-ring (*küeh*). This means that from the kind of visible girdle-ornaments which a man wears an inference on his abilities can be drawn."

The *Ku yü t'u p'u* (Chs. 55 and 56) contains the following seven girdle-rings (Figs. 110–116). Figure 110 is a ring decorated with "connected clouds" as they occur on the *Ju-i* sceptres (Plate LXVIII); a ring of this design was apparently given as a present implying every good wish (*ju i*). The design in Fig. 111 is the same, but treated in another technique, the whole being carved in open-work (*lou k'ung*). Both rings are made pre-Han. Figure 112 is designated as a girdle-ring with "coiled clouds" and attributed to the Han period. Figure 113 shows a coiled phenix of which it is said "that feathers, wings, crest and beak are filled with life's motion (*shêng tung*) like in painting (or

---

[1] I cannot find this passage in the *Li-sao*. In stanza 21, K'ü Yüan mentions that he wore a belt consisting of two aromatic plants symbolic of moral qualities (D'HERVEY DE SAINT-DENYS, Le Li-Sao, p. 15). In another stanza, he plucks a branch of the fabulous tree *k'iung* (MAYERS, Chinese Reader's Manual, p. 99) to enrich his belt (*l. c.*, p. 39). He further alludes to the honorary decorations conferred upon him to wear in the girdle, in remembrance of his former high position (p. 20), and if he asks himself whether his girdle will be again ornamented (p. 21), he hints at the hope of future splendor. The fragrant girdle is to K'ü Yüan the emblem of his virtues and merits (p. 58).—As we have popular traditions of gold and pearls foreshadowing calamities, gold and jade may also in China augur bad luck. Hiao-Ch'êng, king of Chao (B. C. 265–245), had an extraordinary dream in B. C. 262; he saw himself clad in a costume the two halves of which cut lengthwise on the back showed two different colors; in this garb, he mounted a chariot drawn by dragons taking their flight towards Heaven; but he soon fell from the clouds to land on a heap of gold and jade. The interpretation of this dream by the soothsayer was as follows: the grotesque costume foreshadows distress; the flight towards Heaven and the fall indicate a deceitful phantom void of reality; and the hillock of gold and jade announces great grief (A. TSCHEPE, Histoire du royaume de Han, p. 135).

[2] GILES No. 6521 "a precious stone resembling a pearl." I do not know on what foundation this symbolism rests.

環 帶 意 如 雲 連 玉 古

FIG. 110.
Jade Girdle-Ring with Cloud-Band (from *Ku yü t'u p'u*).

古
玉
如
意
連
雲
帶
環

FIG. 111.
Jade Girdle-Ring in Shape of Cloud-Band
(from *Ku yü t'u p'u*).

as if it were painted)." This phrase goes to show that pictorial influence is imputed to this design,[1] and I therefore doubt if the piece in question is, as stated, a work of the Han. It must come down from the T'ang or Sung period, judging from its style. The ring in Fig. 114 is engraved with a floral wreath styled "auspicious plants" (*jui ts'ao*) and praised in the text as "breathing in its elegance and beauty the spirit of the Han." This is all very well, and the elegance of the pattern, nobody will deny; but,

Fig. 112.
Jade Girdle-Ring in Shape of Coiled Clouds
(from *Ku yü t'u p'u*).

thus far, it cannot be pointed out earlier than on works of the T'ang period and breathes too, according to my feeling, the spirit of the

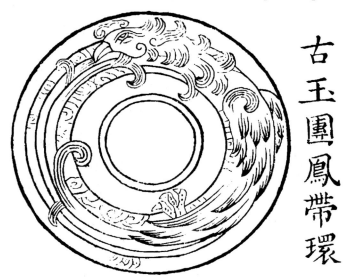

Fig. 113.
Jade Girdle-Ring with Design of Phenix (from *Ku yü t'u p'u*).

T'ang. If this plant of good luck is to be identified with the tea-plant (Pétillon, *l. c.*, p. 247), it is certainly out of the question that the

---

[1] For a full discussion of this subject see below.

pattern could have been devised under the Han, as tea was not yet cultivated.   The ring in Fig. 115 is interesting in showing a coiled

古玉瑞草帶環

FIG. 114.
Jade Girdle-Ring with Design of Auspicious Plants (from *Ku yü t'u p'u*).

古玉蟠螭帶環

FIG. 115.
Jade Girdle-Ring with Coiled Hydra
(from *Ku yü t'u p'u*).

hydra with projecting fangs and long sharp claws, which is an unusual representation; "it is a genuine Han work," remark the editors.   The last ring in Fig. 116 of a jade "purple like vine-grapes" is carved in a band of clouds and "an object posterior to the Han time."

We dwelt above on the peculiar symbolism relative to the incomplete rings called *küeh*. Also Wu Ta-ch'êng alludes to it in figuring a specimen of this kind in his collection (Fig. 117) in which I believe the oldest type of these rings may be found. It is carved from green jade with a black zone and has a double dragon (*shuang lung*) [1] engraved on the one face and "the scarlet bird" (*chu kio* or *chu niao*), the bird of the

FIG. 116.
Jade Girdle-Ring in Shape of Cloud-Band
(from *Ku yü t'u p'u*).

southern quarter on the other face. The form of this bird as here outlined exactly agrees with that on a tile disk of the Han period ("Chinese Pottery of the Han Dynasty," Plate LXVII, Fig. 4). It is not known what its proper significance is on the tile nor in this connection on the ring. The break in the ring is effected by a narrow strip sawn away between the two dragon-heads which cannot touch each other; it symbolically indicates the rupture or the breaking-off of cordial relations between two people.

If we adopt for the scarlet bird the interpretation proposed by L. de Saussure (*T'oung Pao*, 1909, p. 264; 1910, p. 614) who regards it as identical with the quail symbolizing the summer, because it is born from the fire of the summer, we should arrive at the conclusion that the design on this side of the ring represents the summer, and be bound to infer that the dragon on the opposite face should be identical with the Green Dragon corresponding to the spring. Both symbols further

[1] Compare the dragon-images in Figs. 91–93.

correspond to two quarters, the latter to the east, the former to the south; so that in the case of an exiled official his wanderings east and south may be symbolized, or that the break of the ring may indicate a sorrow particularly felt in the spring and summer, or in a more abstract way, that there is a gap in the spring and summer of life.   We shall come back

FIG. 117.
a   *b*
Incomplete Jade Ring, *küeh*, Upper and Lower Faces.

to the subject of these rings in a consideration of the later development of the girdle-pendants.

It should, however, be added that this explanation can be accepted only provisionally, as it does not agree with the archæological facts. In all representations of "the Scarlet Bird," as far as known to me, the bird intended is not a quail but apparently a kind of pheasant.   In the Chinese art-journal *Shên chou kuo kuang tsi*, No. 11 (published in Shanghai) there is on Plate V the illustration of a very interesting stove of Han pottery showing the animals of the four quarters, one on each of the four sides.   The unusual feature of this design is that "the Green Dragon" and "the White Tiger" are here associated each with the figure of a man; and "the Scarlet Bird" is delineated as a large heavy strutting bird with a high crest and a big bunch of curved tail-feathers, almost in the style of a peacock; it may be any species of a large bird, but it is certainly not a quail.

2. THE DEVELOPMENT OF THE GIRDLE-PENDANTS

The ancient girdle-pendant composed of seven jade carvings, a characteristic feature of the culture of the Chou period, is no longer in existence in China. For reasons which escape our knowledge but which we shall try to develop hereafter, it did not possess vitality enough to survive for any great length of time after the downfall of the house of Chou. While possibly still alive during the time of the Han as a result of the revival of the ancient classical traditions, it must have sunk into oblivion shortly after this period, being already displaced by other fashions during this transitional epoch from antiquity to the middle ages. The overthrow of a dynasty and the establishment of a new régime usually was in China also the signal for a change of culture, not always radical, but ushered in by modifications of costume, style, ornament, and subsequent new developments of taste and art. While neither the whole nor any single component of the classical girdle-pendant was perpetuated, a new style of girdle-ornament gradually came into vogue under the Han, doubtless connected with the far-reaching revolution then affecting all domains of taste. This new fashion, curiously enough, developed, according to the views of Chinese archæologists, from that ornament of the Chou dynasty for which we should have predicted the least chance of an extensive popularity, — the gloomy half-ring *küeh* which originally meant separation, banishment, nay, even capital punishment; or, what could not appeal either to the people at large, the decision in literary disputes. But this entire symbolism must have died out during the Han period; for then these objects seem to come into general use, carved into graceful designs not pointing to any serious disaster for the wearer. It is useless to raise here a question of terminology, and to argue that these ornaments differ from the ancient half-rings and may have developed from another type which may have even existed in the Chou period under a different name. This may be, but the brutal fact remains that the long series of these objects is designated *küeh* by the native archæologists, and that in some of them the type, and above all, the designs of the *küeh*, — and these are presumably the oldest in the group of the new *küeh*, — have been faithfully preserved. These ornaments finally end in neat carvings of animal figures, quite in the style of the Japanese Netsuke, purely decorative, with no other object in view than to afford esthetic enjoyment to the wearer and the lookers-on. Also these plastic subjects are styled *küeh*. It is true, in this case the *Ku yü t'u p'u* (Ch. 64, p. 4) objects to this name by saying: "The *küeh* is a broken (or incomplete) sort of disk which is perforated; but the figure of the cicada here illustrated, as well as the following figures of the dragon, fish, tiger, etc. have all solid bodies without perforations. Hence they cannot be designated *küeh*, but should only be styled girdle-ornaments (*p'ei*); merely for the sake of order, they have been arranged among the *küeh*, whereby confusion may arise from this nomenclature, as in the *Po ku t'u* from the name "kettle" (*ting*) which is indiscriminately used for the sacrificial vessels *ts'un* and *i*." Notwithstanding, this work also retains the name *küeh* for some of these objects, while it designates as *p'ei* others which still display a certain relation to the ancient *küeh*.

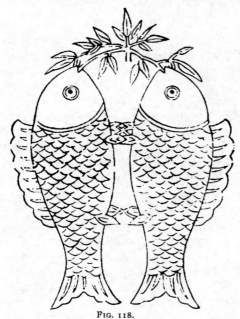

FIG. 118.

FIG. 119.
Jade Girdle-Pendants. Pairs of Fishes (from *Ku yü t'u p'u*).

The twenty-four illustrations of these girdle-ornaments given by the *Ku yü t'u p'u* (Chs. 62–65) present the greatest merit of this book and allow us to trace the development of these decorative objects from the Han to the Sung period, and to connect these more ancient with the modern forms. The majority of these designs, whatever their supposed age may be, have also highly artistic merits and betray an unusual beauty of form and line. These objects are not the result of antiquarian speculation as the jade tablets of rank or the official headdresses, but live affairs which were really made and worn by all classes of people. Because of this intrinsic value, the whole series is here reproduced in its entirety in the same succession as in the original, with the addition of necessary criticism. The first four objects are designated as girdle-ornaments (*p'ei*), all others as *küeh*. Figure 118 represents a pair of fishes standing erect, carved from green jade; their fins are connected, and they are holding in their mouths the leaved branch of a willow (*liu*), according to the Chinese explanation. It should be added that, during the Han period, it was customary to pluck a willow-branch (*chê liu*, see Giles No. 550), and to offer it to a parting friend who was escorted as far as the bridge *Pa* east of Ch'ang-ngan where the branch of separation (*küeh!*) was handed to the departing friend.[1] The double fish is a pattern familiar to us on bronzes of the Han period and symbolizes the mutual harmony between spouses and friends.[2] The significance of this ornament is therefore simple enough: we must part, but we shall remain friends as these two fishes are inseparable. It reveals to us at the same time how the *küeh*, so formidable in the beginning with its message of absolute divorce, was mitigated into a more kind-hearted attitude which made it acceptable to all people, — it became a parting-gift, a farewell trinket. The date of this piece is set at a period covering the Wei and Tsin dynasties, *i. e.* roughly the third and fourth centuries a. d., but I have no doubt that the pattern goes back to the creative period of the Han.

Figure 119 displays a similar design of a pair of fishes, the same carving being brought out on both faces. Also here, the editors explain the plant design as that of a willow. The leaves are represented here on the bodies behind the gills, and a leaf-shaped wreath (with the perforation of the ancient *küeh*) appears between the lower fins. Another diversity is that the tips of the tails here touch each other which seems to hint at a more intimate union of the party concerned, while there is a gap in the previous piece in correspondence with the break in the ancient half-ring. The editors, not unwittily, comment as follows: "Wên-ti (535–551 a. d.), the Emperor of the Wei dynasty, presented the *belle*, Sieh Ling-yün by name,[3] with a girdle-pendant of green jade representing a double fish. Though it is not necessary to assume that our specimen under consideration is identical with just that one of the Wei, still its examination shows that it must be an object from that period (535–554 a. d.)."

The fish, as a symbol of power and rank, came into vogue with the rise of the T'ang dynasty. In the year 618, the first year of the reign of the Emperor Kao-tsu, the T'ang, doubtless for the reason of marking the change of dynasty, abolished the silver badge having the shape of a certain plant and substituted for it the silver tally of the form of a

---

[1] Pétillon, Allusions littéraires, p. 172.

[2] See Paul Carus, The Fish as a Mystic Symbol in China and Japan (*The Open Court*, July, 1911).

[3] She was the emperor's concubine and noted for her skill in accomplishing marvels of needlework in the dark, hence styled the genius of the needle (Pétillon, Allusions littéraires, p. 438).

fish (*yin yü fu*). In 690, the Empress Wu decreed that the form of the tortoise (*kuei*) should replace that of the fish; but in 705, when Chung-tsung mounted the throne, he restored the former fish-symbol which was perpetuated till after the end of the T'ang dynasty and appears again in the epoch of the Kin and the Sung.[1]

Under the Khitan reigning as the Liao dynasty (983–1055 A. D.) mention is constantly made of the gold-fish tallies (*kin yü fu*) worn in the girdle as part of the court dress, an inheritance of the custom of the T'ang. These badges were six inches long, moulded in the shape of a fish, split, as it were, longitudinally into two halves, and the flat surface of each half was engraved with an identical inscription. The left half was kept in the palace when the right half was given to the commander of an army, who had to return it to the treasury when the expedition was over. The halves fitted exactly so that they might be tested whenever necessary.[2]

We must, however, distinguish between such badges conferred upon as a mark of honor or rank, and real tallies of legal force. Thus, we read in the "History of the Liao" that in 1036 A. D. the Emperor Hing-tsung examined a band of scholars and bestowed on them red garments and silver fishes,[3] which, in this case, were mere tokens of recognition, but of no lawful consequence; while in the kingdom of the Liao where military service was compulsory, a cast-bronze tally representing a fish was used for the conscription of troops, and two hundred silver tablets (*p'ai-tse*) were employed to transmit orders for the supply of horses.[4] When the army was mobilized, as ordered by the delivery of one-half tally, it did not march until the emperor despatched a commander with the other half; the two halves were then joined together, and if found to fit, the army began to advance. The "History of the Kin Dynasty" relates that the Princes of the blood used to wear a fish of jade (*p'ei yü yü*) and the officers from the first to the fourth grade, a fish of gold, while a double-fish pendant was reserved for the heir-apparent. These were purely ceremonial badges without giving legal rights, as also the "Sung History" says that they were only worn then as distinctive marks of rank, and not inscribed and tested in the palace as had been the case under the T'ang.

At that period, frequent mention is made also of fish-purses (*yü tai*)

---

[1] Compare CHAVANNES in *T'oung Pao*, 1904, p. 36.

[2] BUSHELL in *Actes du XIe Congrès international des Orientalistes*, IIᵉ section, pp. 17, 18 (Paris, 1898).

[3] H. C. v. D. GABELENTZ, Geschichte der grossen Liao, p. 123 (St. Petersburg, 1877).

[4] *L. c.*, p. 189. Compare also T. DE LACOUPERIE, Beginnings of Writing, pp. 69–70 (London, 1894).

bestowed as gifts upon Turkish princes (CHAVANNES, *l. c.*, pp. 30, 32, 36 etc.). A well illustrated article on these fish-purses will be found in No. 58 of the *Kokka*. Bronze tallies in the forms of a fish, a tortoise, a tiger, and a seal are in our collection.

Figures 120 and 121 are styled girdle-ornaments with double phenixes (*shuang luan p'ei*). The *luan* is a fabulous bird, related in design to the *fêng* and *huang*. The former has been identified by Prof. Newton [1]

<table>
<tr><td>FIG. 120.
Jade Girdle-Pendant, Pair of Phenixes or Peacocks
(from *Ku yü t'u p'u*).</td><td>FIG. 121.
Jade Girdle-Pendant, Pair of Phenixes or
Peacocks (from *Ku yü t'u p'u*).</td></tr>
</table>

with the Argus pheasant of Borneo and Malacca, the latter with the peacock of India. These identifications seem quite plausible, especially as far as the more recent elaborate representations of these birds are concerned. The text explains only that they are holding flowers in their beaks, and refers us to Shih Ts'ung who is known as a practical joker, and who died in 300 A. D. (GILES, Biographical Dictionary, p. 651); he is said to have kept in his seraglio a hundred beauties who all carried in their girdles ornaments of fine jade in the shape of phenixes (*fêng luan*); the present one, thinks the *Ku yü t'u p'u*, is one of that lot. While nothing can force us to make this conclusion our own, it is quite

[1] See GILES, *Adversaria Sinica*, No. 1, p. 9.

evident from this quotation what the Chinese author wants us to understand in regard to the symbolic meaning of this ornament. "In poetry," says MAYERS (Chinese Reader's Manual, p. 41), "many covert allusions to sexual pairing are intimated by reference to the inseparable fellowship of the *fêng* and the *huang*." In Fig. 121, this allusion is undisguised in the osculation of the two birds and allows of the inference that this pendant was plainly a lover's gift to his mistress. This symbolism of the so-called phenix becomes more intelligible, if we say peacock instead of phenix and derive, with the importation of this bird from India into China, also this symbolism from India where (as also later in Europe) the peacock played a prominent rôle in all matters pertaining to love.

In Fig. 120, the cloud-pattern over the head of each bird should be noted, for the poet T'ao Yüan-ming (365–427) says: "The divine *fêng* dances among the clouds, the spiritual *luan* trills its pure notes" (GILES, *Adversaria Sinica*, No. 1, p. 9). The work in Fig. 121 is praised as "clever" (*tsing liang*) in the Sung Catalogue and said to be a beautiful object of the Tsin (265–419 A. D.) or T'ang dynasty (618–905 A. D.).

In Fig. 122 the upper and lower face (*a* and *b*) of the ornament is represented. Two hydras (*shuang ch'ih*) are winding around a perforated jade plaque rounded below and tapering at the upper end. The design is of great elegance, and the editors do not suppress the remark that the curves and wriggles of the monsters are true to life.

The plant design in Fig. 123 is explained as "the fragrant herb" (*hiang ts'ao*) which is designated also as *yü ts'ung*, lit. "aromatic onion." Bretschneider has not identified this name. "This piece is admirable and a beauty in its make-up; it may be a masterpiece of the *San tai*" (Hia, Shang and Chou dynasties). This definition is out of the question; the design, I should say, betrays the style of the T'ang dynasty. The carving is one-sided only. In the next Fig. 124, it is brought out alike on both sides. A butterfly with wings outspread, leaning its antennæ over an oval-shaped ring; alleged to be Han.

Figure 125 shows the design of a *k'uei lung* (GILES No. 6507), a peculiar kind of dragon, coiled in the form of a spiral. The head is winding up into an elephant's nose likewise treated as spiral, and reminds one of the Indian makara (hippocamp); but it is more probable that the tapir is intended, for that animal was familiar to the art of the Han ("Chinese Pottery of the Han Dynasty," p. 152).[1] It is certainly not from the time of the *San tai*, as stated in the text, but not older than the Han period.

[1] It is also familiar, through the medium of Chinese models, to the Japanese netsuke-carvers, as Plate 44 in A. BROCKHAUS, Netsuke, shows; for *baku* is the tapir (written with the same character as the Chinese word for the tapir).

A twisted band carved alike on both sides with meanders ("thunder pattern," *lei wên*) is represented in the ornament Fig. 126 which is called "a work of the Han, elegant and admirable."

a   FIG. 122.   b
Jade Girdle-Pendant, Upper and Lower Faces, Two Hydras (from *Ku yü t'u p'u*).

FIG. 123.
Jade Girdle-Pendant, "The Fragrant Herb" (from *Ku yü t'u p'u*).

FIG. 124.
Jade Girdle-Pendant, Butterfly (from *Ku yü t'u p'u*).

Figures 127 and 128 represent coiled hydras, with the addition of a rodent (*shu*) whisking over the monster's back in Fig. 128. The former is referred to the Han period which may be correct, the latter to the *San tai*, which I think is impossible, despite the reason given for this

FIG. 125.
Jade Girdle-Pendant, Coiled Dragon (from *Ku yü t'u p'u*).

FIG. 126.
Jade Girdle-Pendant, Twisted Band
(from *Ku yü t'u p'u*).

FIG. 128.
Jade Girdle-Pendant, Coiled Hydra with Rodent
(from *Ku yü t'u p'u*).

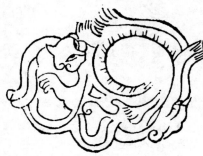

FIG. 127.
Jade Girdle-Pendant, Coiled Hydra
(from *Ku yü t'u p'u*).

dating. In the period *T'ien-chung* (1023–1032), this piece was found in the grave of Kao Ch'ai, a disciple of Confucius, who had lived in the sixth century B. C. (GILES, Biographical Dictionary, p. 651). This news was probably nothing more than a dealer's advertisement and confirms the impression gained from the design that this is a genuine work of the Sung period. The hydra is laid around a spiral explained as the cosmic symbol *t'ai yi*, which occupied the philosophers of the Sung time to a great extent, and the presence of the rat, or whatever species

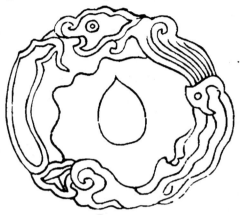

FIG. 129.
Jade Girdle-Pendant, Hydra and Bird-Head
(from *Ku yü t'u p'u*).

FIG. 130.
Jade Girdle-Pendant, Hatchet-Design (from *Ku yü t'u p'u*).

this rodent may be, stamps on this pattern the character of a genre picture as it is characteristic of the impressionistic school of Sung painters.

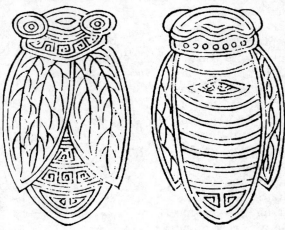

FIG. 131.
Jade Girdle-Pendant, Cicada, Upper and Lower Faces (from *Ku yü t'u p'u*).

Figure 129 is, according to the Chinese explanation, also a single coiled hydra, of the Han period, but the curious bird-head on the right-hand

FIG. 132.
Jade Girdle-Ornament, Cloud-Dragon
(from *Ku yü t'u p'u*).

FIG. 133.
Jade Girdle-Pendant, Single Fish
(from *Ku yü t'u p'u*).

side is not noted, nor is the strip of clouds between it and the hydra-head. There is also a cloud-pattern in the lower portion, and the phenomenon of rain is doubtless expressed in this design.

Figure 130 is interpreted as a *kiao-ch'ih, i. e.* a young dragon whose horns have not yet grown. I am unable to detect any dragon-figure in this design, unless it be on the opposite face which is not represented; but this point is not mentioned in the text. I believe this pattern must be explained as being composed of two axe-shaped implements (of the same type as represented in Figs. 3 and 4 of Plate XXVIII), joined together in the middle, overlaid with and encircled by cloud-ornaments. We remember the so-called cloud-shaped ceremonial dance-axes of the Han (p. 41), and I am inclined to think that a certain share is due to the latter in the conception of this design, which is regarded as pre-Han by the *Ku yü t'u p'u.*

Figure 131 is interesting as showing the full figure of a cicada (see Fig. 168 and Plate XXXVI, Figs. 5–9), the back represented on the upper, and the abdomen of the insect on the lower face of the carving. Nothing is said in regard to its symbolism. It is curious that, despite the manifest tendency to carve the little creature as true to nature as possible,[1] bands of meander scrolls are brought out in the design. We shall hear more of cicadas in the course of this investigation.

The ornament in Fig. 132 carved alike on both faces is again the dragon *k'uei lung*; "fierce and frightful," is the editorial comment. As indicated by the decorative elements, it is doubtless a cloud and thunder-dragon. Its antique elegance, is the editorial conclusion, stamps it as a relic of the Chou or Ts'in.[2]

Figure 133 represents the double-sided carving of a single fish. "The scales and the bristly dorsal fins (*lieh*, GILES No. 7107) are life-like, and it is like an object of the Tsin (265–419 A. D.) or T'ang dynasty (618–905 A. D.)," comment the editors. The scales are conceived of as meander fretwork; but I do not know whether, for this reason, this fish is associated with thunder. The peculiar feature is, at all events, its single-blessedness in distinction from the common fish couples. There is a huge fish in the Yellow River, called *kuan* (GILES No. 6371, PÉTIL-LON, *l. c.*, p. 500)[3] supposed to be a kind of spike, noted for its solitary habits of life, and therefore an emblematic expression for anybody

---

[1] The Chinese text says that it does not differ from living ones, as they appear on paper mulberry-trees after a growth of three years.

[2] The same design as the engraving on an ink-cake in *Fang-shih mo p'u,* Ch. 2, p. 3.

[3] The Chinese theory that this species is not able to close its eyes is certainly mere fancy, as in all fishes the accessory organs of the eye like the lids and lachrymal glands are poorly developed.

deprived of company like an orphan, a widower, a bachelor, or a lonely fellow without kith or kin.[1]   A girdle-ornament of this design was perhaps a gift for a man in this condition.

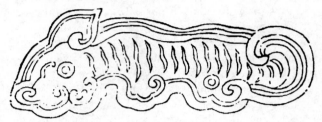

FIG. 134.
Jade Girdle-Pendant, Winding Dragon (from *Ku yü t'u p'u*).

The dragon shown in Fig. 134 is styled a *p'an k'iu* (GILES No. 2346) "a winding young dragon without horns." Its stripes must be understood, as explained in the text, from the natural red veins in the white jade which look like tiger or panther stripes and were skilfully used by

FIG. 135.
Jade Girdle-Pendant, Tiger (from *Ku yü t'u p'u*).

the artist in carving. The editors insist on this piece being a work of the time of the Six Dynasties (*leu ch'ao*), *i. e.* from about the third to the sixth century.

[1] In this sense, it is mentioned as early as in the Shu king. In one poem of the Shi king, No. 9 of the songs of the country of Ts'i, Wên Kiang, the widow of Prince Huan of Lu, is censured for returning several times into her native country of Ts'i where she entertained an incestuous intercourse with her own brother, the prince Siang. The poet compares her to the fish *kuan* who is restless and sleepless at night for lack of a bed-fellow (see LEGGE, Shi king, Vol. I, p. 159, and Vol. II, p. 293).

The carving of a tiger in Fig. 135 is called "ornamented (*lit.* embroidered) tiger-spirit (*siu hu shên*), graceful and yet ferocious, severe and yet majestic, with a flavor of the idea that he is the king of the animals,

a FIG. 136. *b*
Jade Girdle-Pendant. Upper and Lower Faces. Single Hydra (from *Ku yü t'u p'u*).

a curious object of the *San-tai.*" I am under the impression that this design has been influenced by pictorial art (note the word *siu*) and may be of mediæval origin (*T'ang* or *Sung*).

The following ornaments are arranged together in a separate chapter (Ch. 65), and it will be seen that they are, with the exception of Fig. 137, built up on the same principle as we found in Fig. 122. The dragons are, as the *Ku yü t'u p'u* says, laid around the *küeh*; in Fig. 136 it is a single hydra, in Fig. 138 a couple of hydras engaged in "loveplay" (*kiao hi*), with teeth and claws "true to nature," while the finesse of detail is extolled in Fig. 136 and "the admirable life's motion" is emphasized in the coiled hydra of Fig. 137. All three are beyond cavil productions of the Han period.

FIG. 137.
Jade Girdle-Pendant, Coiled Hydra
(from *Ku yü t'u p'u*).

The carvings in Figs. 139–141 represent fine variations of the same motive. The editors justly become enthusiastic over the beauty of these little artworks. The one in Fig. 139, carved out of a pale blue jade, — the two hydras as gentle and genial creatures gracefully playing around the upper and lower edges of the oval ring, — is defined as a "masterpiece of the Han."   In Fig. 140, the two creatures are stretching their paws out to caress each other, and the Chinese interpretation lays stress on their "spiritual life's motion" (*shên shêng tung*) as it

a   FIG. 138.   b
Jade Girdle-Pendant, Upper and Lower Faces, Couple of Hydras (from *Ku yü t'u p'u*).

a   FIG. 139.   b
Jade Girdle-Pendant, Upper and Lower Faces, Playing Hydras (from *Ku yü t'u p'u*).

was possible only for the clever craftsmen (*liang kung*) of the Han time to accomplish.   The piece in Fig. 141 is the most remarkable of all for its asymmetrical arrangement, — the dragons, soaring in the clouds, adjoining the upper and right-hand edges, — and for the bold and unconventional treatment of the design expressed in the wooing of the two happy creatures, — the one biting the tail of the other.[1]   It is

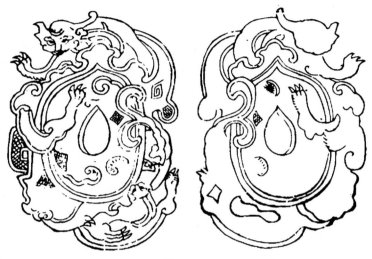

a   FIG. 140.   b
Jade Girdle-Pendant, Upper and Lower Faces, Playing Hydras (from *Ku yü t'u p'u*).

a   FIG. 141.   b
Jade Girdle-Pendant, Upper and Lower Faces, Playing Hydras (from *Ku yü t'u p'u*).

evident that the "loveplay of nature" is brought out with supreme volition in these designs as well as in those of the paired fishes and

---

[1] The Chinese comment tersely utters the two words *kiao yu*, "they get ready for cohabitation."

phenixes, and we shall soon perceive hereafter that also a deeper religious and philosophical idea is underlying this tendency.

First of all, we are now able to recognize the general character of these girdle-pendants: they are related to thoughts of friendship and love, and in this sense, are closely joined to the ancient jewels of the Chou period. But while they are expressive of the same range of emotions, they differ from them in form and design, and this difference was brought about by the tendency to lend a marked expression to these emotional thoughts in the designs themselves. In other words, the stiff and formal traditions of the Chou period were no longer sufficient for the temperament of the people of the Han time who were framed of a different mould. The girdle-ornaments of the Chou, as we saw, were geometrical in shape, cut out in circles, half-circles, squares and rectangles, in conformity with the whole geometric trend of mind ruling at that time, which measured, surveyed and weighed everything; the symbolism of these ornaments did not refer to their designs, but to their designations only by way of a phonetic rebus, an esthetic pleasure merely caught by the ear and eye, and a means of expression for poetry only. But Art had to stand behind with empty hands. The Han people broke with this spoken and written symbolism and created the symbolism of the subject, giving, by so doing, a powerful impetus to the development of art. They dropped all the nice words of their predecessors, the *kü*, the *kiu*, the *yü*, the *hêng* and the *huang* etc., and crystallized their sentiments into the *küeh*. In place of words, they enthroned the artistic motive, and the sound of the verse was exchanged for an enlivened rhythm of material form and line. They were, in fact, productive creators, as we had occasion to admire in several types of Han ceramics and, seeking forms for the expression of their emotions, their art became essentially emotional and, as a consequence, the emotions instilled into their productions of art must become our guides in attempting to understand them. For this reason we must ransack all available sources for tracing any real or alleged symbolism connected with them, for this is the key to the treasury leading to the heart of their art, and not only *their* art, but that of China in general whose ideals are still based on and nourished by that memorable period. Neglecting or disregarding the interpretations of the Chinese would not only result in an absolute failure of a proper understanding and appreciation of their art, but might also lead to such abortive caricatures as have unfortunately been drawn of China's culture-historical development.

But to revert to our subject, — the girdle-pendant of the Chou was the product of the impersonal and ethnical character of the art of that age; it was general and communistic, it applied to everybody in the

JADE GIRDLE-ORNAMENTS OF WOMEN, HAN PERIOD.

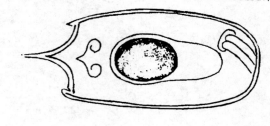

DESIGNS ON THE LOWER FACES OF THE GIRDLE-ORNAMENTS IN FIGS. 1, 3 AND 4 OF PRECEDING PLATE.

community in the same form; it did not spring up from an individual thought, but presented an ethnical element, a national type. Sentiments move on manifold lines and pendulate between numerous degrees of variations. When sentiment demanded its right and conquered its place in the art of the Han, the natural consequence was that at the same time when the individual keynote was sounded in the art-motives, also variations of motives sprang into existence in proportion to the variations of sentiments.

This implies the two new great factors which characterize the spirit of the Han time, — individualism and variability, in poetry,[1] in art, in culture and life in general. The personal spirit in taste gradually awakens: it was now possible for every one to choose a girdle-ornament according to his liking. For the first time, we hear of names of artists under the Han, six painters under the Western Han, and nine under the Eastern Han (GILES, Introduction etc., pp. 6, 7), also of workers in bronze and other craftsmen (LAUFER, Chinese Pottery, pp. 196, 292, 296). The typical, traditional objects of antiquity now received a tinge of personality or even gave way to new forms; these dissolved into numerous variations to express the correspondingly numerous shades of sentiments, and to answer the demands of customers of various minds.

I am in a position to lay before the reader four burial pieces of girdle-ornaments (Plate XXIX) which in style and technique agree with the last six of the *Ku yü t'u p'u*. They are authentic objects of the Han period originating from graves west of Si-ngan fu. Their character as mortuary offerings will allow us to advance one step further in the understanding of their symbolism. The finest qualities of jade of the most exquisite colors have been selected for this purpose, and the execution of the work in which the three processes of engraving, carving in relief and *à jour* are united, is perfect. The glyptic art of antiquity has reached in these carvings a climax unattained by any later age. The piece in Fig. 1, Plate XXIX, is remarkable for the magnificent coloration of the jade in various shades of brown and red standing out from an apple-green background, and for the freedom and mastery in the treatment of the design. Around a perforated elongated foundation,

---

[1] J. EDKINS, On the Poets of China (*Journal of the Peking Oriental Society*, Vol. II, p. 219) has given a good characteristic of Han poetry. "The Han poets were men who felt within themselves the impulses of poetry, which must find expression in some way. The old Odes were like the pleasant murmuring of the brook, the whisper of the pines in mountain hollows, the tinkling of the sheep bell heard from afar. The compositions of the Ch'u poets were marked by the depth and dashing speed of the river which forces its way through rocks attended by deafening sound and distinct contrasts of light and shadow. There was more art in these compositions than in those of earlier date, and it was accompanied by profounder feeling. Consequently, the Han poets could adopt no other course. In short, they made poems of the same kind" etc.

the survival of the ancient *küeh*, there is on the right the figure of a phenix standing on clouds and looking towards the slender-bodied hydra (*ch'ih*) below, which has the bearded head of a bird with pointed beak very similar to that of the phenix.   The left hind-foot of the monster terminates in the figure of a bird's head, presumably symbolizing a cloud.   It is rearing the left fore-paw in the direction of the bird, supporting the right on the clouds below.   Large cloud-bands are gracefully displayed along the upper edge.   The reverse (Plate XXX, Fig. 1) is plain except that the central medallion is filled with engraved spiral patterns; the color of the jade is here green-gray interspersed with reddish specks and veins.   This carving (15.1 cm in length, 7.1 cm in width and 4–6 mm in thickness) was discovered in a grave of the Han period in the village *Wan-ts'un* west of the city of Si-ngan fu.

The designs in Figs. 2 and 3, differing only as to their dimensions are identical even in minute details, which goes to show that there were typical patterns available for this purpose.   Two hydras, their heads seen from above, are winding around the lower and left side of the jade plaque, and spiral ornaments in open-work surround the two spear-shaped points.   The jade of the carving in Fig. 2 is grayish, and red in the left and lower right portion; that in Fig. 3 bluish green and gray.

The girdle-ornament in Fig. 4 of the same Plate is of a milk-white, smooth, lustrous jade and has the fundamental nucleus with oval perforation shaped in the same way as Fig. 1.   Four monsters of conventionalized design, carved in open-work are laid around the edges; above a bearded creature with long head similar to the *k'uei lung*, to the left a bird joining it, then an elephant or tapir head, and below a running quadruped the head of which is not clearly outlined.   This find comes from a grave of the Han period in the village *Kiao ts'un*, west of the city of Si-ngan fu.

We have seen that these designs relate to and are emblematic of procreation.   Not only the pairing of the animals, the most conspicuous feature in all of them, but also the minor decorative elements point in the same direction.   The ornamental clouds are emblematic of generative power too, as they send fertilizing rain.   On the lower faces of the four carvings in our collection, phallic designs are engraved with undisguised explicitness (Plate XXX); the spear-like ends of the oval *küeh* are presumably also intended as emblems of that kind.   These four pieces were found in women's graves; apparently they had been worn by these women during their life as girdle-pendants and were, on their death, buried with them.   The opinion of Chinese archæologists in Si-ngan fu, where this type is designated "chicken-heart girdle-ornament" (see below p. 238), is that it was placed in the womb with the idea of preventing the flesh from decay; that they were, accordingly, protecting amulets in the grave. This custom will be discussed at length in Chapter VIII. From this usage we recognize that the symbolism of these ornaments was deep and serious with an ethical reminder of death. It was a love-token given by the husband to his wife to remind her of their happy union and at the same time of their final separation by death; hence the appropriate application of the *küeh*, the ring of separation; and now we understand why all these ornaments, despite the fact that they allude to a union in the duplication of animals, are simultaneously emblems of parting (*küeh*) and death. But death, in the view of the Chinese, does not mean a permanent, but rather a temporary separation. The relations of a husband and a wife did not cease at the moment of death; they continued to be united even beyond the grave, and expected to resume their marital relations in a future life.[1] The custom tending to preserve the flesh by means of the jade substance (even though imaginary) shows plainly that the *post-mortem* relations were not viewed as merely platonic, but also as substantial enjoyments. This abundance of ideas covering the span of life and death in the fundamental human relations imbues the art of the Han with a spiritual tendency and an intrinsic idealistic import. From this point of view, the art of the Han period as embodying ideas and ideals is preëminently idealistic.[2] It should, further, be defined as emotional transcendentalism, as these ideas have their basis in emotions, not in deductions. Whether the conception of transcendental love meant also eternal love, we do not know; but however this may be, this is of all Chinese ideas the most idealistic ever conceived of, one which had a profound bearing on ethical conduct and at the same time a fruitful effect on art. This belief in the resurrection of love somewhat savours of that mysterious symbolism which the greatest poet of the Germanic race, Ibsen, has embodied in his dramatic legacy "When We Dead Awaken."

The four girdle-pendants in Figs. 142–145 are derived from the work of Wu Ta-ch'êng, not being commented upon by him in an explanatory text. For several reasons, they deserve reproduction. Figures 142 and 144 are designated as "dragon-shaped girdle-pendants, both of white jade, the one with additional russet spots, the other with a yellow

---

[1] Also in ancient times as at the present time husband and wife were buried in the same grave, but in different coffins. The grave was considered a dwelling-place, and a widow mourning the death of her husband expresses the desire "to go home to his abode or chamber" on her death (*Shi king*, ed. Legge, Vol. I, p. 187).

[2] Like the poetry introduced by the *Pan* family in which, according to the Chinese critics, human feelings and moral sentiments were involved. Poetry became with them a moral instructor and, at the same time, touched the feelings (Edkins, *l. c.*, p. 238).

mist;'' Fig. 143 is labeled ''hydra (*ch'ih*)-shaped girdle-pendant of green jade with russet speckles.'' All three are flat plaques, showing the mon-

FIG. 142.
Dragon-Shaped Girdle-Pendant of White Jade with Russet Spots.

FIG. 143.
Hydra-Shaped Girdle-Pendant of Green Jade with Russet Spots.

sters freely carved in sharp outlines. All three have that peculiar feature in common that the head of a bird is attached to the ends of the monsters, in Figs. 142 and 143 to the right-hand sides; in Fig. 144

very clearly in the middle portion, and it seems that a feathered crest surrounds the fish-tail of the monster. On the funeral stone sculpture-work of the Han period we meet with representations of atmospheric phenomena exhibiting clouds conceived of as birds or clouds with bird-heads attached, and celestial spirits connected with figures of birds in

FIG. 144.
Dragon-Shaped Girdle-Pendant of White Jade with Yellow Mist.

the act of instigating dragons to send rain. The bodies of the two monsters in Figs. 142 and 143 are, further, filled with spirals suggestive to the Chinese mind of clouds, rain and thunderstorm and therefore known under the name of cloud-pattern. We have, accordingly, in these jade plaques an abridged conventionalized representation of what is depicted on a larger scale in a more realistic manner on the Han grave-sculptures, — the motive of the dragon assisted by birds in moving the clouds and sending down beneficial rain (compare above p. 164).

Figure 145 is styled by Wu "girdle-pendant with designs of aquatic plants, of white jade with russet specks." The word *tsao* is a generic term for aquatic plants and is mentioned as early as in the *Shu king*.

The jade piece in Fig. 146 is grouped by *Wu* in this class, though he expressly states that this is not a girdle-ornament. He simply calls it "jade with dragon-design, made of green jade with clay spots all over," and adds: "To judge from its make-up, it is very old, but it is not known what part of the body it was to adorn." The ornament differs in technique and design from those discussed previously; it is relief-work on a rectangular plaque, divided into two sections, the upper one occupied by a bird's head, the lower one by a dragon's head. There is in fact no connection between the two in the mere technical composition, and the cloud-pattern is wanting here. But on the other hand, there is room for consideration whether, in this case too, a spiritual relation between the two creatures may be intended.

FIG. 145.
Girdle-Pendant, with Designs of Aquatic Plants, of White Jade with Russet Spots.

As the last of his series of personal ornaments Wu Ta-ch'êng illustrates a piece of curious shape, almost like an escutcheon, labeled "girdle-pendant with cloud patterns" (*yün wên p'ei*), the groups of small spirals being understood under the latter (Fig. 147). The lower face is plain; there is a five-sided perforation in the centre. The object is made of a white jade with black stripes. The colloquial name for it, says Wu, is "girdle-pendant in the shape of a chicken's heart (*chi sin p'ei*), but it is a subject not yet investigated." In Chapter VIII we shall meet a series of ornaments for burial purposes, to which the same term is applied in Si-ngan fu.

Two jade implements with which to loosen knots (*hi* or *chuei*) are figured by Wu Ta-ch'êng and here reproduced in Figs. 148 and 149, distinguished as big and small one, both of Shan-yüan jade. In ancient times they were made of horn and ivory, as indicated by the composition

龍文玉

青玉滿身土斑

此非佩

玉也制

作甚古

不知何

所施

FIG. 146.
Green Jade Ornament with Design of Bird and Dragon.

of the character with the radical "horn;" but those of jade which have come down to us, are decidedly rare, observes Wu. On the top of the larger one, a hydra is cut out à jour.

These implements are mentioned as early as in the *Shi king* (*Wei fêng*, VI; LEGGE, Vol. I, p. 103; ed. COUVREUR, p. 72).[1] According to the *Li ki* (*Nei tse*, I, 2; ed. COUVREUR, p. 621), every young gentleman used to wear a small one on the left side of his belt and a large one on the right side of it, in order to unloose large or small knots. The same pair of implements was worn and used in the same way by married women (*Ibid.*, I, 3). So it seems that the large and small specimen of Wu represent such a pair as sanctioned by the rules of propriety in the *Li ki.*

The *Ku yü t'u p'u* (Ch. 56) gives a series of these implements in the order of its usual paradigma. One is reproduced in Fig. 150, alleged to be adorned with a hydra-head, though it rather seems to be a pig's head.[2] The handle is perforated, and a double ring is attached to it for suspension from the girdle. This object is made a pre-Han. The next with a one-horned hydra head is alleged to be Han. Then follows, as could not be expected otherwise, the same implement with a phenix-head in which, again, the expression of life is extolled, and which "is a beautiful object of genuine Han people." The series winds up with another phenix-head specimen, "which is slightly different from the preceding one, but is an object left by the early Ts'in

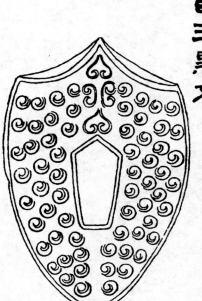

FIG. 147.
Girdle-Ornament, with Cloud-Pattern, of White Jade with Black Stripes.

[1] LEGGE remarks that it belonged to the equipment of grown-up men, and was supposed to indicate their competency for the management of business, however intricate. The youth in the song who is ridiculed had assumed it from vanity.

[2] Here again, the phraseology of pictorial art is used *chêng nêng ju shêng* "fierce as if alive." A pig's head would be a suitable decoration for this implement, as in the beginning boar's teeth were employed for this purpose (compare p. 203).

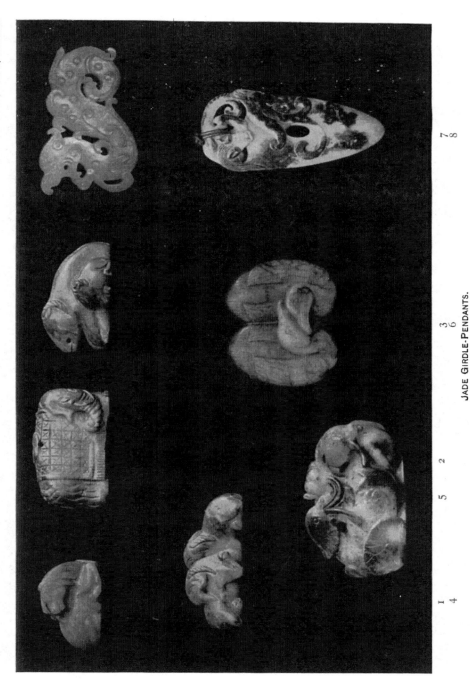

JADE GIRDLE-PENDANTS.

'period.'' Not the faintest reason is given for this chronological definition. For the rest, the forms of all these variants are just alike, and
as much as they are at variance with the two real specimens of Wu,
they show a striking resemblance to the fanciful reconstructions in the
San li t'u and *Leu king t'u* (compare *e. g.* the woodcut in COUVREUR'S
Dictionary, p. 112), so that they cannot receive any credit.

I may be allowed to join here eight jade girdle-pendants (Plate
XXXI) which, though works of the eighteenth century, are in close

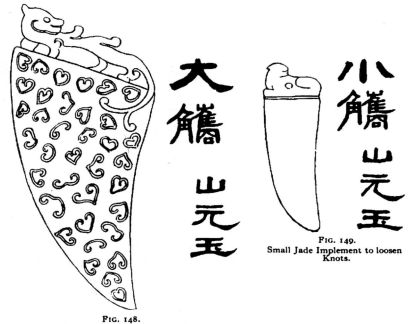

FIG. 149.
Small Jade Implement to loosen
Knots.

FIG. 148.
Large Jade Implement to loosen Knots.

historical connection with the ancient works and partially represent
subjects traceable to the Han period. Therefore, they find a suitable
place here; they further illustrate the development of these ornaments
in modern times. Figure 1 is a carving of gray jade (3.5 cm × 2.5 cm)
showing a reclining horse with head turned back towards a monkey who
is leaning against the horse's neck on the other side. In the illustration,
the monkey is not visible, being covered by the horse's head. One of
the peculiar charms of these ornaments is that their two sides often
present different views. Carried in the girdle, only one face is directed
towards the looker-on whose curiosity is naturally aroused to know
what may be carved on the opposite face, and there he often meets

with something quite unexpected like the figure of a monkey. Thus, a girdle-ornament may attract a person's attention to the wearer and lead to the making of acquaintances. The esthetic enjoyment evinced by the Chinese over the sight of this carving is enhanced by a layer of brown agate-like color strewn only over the opposite side, while the front is of a pure gray-white tinge.

FIG. 150.
Jade Implement to loosen Knots
(from *Ku yü t'u p'u*).

Figure 2, of the same Plate, is the carving of an elephant (5 cm × 2.7 cm) covered with a saddle-cloth[1] on the back of which the dual symbol of Yang and Yin in a circle is brought out. This piece is almost rectangular in shape and creates the impression that the artist's first thought was not to carve the figure of an elephant, but that he chanced on a piece of crude jade of this shape which suggested to him the form of an elephant. Then he set to work to solve his task very ingeniously. The animal is turning its head back, *i. e.* it is carved in high relief on the front side of the plaque; the four feet are cut out on the lower side, as also in the preceding figure of the horse. In this case too, a color surprise is brought in; the surface in front is gray in color framed by brown tinges, while the other side is entirely imbued with a deep red-brown hue.

The carving represented in Fig. 3 of Plate XXXI is a reclining cow with a young calf in the same position at her side (4.5 cm × 2.7 cm), of the same jade with the same color effects as in the two preceding pieces.

In Fig. 4 two goats admirably carved from one pebble are peacefully

[1] Such elephants are frequently used as ornaments in the sense of tribute-bearers with gifts for the imperial court. We have in the Mrs. Blackstone collection several splendid bronze sets of the five sacrificial vessels (*wu kung*) mounted on elephants and lions with an elaborate treatment of this motive.

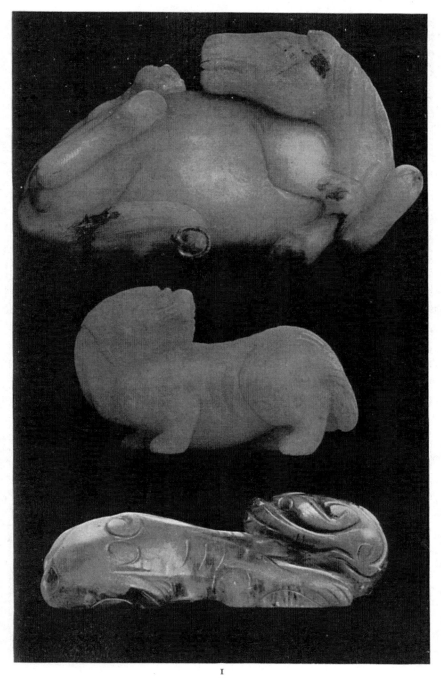

1
2
3
CARVINGS OF ANIMALS.

Explanation of Pl. XXXII.

Fig. 1. Carving of Recumbent Mare and Foal, of White Jade.
Fig. 2. Carving of Lion, of White Jade.
Fig. 3. Carving of Crouching Tiger, of Rock-Crystal.

resting side by side; the jade is also gray and brown (size 5.7 cm × 2.2 cm).[1]

The carving (6.5 cm × 4 cm) in Fig. 5 represents a drake and duck, surrounded by lotus-flowers and leaves and holding the ends of the stems in their beaks. This design has its foundation in a phonetic rebus based on the word *lien* "lotus" written with the phonetic element *lien* "to connect, join." The two ducks are joined (*lien*) in partnership by the lotuses (*lien*), alluding to conjugal happiness.

In the carving in Fig. 6 two mandarin-ducks are represented swimming side by side, holding their heads together; their feet in the act of swimming are brought out in flat relief on the lower side.[2] It is of a jade yellow-brown in tinge interspersed with black patches.

These six pieces represent animal-figures in plastic execution and are all exquisite in the choice and workmanship of the design. They must certainly belong to the sources from which the Japanese received the stimulus for their Netsuke. It is difficult to fix a date for these carvings with absolute certainty; the Chinese range them in the K'ien-lung period (1736–1795 A. D.). They are certainly not modern for two reasons; first, being made of a jade material no longer available, and second, being of a superior workmanship not attained by any article of the present time.

The two pendants in Figs. 7 and 8 of Plate XXXI are of a different technique; the one (7 cm × 4.5 cm) being a flat dragon cut out and sculptured on both sides alike, of gray jade; the other (6.3 cm × 3 cm) a thin heart-shaped plaque of the type *chi sin p'ei* (see pp. 234, 238); on the lower face, the same escutcheon-like figure[3] with the same designs is carved as in Fig. 147. On the upper face, a hydra half in open-work and half in relief holding this part in its mouth gracefully encircles the plaque. It is of fine milk-white jade with a thin layer of brown patches in the upper surface. We note how the traditions of the Han period are still alive and efficient, and how deeply and lastingly they must impress the minds of the Chinese.

The three following carvings, though not girdle-pendants, may conclude these notes, as the subjects embodied in them move on similar lines.

The three carvings of animals grouped on Plate XXXII are used as paper-weights, if employed for a practical purpose, and are good specimens of this kind of work, as turned out in the K'ien-lung epoch.

Figure 1 is a large carving of gray jade with brown spots on the lower

[1] Compare A. BROCKHAUS, Netsuke, p. 438. Brockhaus admits the Chinese origin of the Japanese goat Netsuke, since Japan has neither goats nor sheep.

[2] See LAUFER, Chinese Grave-Sculptures of the Han Period, p. 18, and Plate III.

[3] It exactly agrees with what is called the spade-shield in our heraldry.

side (12 cm × 6 cm) representing a recumbent mare with her foal, both in full figure. The foal is trying to climb on all fours up the back of the mother who has quickly turned her head towards it and in consequence drawn up her right leg; it is just this sudden moment of surprise which

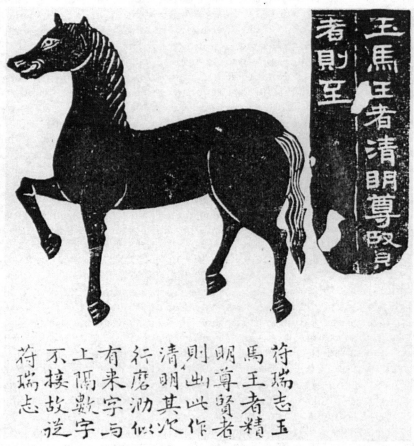

FIG. 151.
"Jade Horse" on a Han Bas-Relief (from *Kin-shih so*).

the artist had in mind to represent, the mare somewhat scared as if she would ask, What do you want? That she had some reason to evince such a feeling of surprise, the artist has not neglected either to express. The young horse has a halter (*lung-t'ou*) around its head with a long rope attached which indicates that it had been fastened to a stake and ran away tearing the rope, so that its visit came somewhat unexpectedly.

The sculptor has bestowed much pains on this little artwork: the mane, ears, nostrils, feet, hoofs, the bunch of hair above the hoofs, and the tails being executed with care and a gleam of realism. It is a classical example of what the Chinese art-philosophers understand by "life's motion" (p. 211). The figure of a horse of jade appears as early as the Han period on the bas-reliefs of Wu-liang among the twenty-two "marvellous objects of good omen" (Fig. 151) accompanied by the

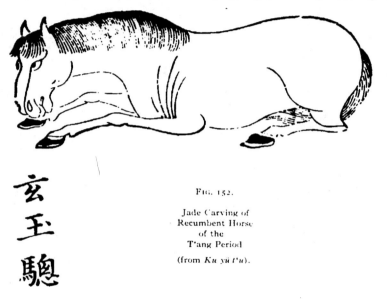

玄
玉
驄

FIG. 152.

Jade Carving of
Recumbent Horse
of the
T'ang Period

(from *Ku yü t'u*).

inscription, "The jade horse: if the reigning sovereign is pure and intelligent and honors the worthies, then it will appear." [1]

Two interesting jade carvings of horses are published in the *Ku yü t'u* (Ch. 2). The one (Fig. 152) is entitled "Piebald of black jade, a type of a horse of the T'ang dynasty," with the following annotation: "In the period K'ai-yüan (713–742 A. D.), Wang Mao-chung offered as tribute five-colored (*i. e.* varicolored) horses for the employment in cavalry. The Emperor Hüan-tsung (713–755 A. D.) ordered them to be represented in sculpture, and his jade sculptors, taking the colors of the five cavalry regiments as basis, took these horses as models and carved them in jade. They were put up on a square table." This is an interesting example of sculpture-modeling after life.

Another full figure of a horse with the separate figure of a man standing behind it is illustrated in the same book (Fig. 153) and is also notable

[1] Compare CHAVANNES, La sculpture sur pierre en Chine, p. 34.

for its curious bit of history.   The horse is stated to be of reddish jade,[1]
the man of yellow jade; the horse's head stands $4\frac{8}{10}$ inches (Chinese)
high, its body $4\frac{1}{10}$ inches, and is $5\frac{4}{10}$ inches long.   In the text it is stated:
"Mane and tail are intact, but the four feet were broken off.   In the
period Chih-chih (1321–1324, Yüan dynasty), the governor of Nan-

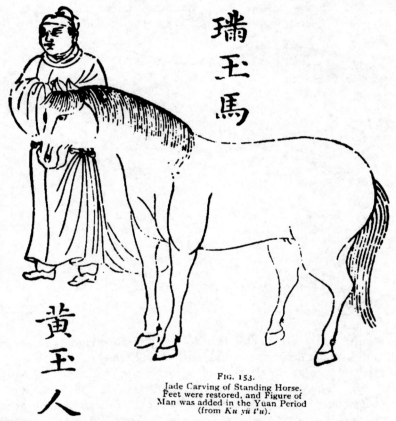

FIG. 153.
Jade Carving of Standing Horse.
Feet were restored, and Figure of
Man was added in the Yüan Period
(from *Ku yü t'u*).

hiung (in Kuang-tung Province), Chao Po-ang [2] possessed an ancient
scroll (with the picture of such a horse) which he exchanged for a porcelain vessel.   The artisan Liu Kia-ming (using this picture as model)
was charged with the task of supplementing the missing feet.   The
master Yao Mu-ngan added the figure of a man of yellow jade to give
the horse an assistant in the way of an equery who has the appearance

---

[1] "Like rouge with spots shining like peach-blossoms, mixed with light green."
[2] Presumably a member of the family of the great painter Chao Mêng-fu (1254–
1322).   He was not an artist himself as several others of the family; his name does
not appear in the Catalogue of Painters *Li-tai hua shih hui chuan*, Ch. 47.

of offering the horse to his lord." [1]   This revelation throws an interesting light on that fatal Chinese custom of restoring and fixing up fragmentary objects which certainly lose a great deal of their archæological value by such acts of violence.   This is practised universally at the present time and admonishes the collector to use great caution; and we now see that it is a time-honored practice too.

In this connection I may be allowed by the way to sound a warning to those who are interested in the mediæval mortuary terra-cotta figurines which have been unearthed from graves in large quantities during recent years.   The complete finds have naturally been rare, and in most cases masses of single heads and parts of the body have come to light which were stuck together haphazardly by inventive Chinamen or even completed with additional clay substances in the hope of an increase in price from the prospective sale to the foreigner. The most grotesque monstrosities have originated in this way, and collectors should be on their guard for such "fakes," and still more archæologists before venturing to base any conclusions on these antiquities.   The addition of the man to the jade figure of the horse is certainly silly and in contradiction to all artistic traditions, as there is no connection between the two, neither by action nor by any technical means; the ancient artwork has been degraded into a toy. [2]

Figure 2 on Plate XXXII represents a standing lion, of white jade (6.5 cm × 4 cm), in the Indian-Buddhistic style, nose, brows, mane and joints of feet being represented by spirals.   The head is elegantly curved back, which has become a favorite position in animals ever since the Han period.   In the open jaws, the tongue, four fangs and the incisors are cut out with minute care.

Figure 3 shows a squatting tiger (9 cm × 3 cm) looking backward carved from a pure clear rock-crystal without flaw.   The spiral

---

[1] Another horse, belonging to the Sung period, is shown in the *Ku yü t'u p'u* (Ch. 37, p. 7).   It is the full figure of a standing horse tied to a post, forming the handle of a jade seal used by the emperor Hui-tsung (1101–1125 A. D.).   The carving is ascribed to a certain Wang Yu, and the editors express a deep admiration for this work.

[2] In the Bishop collection (Vol. II, p. 145) there is a jade-carved horse called *t'o shu ma* "the horse carrying books" which is explained by BUSHELL as "the horse emerging from the Yellow River with the nine volumes containing the nine (?) mystic diagrams."   This is due to a confusion with the legendary dragon-horse (*lung ma*) carrying a tablet (not books which were not then invented) on which the eight diagrams were inscribed.   But that horse is the famous white horse (*pai ma*) of the Buddhist pilgrim and traveller Hsüan Tsang who is supposed to have carried on its back his Sanskrit books from India to Lo-yang; the temple *Pai-ma se* east of the city of Ho-nan fu is named for this horse who plays a great rôle in the fantastic novel *Si yu ki* and is familiar to every visitor of the Chinese theatre.   The drawing in Bishop's work conspicuously shows nine Buddhistic volumes of the peculiar oblong size of the Tripitaka bound in a wrapper and covered with a silk cloth, — carried on the back of the horse.

visible in the illustration is attached to the tail hanging over the back in relief.

The inability of the Chinese for sculpture-work has often been noted, and it is true that generally speaking great sculpture and statuary is lacking in China. Buddhism is largely responsible for anything that exists in this line. The spacious and the majestic is not congenial to the Chinese way of thinking, while they approach perfection in things minute and pretty. This is the element undubitably their own and, to be just, we must judge them from their merits and accomplishments, not from supposed drawbacks or from what does not come up to our expectations. The lack of imposing qualities is fully compensated by their ingenious technical skill, by the wonderful thoroughness and solidity of execution, by an immense adaptability of their work to decorative intentions, by all this beauty of delicate form and line resulting in a microcosm of quaint grace and taste. And this beauty is not cold, but animated by a depth and warmth of colors, — artificial in painting and embroidery, and natural in stone. Their talent in the utilization of the natural coloring in jade and other stones is the best proof of their highly developed color-sense and their innate love of nature, — a phenomenon presumably unique in the history of art. The subjects chosen for their dainty carvings betray a sympathetic insight into the life of the animal and plant world, their power of natural observation, and their faculty of evincing and expressing good and noble sentiments, which rank supreme in their artistic aspirations, whereas striving for naturalness is always subordinate to emotional powers. And that these are not slight or superficial, but of a wide capacity and mental depth, we recognized in the love emblems of the Han period portraying the joy of life, the horror of death and the hope of a hereafter. Idealistic as this glyptic art began, it ended in impressionism, in the still-life, in the *Stimmungsbild*. The artist's sentiment and the expression of his impressions has become the leading motive of art ever since the days of the Sung period. The ancient philosophical and religious emblemizing was destroyed, and the purely personal artwork arose with the sole object to impress and to please.

3. ORNAMENTS FOR HEADGEAR AND HAIR

Ceremonial headgear was utilized to a large extent during the Chou period. The ceremonial cap *mien* was surmounted by a rectangular board (*yen*) in horizontal position from which twelve pendants of globular jade beads, six in front and six behind, were suspended, strung on varicolored silk threads. This headdress was worn in connection with the robe embroidered with dragons by the Son of Heaven, when he sacrificed in the ancestral temple.[1] These jade pendants are called *yü tsao;* the word *tsao* is a general term for aquatic plants which is frequently mentioned in classical literature (BRETSCHNEIDER, Botanicon Sinicum, Part II, p. 224) and appears also as an ornament embroidered on the robes of the emperor and the officials. In a similar manner, as the word *tsao* denotes also an elegant composition, a meaning derived from the fine shape of the leaves of the plant, so it became emblematic of the jade pendants gracefully hanging down like the leaves waving over the pond.

Not only the emperor, but also the officials were entitled to this ceremonial headdress, yet the number of pendants was graduated according to their ranks. "The cap of the Son of Heaven had twelve pendants of jade beads set on strings hanging down, of red and green silk; that of feudal princes (*chu hou*), nine; that of the great prefects of the first class (*shang tai fu*), seven; that of the great prefects of the second class (*hia tai fu*), five; and that of the ordinary officers (*shih*), three. In these cases, the ornament was a mark of distinction."[2] None of these pendants, as far as I know, have survived to the present day.

But another jade ornament is preserved which WU TA-CH'ÊNG presumes served for the decoration of a headdress. This is a round flat button called *k'i* (GILES No. 1048) which was sewed on to the front of the conical leather cap *pien*, worn by the emperor and all officials (BIOT, Vol. II, pp. 152, 234).

Besides the two specimens here figured (Figs. 154 and 155) WU gives two plain ones. All four have small holes on the lower face, three of

---

[1] LEGGE, Li Ki, Vol. II, p. 1. COUVREUR, Vol. I, p. 677. In the illustration on p. 676 and repeated in his Dictionary on p. 541 which is derived from the *Leu king t'u*, the beads appear as strung on threads; nevertheless COUVREUR (Dictionnaire, p. 927) speaks of bands ornamented with embroideries in five colors and provided with pearls of jade.

[2] LEGGE, Li Ki, Vol. I, p. 400. COUVREUR, Vol. I, p. 549. BIOT, Vol. II, p. 235. According to the regulations of the ceremonial of the Han, the Emperor wore a jade hat with nine strings of jade beads hanging down in front and behind (*yü kuan kiu liang*).

them two, and only the one in Fig. 154 three holes which Wu compares with an elephant's nostrils. These holes do not penetrate the surface, but intercommunicate beneath it, as is also done with buttons among us. It will be seen that in shape and design there is a certain resem-

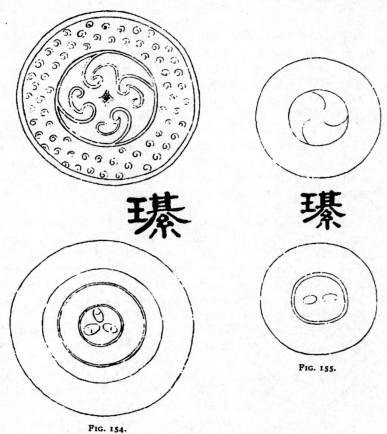

FIG. 154.
Jade Buttons, Upper and Lower Faces.

FIG. 155.

blance between these objects and the central pieces of the girdle-pendant. The jade in Fig. 154 is "white with a black mist," that in Fig. 155 green.

In regard to the symbolism of these buttons, the commentator Chêng K'ang-ch'êng, who likens them to the jade disks *pi*, is of opinion that the large ones symbolize the sun and the moon, the small ones the stars, and that hence the single leather strips of which the cap *pien* was sewn together were called also "stars." This explanation is approved as correct by Wu.

The same cap *pien* was held by a hat-pin of jade stuck through it, as we are informed by the *Chou li*. The *Ku yü t'u p'u* illustrates a number of such jade pins (*pien ch'ai*), but for lack of other comparative material I am not prepared to judge whether, or in how far they are authentic. It may be sufficient to reproduce one of them in Fig. 156, which is alleged to come down from the Shang or Chou period. Only the handle is decorated. The second specimen in the

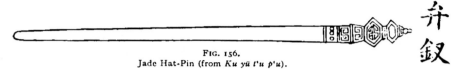

FIG. 156.
Jade Hat-Pin (from *Ku yü t'u p'u*).

above work has a similar handle, and the pin, hexagonal in section, is engraved with a band of cloud-ornaments. The third is surmounted by a dragon-head, and the body of the pin consists of dragon-scales carved in open-work. It is recorded as a special peculiarity of this piece that the jade exhales a natural perfume like garoo-wood, and that, if placed on the head, the whole house will be filled by this odor. Another pin is surmounted by the full figure of a swallow "whose shape, wings and feathers show life's motion and are impressing, as if alive; but to

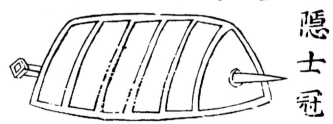

FIG. 157.
Cap showing Wear of Jade Pin, after Sketch by the Painter Li Kung-lin (from *Ku yü t'u p'u*).

judge from this bird, it is not a Han work." The next is shaped into the appearance of a bamboo stem, hollow, and stated to be a work of the Tsin or T'ang period. The last is surmounted by a phenix-head and alleged to be Han.

In Fig. 157 for two reasons, I reproduce from the same work (Ch. 51) a cap designated as "head-dress of retired scholars" (*yin shih kuan*); first, because it gives some idea of how the hat-pin was supposed to have been used, and secondly, because this design is associated with the name of the famous painter Li Kung-lin. I shall not waste time in proving in detail that all the official head-dresses pictured in the *Ku yü t'u p'u* with an elaborate decorative material and alleged to be entirely of jade have never existed in this form, but are just imaginary sketches. This one case may suffice for all. The text has it: "I have formerly seen a painting by Li Kung-lin representing Huang-ti, how he asked for the road on the mountain K'ung-t'ung (*Huang-ti K'ung-t'ung wên tao t'u*),[1] and the cap worn by Huang-ti on this painting is identical with the present one here illustrated." This confession warrants the suspicion that this cap is simply drawn from Li's painting of Huang-ti. The *Ku yü t'u p'u* was compiled between 1174–1189; Li Kung-lin or Li Lung-mien died in 1106 (GILES, Introduction etc., p. 108), and in the collection of the imperial house of Sung one hundred and seven of his works were preserved. In the *Süan ho hua p'u* (Ch. 7), the descriptive catalogue of this collection, the title of the picture in question is not given; it may have been one in a series.

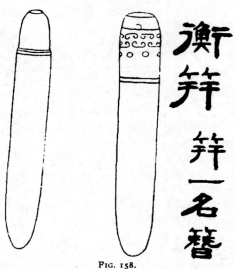

FIG. 158.
Jade Bonnet-Pendants worn by the Empress.

Another question would be how far Li Kung-lin was able to represent faithfully the cap of the ancient Huang-ti, whether he followed some tradition or was merely guided by imagination. We know that he had a faible for antiquities, and that he well understood how to decipher antique inscriptions (CHAVANNES, *T'oung Pao*, 1904, p. 496). So he may be credited, at least, with the earnest intention of aspiring after historical truth to the best of his knowledge, though we may justly doubt that he was in possession of any palpable authentic material to serve him as a correct standard. At all events, this crude sketch is interesting as a scrap from the workshop of that great artist.

The *Chou li* (BIOT, Vol. I, p. 167) mentions the office of the jeweller [2] at the court of the Chou dynasty; he was in charge of all the head-ornaments used by the empress: he made the bonnets worn by her, when she accompanied the emperor on the occasions of sacrifices, and the wigs

[1] Compare CHAVANNES, Se-ma Ts'ien, Vol. I, p. 30.

[2] *Tui* (GILES No. 2801) *shih*, which means as much as carver or engraver of jade, the explanation usually adopted. But the commentator Chêng Se-nung interprets the word (in this case read *chui*) as the designation of a head-dress (*kuan*), which seems to be quite appropriate with the functions of this office.

which she wore in gathering the leaves of the mulberry-tree, — a custom that in view of this artifice loses much of its natural charm; then he made for her the toupets necessary on her visits to the emperor (false locks are alluded to also in the *Shi king:* Legge, Vol. I, p. 77) and the pins for the bonnet and the hair-pins, both of jade. The former were called *hêng ki* (Giles No. 888) and were suspended from the two sides of the bonnet over the ears, while the hair-pin held the hair-dressing together. In Fig. 158, two such bonnet-pendants of white jade are illustrated after Wu Ta-ch'êng who believes he is justified in this identification; as added in the legend to this Figure, he gives the word *tsan* as another name for *ki* which is a definition derived from the *Shuo wên*. Wu interprets also the word *ki* occurring in the *Shi king* (Legge, Vol. I, p. 76): *pi ki leu kia* in the same sense; they are, accordingly, not hair-pins, as translated by Giles (No. 888). This passage intimates that this pin-like pendant was adorned with six gems (*kia*, Giles No. 1146), the character being formed with the verb *kia* "to add" and therefore explained as "gems attached." The *Ku yü t'u* (Ch. 2, pp. 3 b,

Fig. 159.
Earring of White Jade. Upper and Lower Sides.

4 a) pictures two of these jade gems, the one in the shape of a crescent, the other an hexagonal short tube with two perforations going through the axis.

A curious ornament called *t'ien* or *ch'ung êrh* (*i. e.* filling the ears) is mentioned in several passages of the *Shi king* (Legge, Vol. I, pp. 77, 92, 152; Vol. II, p. 410). This was simply an earring, leaf-shaped in form (see Fig. 159 from Wu Ta-ch'êng) as it is still made of jadeite in a similar shape in Suchow for the use of women. In ancient times, such earrings were worn by men,[1] suspended from the cap by means of threads of white, green or yellow silk. Mao, the learned commentator of the *Shi king* annotates that they were made of jade for the Son of Heaven, of stone for the feudal princes, but of jade again for all the gentlemen of old age. The symbolical idea underlying this custom was, as it is put, *sai êrh* "to obstruct the ears," *i. e.* they should be a reminder not to listen to bad discourses, and to shut off the voice of evil. Under the same name, Wu gives, without further explanation, four other

In one case, they are attributed to a woman (*l. c.*, p. 77).

pieces of a different type, shaped like nails or spikes with an ear in the top (Fig. 160). There is nothing on record in regard to the shapes of these earrings, and WU simply makes a plausible guess, but one more satisfactory indeed than the fancifully decorated figure invented by the Chinese draughtsmen of later ages to supply an illustration of this object (figured in COUVREUR's Dictionnaire, p. 883). We see more and more that the ornaments of the earliest times were plain and dignified,

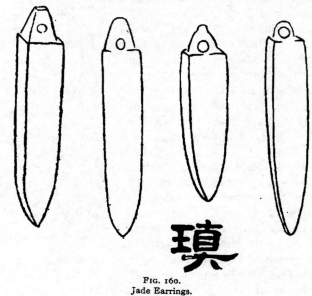

FIG. 160.
Jade Earrings.

and that all the complicated pretentious designs are products of later ages. Also the *Ku yü t'u* (Ch. 2, pp. 3 b and 4 a) offers two designs of such earrings, the one heart-shaped, the other a short tube perforated longitudinally.

Instruments to scratch the head, worn by women are mentioned as early as in the *Shi king* (LEGGE, Vol. I, p. 77). It was a rudimentary comb, explains Legge, consisting of a single tooth, and is said therefore to correspond to the modern comb. Being elegantly made of ivory, it was worn in the hair as an ornament. Men wore the same implement suspended from the girdle (*Ibid.*, p. 164). A boar's tooth is still used for this purpose, also among the Tibetans, and forms one of the five articles of the chatelaine mentioned above, p. 203. There was certainly a deep-felt necessity for this instrument in ancient times as nowadays.[1]

[1] The ancient Chinese also scratched their heads when in perplexity, as seen from a verse in the *Shi king* (LEGGE, Vol. I, p. 68).

The modern Chinese back-scratcher of wood, bone, or ivory, a plain stick surmounted by a hand, is a good example of an implement grown out of the projection of a bodily organ. It has obtained naturalization in this country, and judging from the large sales in New York China-town, it meets the demand of a favorite household-article. The Sung Catalogue of Jades (Ch. 76, p. 7) figures two jade back-scratchers (*sao-lao*), one plain, the other with stem of the appearance of a bamboo and surmounted by a plaque carved into a cloud-pattern in the place of the hand. In the text, such implements of jade and bronze are ascribed to the Emperor Yang of the Sui dynasty (605–618 A. D.), and the two pieces in question are stated to be antiquities of the Sui or T'ang period. This may be true, as far as jade carvings are concerned; but the implement itself is much older and doubtless as old as the Chinese nation itself.[1]

[1] The Eskimo make use of a back-scratcher called *kumakssium* (*lit.* "instrument used against lice"), a long, slightly curved piece of bone with a piece of bear-fur on the end (A. L. KROEBER, The Eskimo of Smith Sound, *Bull. Am. Mus. Nat. Hist.*, Vol. XII, 1900, p. 289).

4. Jade Clasps

In the *Ku yü t'u p'u* (Ch. 59) a number of jade ornaments are illustrated, rectangular in shape, and with a long loop, rectangular in section, attached to the back (Fig. 161); these are interpreted as *wei*, *i. e.* as decorations placed on the top of a sword-scabbard to fasten it to the girdle by this means. It is not difficult to see that this mode of use is impossible, and that this explanation, also given in the *Ku yü t'u* (Ch. 2, p. 9), is erroneous. Wu Ta-ch'êng has not failed to correct this mistake in his book, and to define these objects justly as plain girdle-clasps (*suei*, Giles No. 10407). But even without this authorization of a competent Chinese critic, commonsense could lead us to no other result. The word *suei* is very old and occurs as early as in the *Shi king;*[1] it is defined in the *Êrh ya*, but not in the *Shuo wên*. Wu is presumably also right in making out that this clasp was employed in connection with leather belts which passed through the loop on the back, — I suppose in such a way that the end of the belt was drawn through the loop and then tightly sewed on in a seam running parallel with the long side of the loop. The clasp was accordingly closely attached to this end of the belt and remained in the centre hanging down in a vertical position; while the other end of the belt was fastened, probably by means of metal hooks, to the upper and lower ends of the clasp curved inward and grooved.[2] Or, it could be imagined that to this end of the belt two metal plaques were attached by means of hinges, one above, the other below, corresponding in size to the spaces above and below the loop on the clasp, and that these plaques were slipped in there and held between the grooves. At all events, this can be supposed only, as there is no account extant illustrating the mode of wear. Certain it is — and also Wu expressly insists on it — that these clasps were worn pendent in a vertical position as they are here reproduced. This is evidenced by the loop and further by the way in which the ornaments are arranged; note the animal-heads on the top looking downward on Plate XXXIII.

The most interesting point with reference to these clasps is that they

[1] Legge, Vol. II, p. 355. Legge missed the right meaning of the word by translating: "If we give them long girdle-pendants with their stones, they do not think them long enough," which does not make sense. The meaning is, if we give them long-shaped jade clasps for their girdles, they find fault with them as to the length of their loops, since indeed this clasp was useless, unless it fitted the belt. Couvreur (p. 265) has the same mistranslation. Giles: jewels or ornaments hung at the girdle.

[2] In one specimen delineated in the *Ku yü t'u* (Ch. 2, p. 12 a) the upper and lower ends are provided with angular projections to which it would have been possible to fasten the belt by means of cord or hooks.

represent original jade types and are not imitated in jade from models in metal. No types corresponding to them in bronze have ever been found, as is the case with the jade buckles to be discussed hereafter.

a FIG. 161. *b*
Jade Girdle-Clasp, Front and Back (from *Ku yü t'u p'u*).

A priori we might infer from this that it is a type of considerable antiquity; but from an historical viewpoint this is not the case. Wu does not discuss the question of age; the older works make these objects

Han.   The specimens in my collection are all of the Han period, judging from material, technique and ornamentation.   So I am inclined to

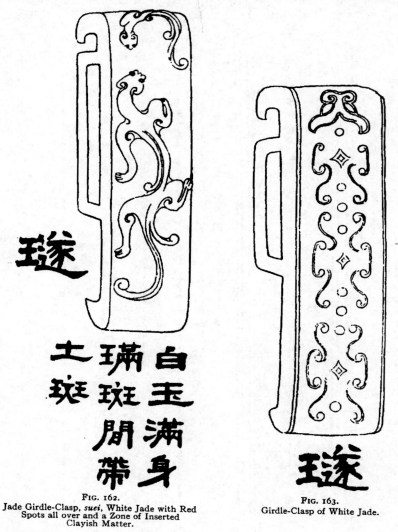

FIG. 162.
Jade Girdle-Clasp, *suei*, White Jade with Red Spots all over and a Zone of Inserted Clayish Matter.

FIG. 163.
Girdle-Clasp of White Jade.

believe that the type itself is not older than this epoch, and that its formation may be credited to the Han.

The colloquial name by which this ornament is still known in Si-ngan fu is *chao wên tai*, a term mentioned also by Wu Ta-ch'êng with the note that it is presumably an old designation whose origin, however,

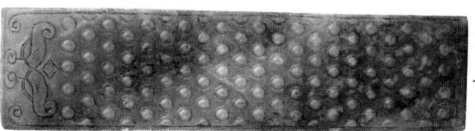

JADE CLASPS OF THE HAN PERIOD.

is not known. Prof. GILES (No. 473) registers this phrase with the translation "a pouch, slung over the shoulders, for carrying despatches etc." (compare *chao hui* "official despatch"). It may be that such or a similar idea ("a girdle ornament used, or authorizing the bearer to carry official documents") was instrumental in causing the people to transfer this expression to the ancient girdle-clasps the proper use of which they no longer understood. Or the name may have arisen out of a confusion with the *ch'ao tai*, the court girdles of jade of the T'ang dynasty.

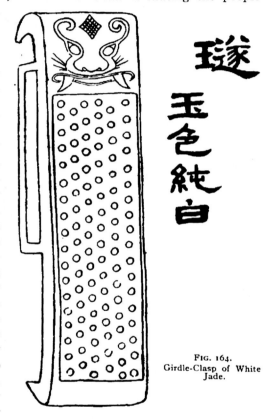

FIG. 164.
Girdle-Clasp of White Jade.

From WU's collection of eight specimens, four have been selected (Figs. 162–165) to show the identity in style and make-up with those in our collection. It should be remarked that these drawings are incorrect from our point of view; if, as here, the side-view with the loop is shown, the full view of the surface cannot appear at the same time, but it must naturally be shortened according to the laws of perspective. If, on the other hand, the surface is illustrated as on our Plate XXXIII, the loop on the back naturally remains invisible. Despite this obvious misrepresentation, I believed I should retain these sketches, because in their method they are so characteristic of Chinese drawing, and because they will afford some idea of the appearance of the sections of these pieces. And, as will be seen presently, I have reason to insist on the perfect identity of Wu's specimens with my own.

In our collection there are six such jade clasps of the Han period, grouped on Plate XXXIII, representing five different kinds of jade, 2 and 3 being of the same light gray. The clasp in Fig. 1 (11.7 cm × 3 cm) is light green in color with layers of brownish-red. It is decorated with what is called the millet-pattern, — twenty-one rows of alternately five and six knobs forming horizontal and diagonal lines.   The figure in slight relief on the upper end may be a conventionalized *t'ao-t'ieh* head. The pieces in 2 and 3 (7.3 cm × 2.2 cm) differ only slightly except that No. 3 is filled to a greater extent with clay matter.   On both, the

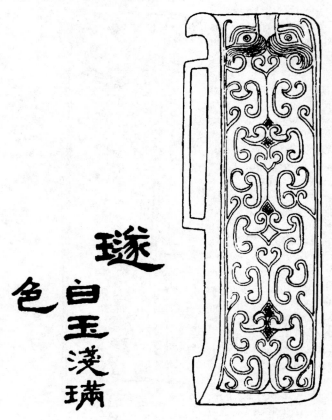

FIG. 165.
Girdle-Clasp of White Jade with Light-Red Tinge.

motive "the hydra watching its young one" is brought out in high undercut relief.[1]   The three following pieces (8.2 cm × 2.2 cm, 8 cm × 1.9 cm, 6.8 cm × 1.7 cm) are all engraved with spiral bands and some conventional design at the top which is in Fig. 5 a well outlined monster's head with two projecting fangs. (Compare Fig. 164.)   The jade

[1] The same motive is represented also on two girdle-clasps illustrated by Wu Ta-ch'êng (one of them reproduced in Fig. 162) and on one in the *Ku yü t'u*.

in Fig. 4 is light-green with purple clouds; that in Fig. 5 is white with brown-red patches in the lower end and on the back; [1] that in Fig. 6 is black with a few white spots.

The various dimensions of the loops on the backs of these pieces show that belts varied much in width. As the length of these openings indicates the width of the girdle, we have girdles of 4.7, 2.6, 2.5, 2.8, 3.1, and 1.8 cm in width, which may be accounted for by assigning them to men, women and children according to size.

If we now glance back at Fig. 161 derived from the *Ku yü t'u p'u,* we shall notice that this piece is in many respects at variance with those published by Wu and me. First, technically, the loop occupies only a small strip in the centre of the back, while in our pieces it extends over the whole width of the clasp; the two projections at the ends, as here drawn, do not occur in any of our pieces. There is not only no sense and purpose in them, but also, as they are visible on the front, they disfigure the artistic unity and impression of the object, and it may be boldly stated that no artist of the Han period could have been guilty of such an absurd breach of good taste. In regard to decoration, it is curious that it is divided into two fields by a central zone, and that each field winds up with an animal's head, the upper one looking upward, the lower one downward, the two looking away from each other. This arrangement is so inartistic that certainly no Han artist has conjured it up. Now if all these absurdities would occur in just this one piece, one might pass over the matter in silence with a forgiving spirit; but exactly the same folly is repeated in five other pieces, — the same loop, the same prongs, the same division of ornaments, and the same heads, while not one normal specimen is reproduced. And in this creation the *Ku yü t'u p'u* stands alone. The seven specimens illustrated in the *Ku yü t'u,* the appendix to the *Po ku t'u,* are of exactly the same style as those of Wu and my own. Since that work is known to us as unreliable and teeming with fictitious matter and late productions, not to use such a harsh word as forgeries, we shall not err in declining to see in those girdle-clasps works of the Han period for which they are given out; they may have been made under the Sung, if ever made at all, — as such a specimen has never turned up, — unless simply drawn from hearsay, or as an attempt at traditionary reconstruction.

[1] This jade is called "red jade of the Han dynasty" (*hung Han yü*). The specimen in question was found in the village Wan-ts'un west of Si-ngan fu.

5. JADE BUCKLES

Among the jade relics of the Han period, there are also buckles (*kou*) for the belt of a gracefully curved shape, at the top always provided with some animal-head turned over and looking downward,[1] and at the back with a projecting stud which was either stuck like a collar-button into a slit of the belt or slipped under a metal ring fastened to the belt. An idea of their general appearance will be best gained from Figs. 166 and 167 reproducing two specimens in WU's collection; the one is of white jade with "a yellow mist," the other of uniformly white jade. WU has appended no discussion to this type, since it is generally known to Chinese archæologists. It has survived until the present day, and pieces similar to the Han models, though much clumsier and far from reaching their beauty, aside from the inferior glassy modern material, are still turned out. The gulf separating the Han from the modern buckles is so deep that, at first sight, the two can be distinguished.

Two jade girdle-buckles of the Han period are represented on Plate XXXIV, Figs. 1 and 2. The one (8.5 cm × 1.8 cm) of gray jade with moss-green and black layers is surmounted by a well carved horse-head looking downward, while from below, a monkey seizing a bee in its right forepaw is crawling upward towards the horse.[2]

The other piece in Fig. 2 (7.8 cm × 2.3 cm) of white and brown-red jade terminates in a dragon's head with two long horns. A hydra is cut out in high relief on the surface.[3]

A number of bronze buckles of the same type are in our collection, one of which is selected here for comparison with the jade types, in

---

[1] The lower end opposite this head is called "the tail" (*wei*).

[2] The monkey with the bee (*fêng hou*) is usually a rebus with the meaning "to bestow on one the investiture (*fêng*) of a vassal prince (*hou*)" (compare GRUBE, Zur Pekinger Volkskunde, p. 95). Neither the cause for the association of the monkey with the horse, nor the meaning of the word *ma* "horse" with reference to the rebus, are known. The proverb quoted by GILES (under *ma*) may be called to mind: "The heart is like a monkey, thought like a horse," — for restlessness and speed. Pictures of monkeys are sometimes presented to candidates where *hou* "the monkey" is read *hou* "to expect" *scil.* an office. Perhaps the significance of the above rebus was: "May you obtain office or rank with the speed of a horse!" or "Quick promotion in the race for office!" or something like this.

[3] The same type is pictured in the *Ku yü t'u* (Ch. 1, p. 11) and ascribed to the Han period. The following buckle, also with a dragon-head, is without any reason dated in the Shang dynasty, which is absurd. The third is placed in the *San-tai* (*i. e.* Hia, Shang, Chou) period; it has a hydra's head, and is further described to bear on the "tail" (end) a tiger's head and body dragging along its feet a monkey, which is not shown in the illustration. The fourth buckle is even made an object prior to the *San-tai*; it is provided with a horse-head like ours in Fig. 1, Plate XXXIV, and it is curious in that it terminates in a hydra's head at the other end,— the only piece with double head that is known to me.

EXPLANATION OF PL. XXXIV.

Fig. 1. Jade Buckle with Horse-Head.
Fig. 2. Jade Buckle with Dragon-Head.
Fig. 3. Bronze Buckle incrusted with Silver Wire.

BUCKLES OF THE HAN PERIOD.

Fig. 3 (11.3 cm × 2.1 cm). It is cast of solid bronze, surmounted by a hydra-head; the surface is incrusted with silver wire forming a geometric decoration of lozenges and spirals. There are others of exactly the same shape as the jade pieces, and there can be no doubt that the latter have been derived from the former. In their elegant curve, they were

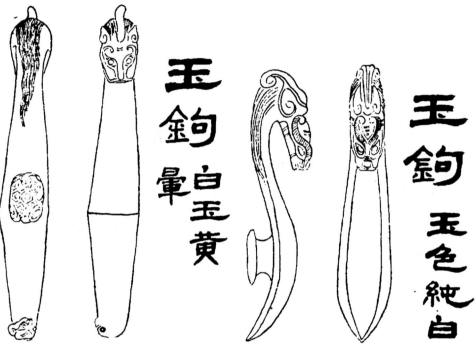

FIG. 166.
Buckle of White Jade with Yellow Mist,
Back and Front.

FIG. 167.
White Jade Buckle, Side and Front.

well adapted to metal casting; carving and grinding them in jade was a task of toil and trouble. This technical consideration is confirmed by the written symbol of the word *kou* which is united with the radical *kin* "metal." In case of a real jade type, we should certainly expect to see the radical *yü* "jade."

In early times, this girdle-buckle does not seem to have been in existence. The word *kou* occurs three times in the *Shi king* (LEGGE, Vol. II, pp. 285, 538, 547), but there it denotes in each case a hook for the trappings of the breast-bands of a horse. This would have an interesting bearing on the question why a horse's head is represented in several of these buckles, if it could be unquestionably established that it was only an equine ornament in its origin and later transferred from the horse to the man, with the idea perhaps of imparting to him the strength and alacrity of the animal. This supposition, however, remains hypothetical, though it may be suggestive. It would form a notable analogy to the wearing of nose-rings by men, supposed to be derived from buffalo's nose-rings to instill the buffalo's strength into

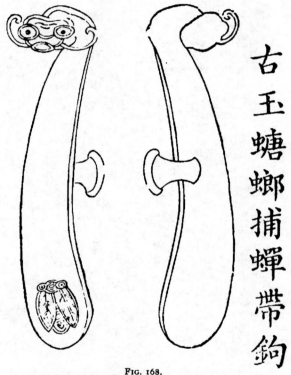

FIG. 168.
Ancient Jade Buckle with the Motive ''Mantis catching the Cicada'' (from *Ku yü t'u p'u*).

the body of man (see LAUFER, Anneaux nasaux en Chine, *T'oung Pao*, 1905, pp. 321–323).

The *Ku yü t'u p'u* (Chs. 57 and 58) offers several interesting jade buckles which may presumably lay claim to authenticity. The one represented in Fig. 168 in two views is explained to symbolize the motive "the praying-mantis (*mantis religiosa*) catching the cicada." The mantis [1] forms the head of the buckle; the cicada is carved on the tail in

[1] *T'ang-lang.* The character here written *t'ang* (GILES No. 10774) should properly be written with No. 10762.

relief (one tenth of a Chinese inch high). The text to this illustration recalls the story of Ts'ai Yung (133–192 A. D.) [1] of the Han dynasty who was once invited to a party, and on reaching the house, heard the sound of a lute played inside. It was a tune to a war-song expressing a desire for murder. Ts'ai, for fear of being killed, at once returned. The host and his guests pursued him, and when questioned, Ts'ai gave the reason for his retreat. The guests said: "When you approached, we seized the lute, as we noticed on a tree in the courtyard a mantis trying to catch a cicada; three times the mantis had reached it, and three times it failed in its attack. We feared that the mantis might miss the cicada (and therefore played the warlike tune)." Ts'ai was thus set at ease. The author of the *Ku yü t'u p'u* adds that the significance of this story escapes him, but supposes that the application of this design to the buckle is derived from it, the meaning underlying it being "murder." This may be correct in general, though I do not believe that the design in question must be traced back to just that particular story.

This story as well as this design are the outcome of popular notions regarding the mantis which is looked upon as a formidable warrior and endowed with great courage. The habits of the mantis are well known: the so-called flower-mantis in tropical regions resembles the flowers of certain plants, and in these flowers it lurks awaiting smaller insects upon which it feeds. What we term the "praying" attitude of the mantis in which its knees are bent and the front-legs supported on a stem, is nothing but this lying in ambush for other insects. Good observers of nature, the ancient Chinese were very familiar with its peculiar traits; they called it "the insect-killer" (*sha ch'ung*) or "the heavenly horse" (*t'ien ma*) from its speed, and greatly admired its bravery.[2] Its eagerness to catch cicadas is repeatedly emphasized, and above all, immortalized by the famous story of the philosopher Chuang-tse.

"When Chuang-tse was wandering in the park at Tiao-ling, he saw a strange bird which came from the south. Its wings were seven feet across. Its eyes were an inch in circumference. And it flew close past Chuang-tse's head to alight in a chestnut grove. 'What manner of bird is this?' cried Chuang-tse. 'With strong wings it does not fly away. With large eyes it does not see.' So he picked up his skirts and strode towards it with his crossbow, anxious to get a shot. Just then he saw a cicada enjoying itself in the shade, forgetful of all else. And he saw a mantis spring and seize it, forgetting in the act its own body, which the strange

[1] GILES, Biographical Dictionary, p. 753.
[2] Compare the Chinese drawing of the mantis in Fig. 169.

蜋   螳

FIG. 169.

The Praying-Mantis *t'ang-lang*   (from *San ts'ai t'u hui*, edition of 1607).

bird immediately pounced upon and made its prey. And this it was which had caused the bird to forget its own nature. 'Alas!' cried Chuang-tse with a sigh, 'how creatures injure one another. Loss follows the pursuit of gain.' " [1]

Surely, this pretty allegorical story has impressed the minds of the Chinese people deeper than the insipid account regarding Ts'ai Yung; and the Han artists, it is more credible, drew on Chuang-tse as the source for the motive of the mantis struggling with the cicada. Also GILES comments in his translation: "This episode has been widely popularised in Chinese every-day life. Its details have been expressed pictorially in a roughly-executed woodcut, with the addition of a tiger about to spring upon the man, and a well into which both will eventually tumble. A legend at the side reads, — All is Destiny!" And in this thought, I believe, we should seek also the explanation of the motive on the Han jade buckle. Certainly, it does not mean such a banality as that frigid "kill!" intimated by the philistine scribbler of the *Ku yü t'u p'u*, but it was a *memento mori* to admonish its wearer: "Be as brave as the mantis, fear not your enemy, but remember your end, as also the undaunted mantis will end!"

In another passage Chuang-tse exclaims: "Don't you know the story of the praying-mantis? In its rage it stretched out its arms to prevent a chariot from passing, unaware that this was beyond its strength, so admirable was its energy!" [2] This is an allusion to another famous story contained in the *Han shih wai chuan*, a work by HAN YING who flourished between B. C. 178–156. It is there narrated: "When Duke Chuang of Ts'i (B. C. 794–731) once went ahunting, there was a mantis raising its feet and seizing the wheel of his chariot. He questioned his charioteer as to this insect who said in reply: ' This is a mantis; it is an insect who knows how to advance, but will never know how to retreat; without measuring its strength, it easily offers resistance.' The Duke answered: ' Truly, if it were a man, it would be the champion-hero of the empire.' Then, he turned his chariot to dodge it, and this act won him all heroes to go over to his side." [3]

Figure 170 represents a jade buckle adorned with the head of what is apparently a wild sheep or antelope. The *Ku yü t'u p'u* entitles it with the curious name *shang yang* [4] "which in form is like a sheep and pos-

[1] GILES, Chuang Tzŭ, Mystic, Moralist, and Social Reformer, p. 258 (London, 1889).

[2] GILES, *l. c.*, p. 49.

[3] Compare *Ko chih king yüan*, Ch. 100, p. 2. PÉTILLON, Allusions littéraires, p. 385 (Shanghai, 1898). LOCKHART, A Manual of Chinese Quotations, p. 335. PFIZMAIER, Denkwürdigkeiten von den Insecten China's, p. 373 (*Sitzungsberichte der Wiener Akademie*, 1874).

[4] GILES (No. 9738) explains the word "a one-legged bird, said to portend rain, and imitated by children hopping about."

sessed with one horn; when it appears, there will be a heavy rainfall."
And therefore the inference is that the ornament of this buckle implies
the idea of "moisture." In my opinion this *shang yang* is identical
with the *ling yang*[1] (*antelope caudata*) and means "sheep of *Shang-chou*,"
the city in Shensi; for in the *T'u king pên ts'ao* by Su Sung of the Sung
period it is expressly stated that this antelope (*ling yang*) occurs in the

mountains of Shang-chou.[2] Transparent lan-
terns are still manufactured in Peking from the
horns of this antelope sliced into thin pieces
which, after having been soaked in water for
some weeks, are joined together. Evidently,
in the application of the *ling-yang* on the
buckle, a rebus is intended by way of punning
with *ling* (GILES No. 7218) "old age," for
which at present the more popular fungus of
immortality (*ling chih*) is used; this seems
more plausible than the suggested allusion to
rain.

In Fig. 171, the head is explained as that
of the fabulus animal *p'i-sieh*; this term means
also "to ward off evil spirits" in the sense of a
talisman and "to avoid evil thoughts," so that
its presence here might express an admonition
and protection to the bearer (compare Ch. X
and Fig. 195). A coiled hydra is engraved on
the body. The date is given as that of the
Han period.

FIG. 170.
Ancient Jade Buckle with
Head of Antelope
(from *Ku yü t'u p'u*).

The head in the buckle of Fig. 172 is said
to be that of the *t'ien-lu*, an animal of lucky
foreboding in the Han palace.[3]

Three other illustrations may be added for the sake of the art-historical interest in the types of the dragon and phenix. The belt-buckle in
Fig. 173 is surmounted by a peculiar dragon-head. "All over the
body it is covered with fish-scales, and the dragon's form in its weird,
wriggling motion is as sublime as the kind of dragons painted by Chang
Sêng-yu." The latter was a famous painter of Buddhist subjects,
living under the Liang dynasty in the first part of the sixth century,

[1] GILES No. 7208.

[2] Quoted in *Hing-ngan fu chi*, Ch. 11, p. 16.

[3] *T'ien-lu* was also the name of a pavilion north of the Han palace *Wei-yang*,
which contained the archives. GILES explains it as a fabulous creature like a deer
with one horn, placed on the top of columns at the graves of officials of the third
rank.

and made a specialty of dragons.[1]   None of his works have survived, but as a number of them were still preserved in the gallery of the Emperor Hui-tsung, the authors of the *Ku yü t'u p'u* were placed in a position to be acquainted with his style.   And it is interesting that they point out to us in the dragon of this jade buckle a type of dragon in the

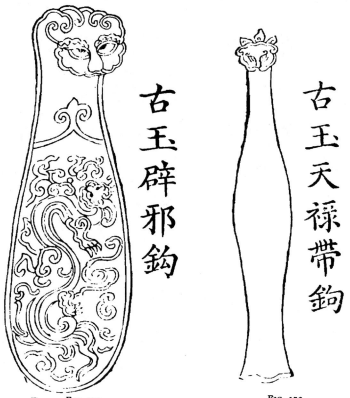

FIG. 171.
Ancient Jade Buckle with Head of *p'i-sieh*
(from *Ku yü t'u p'u*).

FIG. 172.
Ancient Jade Buckle with Head of the
Animal *t'ien-lu* (from *Ku yü t'u p'u*).

style of that great artist.   We doubtless meet here the Indian type of dragon which Chang had received from India with his other Buddhistic motives.   If the authors of the *Ku yü t'u p'u* are correct in their observation, the inference would naturally be to date this jade buckle in the sixth century A. D.

It is therefore a matter of surprise to see the dragon-buckle in the next Fig. 174 dated in the first years of the Eastern Han (*Tung Han*

[1] GILES, An Introduction to the History of Chinese Pictorial Art, pp. 29–31; HIRTH, Scraps from a Collector's Note Book, pp. 59–61.

ch'u nien) which would refer to a period from 25 to about 30 A. D., for also this dragon-head, though not identical with, but differing in style from, the preceding one, savors of Buddhistic influence. Also in this case, the editors intimate a pictorial undercurrent, at least in regard to the pattern of "rolled clouds," of which they say that "it is sublime like painting" (*yen ju t'u hua*). I think they are quite right, for it is

所畫之真龍也　然如張僧繇輩　猙獰蜿蜒之狀儼　周身皆作鱗甲其

古玉脊首帶鉤

FIG. 173.
Ancient Jade Buckle, with Dragon-Head in the Style of the Painter Chang Sêng-yu of the Sixth Century (from *Ku yü t'u p'u*).

indeed the cloud-pattern as it occurs in the Buddhist painting of the T'ang period. But if this style has sprung from pictorial art, it seems also likely that the dragon-head has the same origin, for it is a uniform composition: the dragon soaring in the clouds, as the motive of the preceding composition is the scaly dragon. I should therefore make the piece in question not earlier than the age of the T'ang dynasty, though it must be admitted that it may even go as far down as the Sung period. In such doubtful cases, it is best to indicate the date simply as mediæval.

Such a mediæval object is surely also the buckle in Fig. 175 surmounted by the head of a bearded phenix with cloud-shaped crest. In this case, the date is established as between the Tsin and Sung (*i. e.*

Liu Sung) dynasties, a period between
the end of the third and the end of the
fifth centuries, in the Chinese text. The
geometric feather-ornament on the body
of the buckle is of special interest, as it,
too, displays pictorial influence; the Chinese editors make a very slight allusion to
this point readable between the lines by
remarking that the head, crest and
plumes of the phenix are represented "as
if alive" (*ju shêng*). Now we know that
this is one of the stock-phrases of Chinese
esthetics of painting, and the artists who
worked on the drawings of the *Ku yü t'u
p'u* were fully impregnated with this
phraseology. I have, for this reason,
introduced Fig. 176, a jade buckle with
a horse's head and a coiled hydra, because they here avail themselves of the
term *shêng tung* "life's motion"[1] with
reference to the wriggling motions of the
monster, the application of this term to
the motive in question being very instructive. The mere fact that the Sung
artists operate on these occasions with
the nomenclature of pictorial criticism,
and their hint at the dragon-motive of
Chang Sêng-yu, sufficiently prove that
the great age of these jade buckles is

FIG. 174.
Jade Buckle with Head of Dragon.
Mediæval (from *Ku yü t'u p'u*).

[1] See HIRTH, *l. c.*, p. 58. This phrase is not identical with our word realism, but
denotes the peculiar live action in which a man, an animal or a plant is represented,
and in which the observation of a particular motion is brought out. This is one of
the most characteristic features of Chinese art and especially painting, which has
been recognized among us, before we became acquainted with the Chinese confession
(see LAUFER, The Decorative Art of the Amur Tribes, pp. 77-78, and E. GROSSE,
Kunstwissenschaftliche Studien, p. 205). Dr. HADDON, in a review of the former
work (*Nature*, 1903, p. 561 b) doubted whether this view of mine would appeal to
all readers, and remarked: "The idea that the bulk of the ornamentation of a
group of people is based mainly upon conceptions of motion is certainly new." But
this idea is that of East-Asiatics themselves and has been inborn in their minds and
working in full play for at least 1500 years; the Chinese art critics ever since the
fifth century have expressed it with full consciousness and made it one of the "Six
Canons" to be observed in painting: "life's motion," *i. e.* the specific action, posture,
movement peculiar to any living being or plant at a certain given moment. *E. g.*
bamboo-leaves drooping under the load of heavy rain-drops or agitated by the wind
is, first, an observation made in nature which the artist attempts to outline on paper;
others follow in his footsteps, the motive is copied over and over again, until it finally
degrades into a stereotyped ornament with a stereotyped name.

doubtful, and that the designs applied to them arose, not in the studios of lapidaries, but in the schools of painters.

In view of these self-made confessions, and considering the entire pictorial character of the drawings in the *Ku yü t'u p'u* executed by

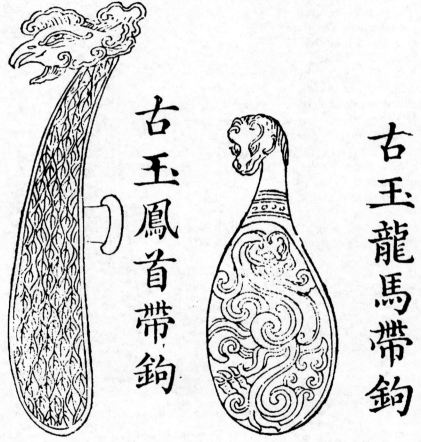

FIG. 175.
Jade Buckle with Head of Phenix, Mediæval
(from *Ku yü t'u p'u*).

FIG. 176.
Ancient Jade Buckle with Horse-Head and Hydra
(from *Ku yü t'u p'u*).

well-known artists of the brush of the Sung period, I cannot banish the thought that these artists themselves may have been the agents in transmitting their own designs to the sculptors who forthwith executed them in jade. This theory — and it has no pretention to be more than that — would possibly furnish a reasonable explanation for the peculiarly systematic and schematic character of this singular and anomalous book, and further, expose the psychological motive by which those Sung artists have been guided in perpetuating their own productions to posterity under the disguise of a pretended collection of ancient jades. By the systematization prevailing in this book, I mean to say that every class of objects is *quasi* conjugated according to a definite scheme, that the same series of motives occurs again in every series of types and is accordingly badgered through every chapter.

We could but reluctantly yield to the assumption that it was ever possible to gather such a seemingly complete collection, so fully representative of each and every variant. It is too beautiful, too obtrusively complete to deserve full credence. It is more credible and plausible that in the same way, as in the court-atelier (*shang fang*) of the Sung emperors the bronze vessels of the Shang and Chou dynasties were reproduced,[1] as the Sung painters imitated the style of the T'ang masters, the ancient works of jade were revived in new forms and with motives partially modified under the suggestive influence of this general renaissance movement in art. And therein the whole secret of the *Ku yü t'u p'u* may be regarded as unveiled. It is, to a large extent, a collection of art-motives, a grammar of ornaments, characteristic of the national awakening of art in the Sung period. Therein lies its importance, but there, also its value ends; it is not a safe guide for the study of jade antiquities.

[1] Still known to the Chinese under the terse label *Shang fang tso ti* "Made in the imperial atelier." There are many beautiful bronze specimens of this class in our collection, many of them being not only just as good, but even better than the ancient models.

6. SWORD ORNAMENTS OF JADE

Wu Ta-ch'êng reports that he owns in his private collection an ancient two-edged bronze sword having a sword-guard of jade and a hilt of the same material, the bronze and jade parts being cleverly joined together. The age of such swords, he adds, cannot be made out definitely, as sword-guards occur both in bronze and in jade. An allusion to a sword-guard of jade is met in the "Annals of the Former Han Dynasty" (*Ts'ien Han shu*) in the chapter *Hiung-nu chuan* "Memoirs of the Hiung-nu (Huns)" where it is on record that the *Shen-yü* (sovereign) of the Huns received as a gift from China a two-edged sword with jade fittings.[1] This passage is explained by the commentaries that the hilt, the ring-shaped knob surmounting the hilt, and the guard were made of jade, a statement which agrees with the specimen of *Wu*. On this occasion the famous Yen Shih-ku (579–645 A. D.) imparts to us three words then current for the designation of a sword-guard: *pi* which means nose, *wei* which means protector, accordingly corresponding to our word guard, and a complicated character composed with the radical for jade (*yü*) and likewise reading *wei*.[2] In all probability the latter word and character have been coined *ad hoc*, in order to designate the special sword-guards made of jade. In the "Biography of Wang Mang" contained in the same annals, it is noted that odd bits of jade (*sui yü*) were chosen for their manufacture.

In our collection there is a beautiful specimen of a jade sword-guard of the Han period, remarkable for its artistic design and splendid workmanship. It is shown on Plate XXXV, Fig. 1 in three views; 1a exhibits its front on which one hydra is carved. Fig. 1b shows the opposite side on which two hydras are displayed, while Fig. 3c represents the lower side with the perforation in which the hilt and blade of the sword join. In Fig. 2 of the same Plate is illustrated an ancient bronze sword-guard acquired by me in Si-ngan fu covered with a thick layer of fine patina, plain and unadorned, but in the rhomboidal outline and in the

---

[1] According to the Annals of the Wei dynasty (*Wei shu*) the emperor Wên possessed a two-edged sword (*kien*), the head of which was mounted with brilliant pearls, and the hilt adorned with jade of Lan-t'ien (*Lan yü*). It was customary for him to hand it to his followers in order to ward off the evil influences of spirits.

[2] The character will be found in Figs. 177–181 representing jade sword-guards. It is omitted in GILES's Dictionary. *Yen Shih-ku* defines also the word *sün* (GILES No. 4906 "knob on the guard of a sword"); it is, according to him "the horizontal part overlapping the mouth of the sword." This cannot be the guard, as the latter is explained by three other words, but only the flat projecting rim occurring on the top of the two-edged bronze swords. The editors of the *Kin-shih so* are of the same opinion; they give the drawing of an ancient sword with the nomenclature of the single parts and denote the part in question by the word *sün*. The Japanese use the character for this word in writing their word *tsuba* which means sword-guard.

EXPLANATION OF PL. XXXV.

Fig. 1. Jade Sword-Guard, *a* Front, *b* Back, *c* Lower Side.
Fig. 2. Bronze Sword-Guard.

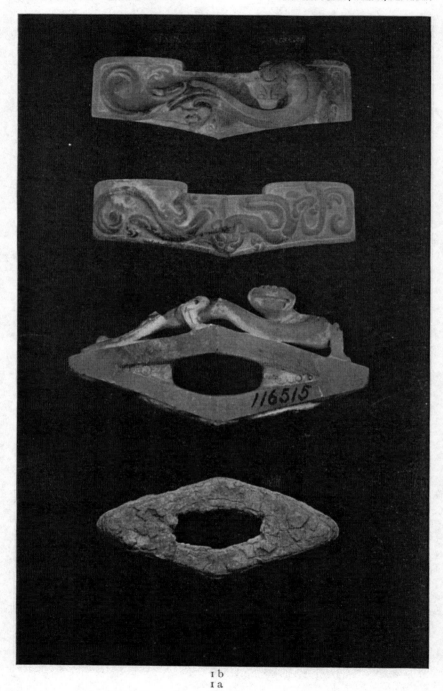

1 b
1 a
1 c
2

SWORD-GUARDS OF THE HAN PERIOD.

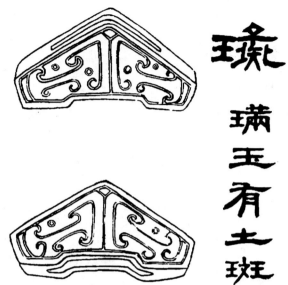

璚
瑀
玉
有
土
斑

FIG. 177.
Sword-Guard of Red Jade with Clayish Substances.

璚
玉色純白下邊
有紅暈一縷

FIG. 178.
Sword-Guard of White Jade.

shape and treatment of the perforation perfectly identical with the specimen in jade.

While it becomes obvious from the peculiar character and technique of these perforations that those objects could have served no other purpose, this fact is set beyond any doubt by two actual finds of Han cast-iron swords in our collection which are provided with bronze guards of exactly the same description.   As a minor evidence, I may mention that also the antiquarians of Si-ngan fu recognized and defined these bronze and jade specimens as sword-guards (called *hu-shou* "guarding the hand").

Figures 1a and b of this Plate XXXV exhibit, as stated, the front and back of the jade sword-guard (5.6 cm × 1.8 cm).   On the one side (b), a hydra is shown in full figure, undercut and almost standing out freely from the surface, raising its head with open jaws, as if ready for attack.   On the other face (c), a mother hydra and its young one are laid out in flat relief.[1]   The jade is light-gray in color and filled with white clay matter; in the middle of the hydra's body in Fig. 1a, we notice a black spot in the illustration which is bright brown in the jade and surrounded by ivory-colored patches.

Five specimens of jade sword-guards are in Wu Ta-ch'êng's collection and here reproduced in Figs. 177–181; each being represented twice as seen from above and below.   While the outward shapes of these somewhat differ from mine, the way of cutting out the perforation is the same.   Figure 177, as the legend attached to it says, is a piece of "reddish jade with earth spots;" that in Fig. 178 of "uniformly white jade with a spray of red mist on the lower side;" those in Figs. 179 and 180 of "white jade with reddish dots;" that in Fig. 181 of "green jade with a black section in which reddish dots are interspersed."   As regards ornamentation, our author has not commented on it; it consists, on the whole, of spiral formations in low relief equally brought out on both faces, except in Fig. 181 where the back is unadorned.   The decoration on the front is curious; the drawing must be held upside down to recognize in it a hydra with raised right fore-paw and head looking backward. The double spiral forms the monster's tail.   This design is perhaps symbolic of the action of the sword.

In the "Book of Songs" (*Shi king*), this verse occurs: "The jades at his scabbard's mouth all gleaming" (LEGGE, Vol. II, p. 383; compare also p. 485).   Two kinds of jade ornaments used for the decoration of scabbards are pointed out, — the one called *pêng* (GILES No.

---

[1] A frequent motive called either "pair of hydras, child and mother," or "the mother watching the cub."   There is, accordingly, an appropriate symbolism brought out on this sword-guard, attack on the one side, and defence on the other.

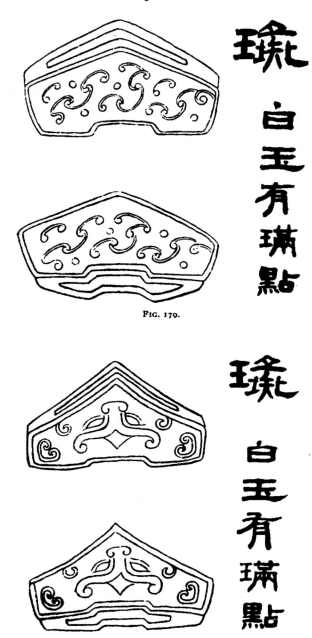

FIG. 179.

FIG. 180.
Sword-Guards of White Jade with Russet Dots.

8870) defined by the dictionary *Shuo wên* as an ornament for the upper part of a knife sheath made of jade for the emperor and of gold (or metal) for the vassal princes; the other called *pi* (GILES No. 8929)

FIG. 181.
Sword-Guard of Green Jade with Black Stripes
and Russet Dots in them.

for the adornment of the lower end of a sheath or scabbard. We know also that this class of objects was buried with the dead, for we read in the *Ku yü t'u* (appended to the *Po ku t'u*, Ch. 2, p. 14) where three sets of them are figured, that in regard to the first of them, Liu Yen-siang had obtained it in the Western Capital (Si-ngan fu), and that according to tradition it was found in the grave of Kao Sing (see below).

It will be seen from the two Figures 182 and 183 derived from WU TA-CH'ÊNG and representing two jade ornaments (the one of white jade, the other of green jade with "earth spots") for the decoration of the mouth of scabbards (*pêng*) that these bear in their forms a striking similarity with the sword-guards of jade, and that their forms are apparently derived from the latter.   Also the prancing hydra on the back of

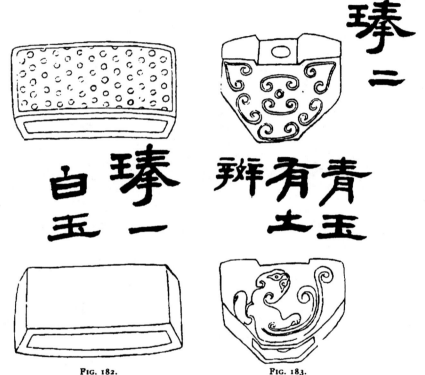

FIG. 182.   FIG. 183.
Jade Ornaments, *pêng*, for the Mouth of Scabbards, White Jade (182) and Green Jade with Clayish Substances (183).

the ornament in Fig. 183 is related in style to that on our jade sword-guard in Plate XXXV, Fig. 1.   Each of the two *pêng* is figured twice, as seen from above and from below, the front being chosen for the former view, the back for the latter.   In Fig. 182, the back is plain; in Fig. 183, two different decorations are brought out on the two faces.

In Figs. 184–187 showing four jade ornaments used as adornment for the extreme ends of scabbards (*pi*) from the collection of WU TA-CH'ÊNG, the double views have been retained only in those cases where the ornaments of both faces are at variance.   In regard to the material, it is remarked of the pieces in Figs. 184 and 185 that they are made of "white jade with reddish dots spread over the entire body;" of the piece in Fig. 186, "of white jade with two black strips;" and of that in Fig. 187, "of white jade with a small zone of black mist, this specimen being the largest of its kind." In form, these *pi* are bell-shaped in general. The small oval perforations show that they were destined for scabbards terminating in a curved narrow tip, accordingly holding one-edged knives or swords. The decorations are self-explanatory,

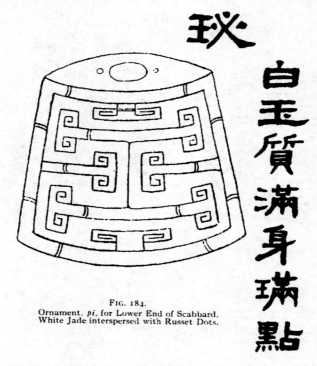

FIG. 184.
Ornament. *pi*, for Lower End of Scabbard.
White Jade interspersed with Russet Dots.

pointing to the style of the Han period. The hydra ready for attack is notable in Fig. 185. In Fig. 186, a bronze vessel with projecting ridges is imitated. Figure 187 offers perhaps a strongly conventionalized form of the monster *t'ao-t'ieh*.

In our own collection there is a carving of cloud-white and greenish jade (3.8 cm × 4.5 cm) which I think may be identified with this type (Plate XXVIII, Fig. 6). The petals of an opening flower seem to be intended in the upper portion, and on one of the lateral sides a veined leaf is engraved, part of which is visible in the illustration. Also this piece is doubtless a work of the Han period.

The *Ku yü t'u* offers also several of these scabbard ornaments. Its first set is dated in the time before the Shang dynasty (B. C. 1766) on the ground that it was found in the grave of Kao Sing, a pretended

FIG. 185.
Ornament, *pi*, for Lower End of Scabbard, White Jade interspersed with Russet Specks.

contemporary of the legendary Emperor Shun (alleged B. C. 2258–2206). Alleged finds from his grave are alluded to also in the *Ku yü t'u p'u*, but this grave must be looked upon as legendary as well as the personage.

Figure 188 represents a *pêng* and a *pi* illustrated in the *K'ao ku t'u* by Lü Ta-lin (Ch. 8, p. 5), and accompanied by an engraving of an ancient king (Fig. 189) holding a sword adorned with these ornaments. This picture has a special interest to us, as it is stated in the accompanying text that it is derived from the series of Virtuous Women (*Lieh nü t'u*) of the famous painter Ku Ch'ang-k'ang, or Ku K'ai-chih of the fourth century A. D., and that it represents Wu, king of Ch'u (B. C. 740–690) carrying a sword in his belt which is ornamented at the upper and lower ends with the jade carvings called *pêng* and *pi*.[1] The series of pictures

FIG. 186.
White Jade Ornament, *pi*, for Lower End of Scabbard.

here alluded to was first published in print in 1063 A. D., and republished in facsimile in four volumes at Yang-chou in 1825 (under the title *Ku lieh nü chuan*).[2] There we find the same figure in the second volume No. 2, with the same designation; the king is engaged in conversation with the queen, the princess Man of Têng. It is the illustration of the scene described by TSCHEPE (Histoire du royaume de Tch'ou, p. 22). The drawing of the sword is much plainer there than in our picture and lacking in those ornamental characteristics for the sake of which it is

[1] On KU K'AI-CHIH see GILES, Introduction etc., pp. 17–21; HIRTH, Scraps, pp. 51–53; CHAVANNES, *T'oung Pao*, 1904, pp. 323–325. The biography translated by Chavannes from the *Tsin shu* is certainly not the life of the painter, but a collection of merry tricks in the style of *Eulenspiegel* which tradition has centralized around his person; such types have been created by popular tradition everywhere, and any good jokes are finally ascribed to them. Most of these anecdotes concerning the painter have a world-wide currency; the last, *e. g.*, occurs in the Turkish stories of Nassr-eddin.

[2] A copy procured by me in China is in the John Crerar Library (No. 673).

reproduced in the *K'ao ku t'u*. The object held by the king in his right is drawn there as a palm-leaf fan.

In this connection, also the archer's thumb-ring of jade may be noticed, as there is an ancient specimen of yellow-reddish tinge in our

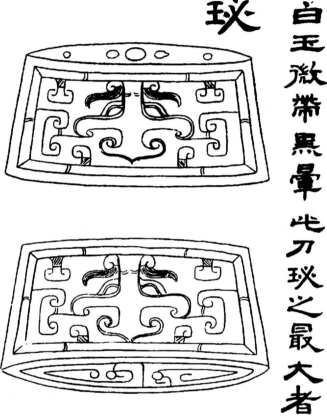

玔

白玉激帶黑暈此刀玔之最大者

FIG. 187.
White Jade Ornament, *pi*, for Lower End of Scabbard.

collection (Plate XXVIII, Fig. 7). The illustration shows the side-view. The ring is 3 cm wide and high, 2.5 cm long, the opening having a diameter of 1.8 cm. It is shaped like an arch in the upper part and flat in the lower.

Dr. BUSHELL has likewise illustrated a piece of this type in the Bishop collection No. 330 (Vol. II, p. 109) which he describes as being of a jade gray clouded and veined with very dark brown, and which he dates likewise in the Han dynasty.

Archery is a very ancient practice in China, and these thumb-rings as also the arm-guards are mentioned as early as in the *Shi king* [1] under two different names, *küeh* (GILES No. 3220) and *shih*. They were worn on the thumb of the right hand to protect it from being injured by the bow-string after the release of the arrow, or as Mr. MORSE [2] explains, the thick edge of the ring is brought to bear upon the string as it is

FIG. 188.
Jade Ornaments for Scabbard (from *K'ao ku t'u*).

drawn back, and at the same time the string is quickly released by straightening the thumb.

WU TA-CH'ÊNG figures also a specimen of pure white jade which it is not necessary to reproduce, as it has the same shape as ours, and arrives at the conclusion that this particular piece was reserved for imperial use, on the ground that such rings of white jade were permitted to the emperor only, while those of the officials were of ivory.

These thumb-rings are still used in archery and manufactured in Peking from the antler of an elk which is there designated also by the Chinese with the Manchu name *hantahan*.

The mode of wearing the ring may be seen in a Chinese illustration given by P. ETIENNE ZI (Pratique des examens militaires en Chine, p. 18, Shanghai, 1896). Father ZI remarks that the most prized rings are those made of jade of the Han period (*Han yü*) of a white gray with red veins and green stripes; those taken from the graves of students who were graduated at the time of the military examinations are reddish in color, and a notion that they afford protection against spirits is attached to them. MORSE (*l. c.*) has made a most careful technical study of this question and gives

---

[1] LEGGE, Vol. I, p. 103, ed. COUVREUR, pp. 72, 208.   Compare also *Li ki*, ed. COUVREUR, Vol. I, p. 621.

[2] EDWARD S. MORSE, Ancient and Modern Methods of Arrow-Release (*Bulletin of the Essex Institute*, Vol. XVII, 1885, p. 17).

an exact sketch of the Chinese method of arrow-release (p. 16); he figures

FIG. 189.
King Wu of Ch'u,
after Painting by Ku K'ai-chih
(from *K'ao ku t'u*).

likewise two modern Chinese thumb-rings (p. 17) and shows that the
Turk and Persians also avail themselves of such rings of similar shapes.

7. THE JADE COURT-GIRDLES OF THE T'ANG DYNASTY

Under the T'ang dynasty (618–907 A. D.), a new fashion in court-girdles came into existence. The *T'ang shih lu* (as quoted in the *Ku yü t'u p'u*, Ch. 52) reports: "Girdles for the loins have existed since times of old, and all used to wear them; but these were leather belts throughout. Kao-tsu (618–627 A. D.) of the T'ang dynasty was the first to institute regulations for the girdles: all princes and nobles, lords, ministers of state and generals above the second rank were allowed to wear jade girdles. The one of the Son of Heaven consisted of twenty-four plaques, all others of thirteen plaques with two additional plaques at the ends ('tails'). There were two kinds, ornamented and plain ones, the former only for the use of the Son of Heaven, while the princes and ministers could wear the girdles with dragon-designs only in case that they were bestowed on them." This was a new departure from the custom obtaining under the preceding Sui dynasty under whose regulations the emperor wore a girdle with twelve metal rings, and the princes and high officials one with nine such rings. We shall soon see by what factors this innovation of Kao-tsu was instigated.[1]

Figure 190 represents the first of these girdles depicted in the *Ku yü t'u p'u* consisting of twenty-four plaques carved from a lustrous white jade, the girdle running all round; the single plaques are stated to be $2\frac{7}{10}$ inches (Chinese) long, and $2\frac{3}{10}$ inches wide and $\frac{3}{10}$ inches thick, the round ones having a diameter of $5\frac{5}{10}$ inches, the square ones being $2\frac{7}{10}$ inches long and $1\frac{4}{10}$ inches wide; the two tail-pieces being $5\frac{9}{10}$ inches long and $1\frac{9}{10}$ inches wide. The two central plaques display each a dragon with four claws soaring in the clouds, while the others are occupied by cloud patterns. The judgment of the *Ku yü t'u p'u* on the work is: "The patterns are in open-work and clever, the carving is fine like down, showing the excellent workmanship of the T'ang." This, as well as the next girdle (Fig. 191) of which only half is reproduced, is an imperial privilege. The jade in this specimen is white and "glossy like mutton-fat." It is carved with designs styled *wan shou* "ten thousand ages," a hyperbolic expression for the emperor (also for his birth-

---

[1] There are many other regulations of the T'ang concerning girdles published in the *T'ang shu* and the *Ku kin chu*, which we cannot treat here in full, as they are not concerned with jade. Kao-tsu bestowed girdles of black tortoise-shell on civil, and girdles of black silver on military officials, "for the reason that they should indicate the unchangeability of color," a curious double symbolism with reference to color and faithfulness of the officials. In the period *Shang-yüan* (674–676 A. D.), an edict conferred girdles of gold and jade on civil and military officials above the third grades, gold girdles on those of the fourth and fifth grades, silver girdles on those of the sixth and seventh grades, brass (*t'ou shih*) girdles on those of the eighth and ninth grades, and assigned copper and iron girdles to the people at large.

day); this name is derived from the svastika on which this pattern is based, and which is also read *wan*, considered as an abbreviated form of this character meaning "ten thousand," and its arrangement in this composite design suggests the character *shou* "long life;" this occurs three times in the well-known form in each of the two central plaques.[1]

The girdle in Fig. 192 is styled "Jade court-girdle of great happiness, equalling heaven, of the T'ang dynasty." It is not of white, but

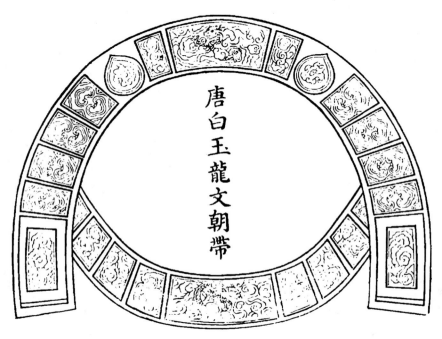

唐白玉龍文朝帶

FIG. 190.
White Jade Court-Girdle with Dragon-Designs of the T'ang Period (from *Ku yü t'u p'u*).

of pale-green jade, because officials are allowed to wear it.[2] The pattern carved on the plaques consists of bats surrounded by clouds (*pien fu yün hia chih wên*) in open-work and cut out in layers "fine like down." The jade of the bats is red like cinnabar (whether natural or artificial is not mentioned), "which is very curious, for this red (*hung*) means the word *hung* (GILES No. 5252) "great," and as the bat *fu* means *fu*

---

[1] For designs of a related character see L. GAILLARD, Croix et Swastika en Chine, Chap. III (Shanghai, 1893).

[2] The *Ku yü t'u p'u* says so, but this girdle consists of twenty-four plaques, while the officials, according to the regulations, as we found, were entitled to a girdle of only 13+2 plaques. I am not able to account for this contradiction.

"happiness," this design has the significance of *hung fu* "great happiness;" the clouds in the sky bear out the meaning "equalling heaven" (*ts'i t'ien*), so that the whole implies a sentence by which blessings are

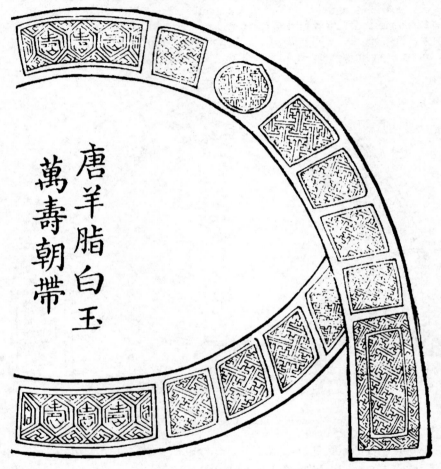

FIG. 191.
Jade Court-Girdle of the Emperors of the T'ang Dynasty (from *Ku yü t'u p'u*).

implored for the imperial palace; *i. e.* when the officials entitled to wear this girdle appeared with it in audience, it was expressive of the wish, "May your Majesty obtain great happiness reaching heaven like the clouds!"

The most interesting point in regard to this rebus is the double symbolism associated with the red color; for the color red is an emblem of luck in itself, but besides this, the color is supposed in this case to be readable with its name *hung* "red" which is punned upon with another word *hung* "great." A pun is also underlying the cloud symbolism, as

FIG. 192.
Jade Court-Girdle of the Officials of the T'ang Dynasty, ornamented with Bats and Clouds
(from *Ku yü t'u p'u*).

the compound *yün hia* "clouds and vapor" is used; and the word *kia*, in the compound *chu kia* "to implore blessings," is written with the same phonetic element as the word *hia* "vapor;" here, accordingly, is a double symbolism, phonetic and one of subject-matter.

While the designs on these three girdles are plainly Chinese, we meet on the plaques on the next girdle in Fig. 193 a floral design of Persian origin occurring in the same form and under the same name (*pao siang hua*, "rose-flower") on the contemporaneous metal mirrors of the T'ang period, of which there are many excellent specimens in our collection.

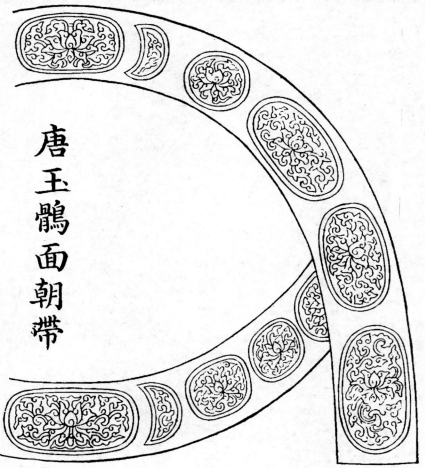

FIG. 193.
Jade Court-Girdle of the Empress of the T'ang Dynasty, with Persian Floral Designs
(from *Ku yü t'u p'u*).

We shall see presently through what channel this elegant plant pattern arrived in China.   The *Ku yü t'u p'u* notices the delicate composition of the design and calls it an unusually precious girdle.   It does not say to what class of wearers it belongs, but I venture to make a guess at a palace lady, — for two reasons, first of all because of the dainty feminine character of the design, and secondly because the color of the jade is stated to be yellow, yellow being the color of Earth, and the empress is mysteriously connected with the deity Earth.[1]　Also the number of twenty plaques in this girdle refers to the female element as an even number, and relates to Earth too, as being 4×5, five being the number

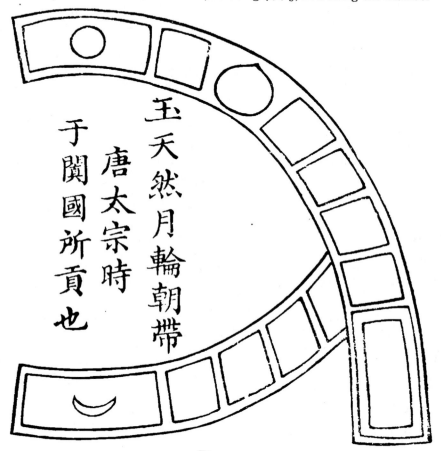

于闐國所貢也

唐太宗時

玉天然月輪朝帶

FIG. 194.
Jade Court-Girdle of the T'ang Dynasty, sent as tribute from Khotan (from *Ku yü t'u p'u*).

of Earth.　Besides, in the legend to the illustration, the girdle is styled court-girdle of Hu-mien which in all probability is a woman's name.

From an historical viewpoint, the most interesting of these girdles is that in Fig. 194 called "Jade court-girdle with the disk of the natural moon." It is remarked in the text that the Emperor T'ai-tsung (627–

[1] Therefore, under the T'ang, the empresses claim the right to participate in the great thanks-offerings to Earth (CHAVANNES, Le T'ai Chan, pp. 185, 206).

649 A. D.) of the T'ang dynasty had obtained it after the pacification of Kiang-nan, and that it had been sent as tribute from Khotan. Now in fact it is on record that a girdle of jade was offered by Wei-ch'ih Wu-mi, king of Khotan, in the year 632 (CHAVANNES, T'oung Pao, 1904, p. 4), i. e. during the reign of T'ai-tsung, so that the account of the *Ku yü t'u p'u* deserves credence, and it is quite possible that this or a similar girdle is the one then sent from Khotan.[1]

This girdle consists of twenty-four green jade plaques unadorned with the exception of the two central plaques, the upper one of which shows a figure of the full moon, and the lower one the crescent. "At the time of the Sui dynasty," adds the Sung Catalogue, "the waxing and waning of the moon corresponding to the aspect of the moon on the fifteenth and last of the lunar month was a symbol of the rise of splendor." This interpretation is unnecessary, for if this girdle was despatched from Khotan, it is likely that the lunar designs had also originated there.[2] In glancing back at the preceding girdles, we now become aware of the fact that they are all modeled on exactly the same principle as this one, and that even the rectangular, square and pear-shaped plaques appear there in the same rotation. Consequently, if this girdle was made in Khotan, the others are simply imitations of it, as girdles of this type had been unknown in China before. Then we are also justified in deriving from Khotan the Persian floral design on the girdle in Fig. 191. Finally, this type of girdle itself goes back to Persia where it is still in use.

Of the last girdle in this book, only the two central plaques (Fig. 195) are reproduced, as this girdle presents exactly the same shape as the preceding one. All plaques are plain except these two filled with a design of hills. It is therefore designated as "a court-girdle of ancient jade with a design of natural hills." The jade is pale-blue and crystal-clear. The *Ku yü t'u p'u* compares these hills with the sacred mountains of China, and remarks that their wonderful summits and superposed

---

[1] The jade of Khotan is mentioned in the Annals of the T'ang dynasty (*T'ang shu*, Ch. 221) where it is remarked in the description of that region that the natives observe during the night the spots where the reflection of the moon-light is intense, and do not fail to find there fine jade, also that they utilize jade for the making of seals (CHAVANNES, Documents sur les Turcs occidentaux, p. 125).—The tribute sent by Khotan consisted in jade, and at certain times, it seems to have been an obligatory tax. At least, we read in the Annals of the Yüan dynasty (*Yüen shih*) under the year 1274 that the people of Khotan were relieved from the burden of collecting jade (BRETSCHNEIDER, Notices of the Mediæval Geography, p. 226).— As a tribute-gift from Turkistan (*Si yü*), the *Ku yü t'u p'u* (Ch. 97, p. 11) figures a neat bird-cage of green jade, and a wine-vessel in form of a dragon sent by a king of Khotan in the period 1023–1031 (Ch. 90, p. 5).

[2] This crescent-design is not Mohammedan in origin, but Sassanidian. For the Persian analogies on textiles after the bas-reliefs of Takht-i Bostān, see J. DE MORGAN, Mission scientifique en Perse, Vol. IV, pp. 325, 327 (Paris, 1897).

peaks are sublime as if painted, and that this piece is the most remark-able of all jades. This design is certainly interesting in that it reflects the style of mountain-painting during the T'ang dynasty, — and it is doubtless copied from some great landscapist of the period, — and in that it offers the oldest known example of a carving of mountain scenery in jade of which the lapidary artists of the Ming and the eighteenth century renaissance period have left us such glorious examples.

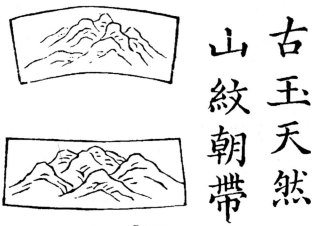

FIG. 195.
Central Plaques from Jade Court-Girdle of the T'ang Dynasty, with Mountain Scenery
(from *Ku yü t'u p'u*).

The emperors of the Sung and Ming dynasties retained the jade girdles of their predecessors, until they were abolished with the rise of the present Manchu dynasty which set up new regulations in this de-partment. The emperors of the reigning house wear a court-girdle of yellow silk adorned with four plaques of gold on which five-clawed dragons are chased. The ornaments worn at the girdle are of lapis-lazuli for the imperial functions in the Temple of Heaven whose cardinal color is blue, of yellow jade for his services at the Altar of Earth, yellow corresponding to the color of Earth, of red coral for the Altar of the Sun, and of pale-white jade for the Altar of the Moon. The jade court-girdles of the T'ang dynasty were preserved until recent times in Korea and Annam.

VIII. JADE AMULETS OF THE DEAD

Dr. BUSHELL has introduced into Chinese archæology the term "tomb jade" on which he remarks (in BISHOP, Vol. II, p. 102; also Chinese Art, Vol. I, p. 145): "The term *tomb* has been adopted as a synonym of the Chinese word *han*, which originally meant "placed in the mouth" of a corpse before burial, was afterward extended to include all jade objects buried in tombs in ancient times, and ultimately employed, with a yet wider signification, to comprise all old jades dug up from the ground, whether lost during floods or earthquakes, or purposely buried in times of famine or rebellion. So our "tomb jade" is to be taken as a synonym of the term *han yü* of the Chinese archæologist in the widest sense of the term."

This terminology, however, is a debatable subject. I should not be so small-minded as to take issue with Dr. Bushell on a mere term-question, but, as it means much more than this and has a large bearing on this whole subject, I am forced to state my opinion as briefly as possible. It seems to me that the definition of Bushell is due to a misunderstanding on his part of his Chinese informants. It is true the modern Chinese concerned with the archæology of jade frequently speak of *Han yü*, but this always and invariably means "jade of the Han dynasty." It never means, however, the word *han* (GILES No. 3821) "to place in the mouth of a corpse" or "the jade amulet placed on the tongue of the dead" which Bushell had in mind. This word is not in colloquial use, is indeed exceedingly rare in literature and only known to men of a thorough literary education. It is, as I may vouchsafe from a long personal experience, entirely unknown to the people of Si-ngan fu who always designate these objects as *ya-shê* ("pressing the tongue"); but these same people have a lot to say about *Han yü*, and this expression exclusively means that a given piece of jade, whether wrought or unwrought, is the peculiar kind of jade particularly used in the Han period. Thus, *e. g.*, the rough water-worn jade pebble illustrated on Plate I, Fig. 1, is a *Han yü*; the bell on Plate LIII, though carved in the K'ien-lung period, is called a *Han yü* because the jade material is identical with that of the Han period; it is not Turkistan or Burmese jade, but a bowlder accidentally found on the soil of Shensi, probably in a river-bed, during the eighteenth century. This one example is conspicuous in showing that Bushell's definition of tomb-jades is beset with grave danger, for the unsophisticated collector receiving such a piece from Chinese hands with the mark *Han yü* would

294

doubtless conclude, on Bushell's authority, that it is a genuine ancient tomb-jade. Certainly what the Chinese call *Han yü* may come down from the Han period, but it must not; it may come out of a Han grave, but it must not; it may have incidentally been found also underground in a field or in a river, outside of a grave.

Nothing is gained for scientific purposes, but on the contrary great harm is caused to a correct understanding of these objects, if we indiscriminately designate as tomb jades all jades found in ancient graves. In the Chinese literature on the archæology of jade, there is in fact no term corresponding to Bushell's tomb jades, and I feel quite confident in so saying, as I discussed this subject with able Chinese scholars. The extension which Bushell lent to the term *han yü* "jade placed in the mouth" is arbitrary and not justified by any Chinese text nor by the opinion of any Chinese antiquarian. We find a great number of jade objects in the ancient graves which could never be called by that name. We became acquainted with the six jade images of the cosmic deities placed in the grave according to the *Chou li;* they have nothing to do with *han yü.* We saw also that jade girdle-ornaments, badges of rank, and even implements like chisels, hammers and knives have been entombed. There are figures of animals in addition; all these are not *han yü.* We consequently recognize that there are different groups of jade objects (as there are also of metal and pottery) surrounding the corpse in the grave, and that these groups emanate from quite different psychical phenomena and must accordingly receive different explanations. There is not one large comprehensive class of tomb-jades, as it appears from Bushell's deductions, but there are several groups widely distinguished one from another. It is therefore preferable to drop entirely this unfortunate term of tomb-jades which is superfluous, apt to lead astray, and has no other meaning than that of a certain ticklish sensation to the collector.

There is, first of all, the group of jade objects in the grave which have a purely social significance. At all times, in China, as everywhere else, people had belongings buried with them which they cherished during life, and from which they did not want to part during the slumber of death. The idea of profession comes in here easily, as people naturally take the greatest liking to the things pertaining to their vocations. The soldier is fond of his sword and takes it along into the grave, the official of his insignia of rank, woman of her jewelry.[1]

---

[1] The burial of uch favorite objects was always left to individual liberty. They were not subject o obligatory regulations. Hence the great diversity of objects discovered in the graves. Also at the present time everybody takes with him whatever he likes. This may also account for the reason why Chinese sources are so reticent about these objects and their arrangement in the grave, while all ceremonies of the burial are minutely described.

There is, secondly, a purely religious group of mortuary jade objects composed of the six cosmic deities and implements of primitive forms originally connected with solar worship. This group belongs to the culture of the Chou era. In the Han period, we meet with jade carvings of animals acting as protectors of the grave.

A third group of burial objects is formed by jade amulets worn by the corpse, as the belief prevailed that jade possessed the property of preserving the flesh of the body and keeping it from decay. These amulets will form the subject of this chapter.

In the following chapter, we shall deal with a fourth group of grave objects used in dressing the corpse for burial.

In the days of the Chou dynasty, jade was taken internally as food. "When the emperor purifies himself by abstinence, the chief in charge of the jade works (*yü fu*) prepares for him the jade which he is obliged to eat," says the *Chou li* (BIOT, Vol. I, p. 125). Jade, add the commentaries to this passage, is the essence of the purity of the male principle, the emperor partakes of it to correct or counteract the water which he drinks (as water belongs to the female principle); the emperor fasts and purifies himself, before communicating with the spirits; he must take the pure extract of jade; it is dissolved that he may eat it. And in another passage of the *Chou li* (BIOT, Vol. I, p. 492), we read that jade is pounded to be mixed with rice to be administered as food to the corpse of an emperor before burial (*tsêng yü*).

In later Taoism, we meet the belief highly developed that jade is the food of spirits and tends to secure immortality (DE GROOT, The Religious System of China, Vol. I, pp. 271–273; Vol. II, p. 395). We remember from a consideration of the symbolism underlying the girdle-ornaments of the Han period that a belief was then dominant in a revival of the corpse, and the hill-censers and hill-jars of Han pottery interred with the dead have taught us how deep the longing for immortality was among the people of that age. Two ideas are, therefore, prominent in the burial of certain jade ornaments with the corpse during the Chou and Han periods,—the preservation of the body by the effect of the qualities inherent to jade, and the hope of a resurrection prompted by this measure.

The idea of jade being apt to prolong life seems to have originated at the same time in connection with the notions and practices of alchemy then coming into existence. A marvellous kind of jade is called *yü ying* "the perfection of jade." It is represented among "the wonderful objects of good omen" (*fu jui*) — there are twenty-two altogether — on the bas-reliefs of Wu-liang of the Han period in Shan-tung where it is pictured as a plain rectangular slab accompanied by the inscription,

"The perfection of jade will appear, when the five virtues are cultivated."[1]   Vessels, it was supposed, could be made of this supernatural substance; in B. C. 163, a jade cup of this kind was discovered on which the words were engraved, "May the sovereign of men have his longevity prolonged!"   The then reigning Emperor Wên took this joyful event as a suitable occasion to choose a new motto for the period of his reign, and to count this year as the first of a new era, celebrated with a banquet throughout the empire.[2]

It was believed that immortality could be obtained by eating from bowls made of this kind of jade.   Thus, the phrase "to eat in the perfection of jade" came to assume the meaning "to obtain eternal life."   In the form of a wish, it appears in prayers cast as inscriptions on certain metal mirrors of the Han period connected with the worship of Mount T'ai in Shan-tung (CHAVANNES, Le T'ai Chan, p. 425).

While the Steward of the Treasury of the Chou dynasty who in the main was the superintendent of the jade insignia of rank was in charge of the mortuary mouth-jade and responsible for its proper delivery when occasion arose, it was not he who was concerned with its manufacture.   This was the duty of the chief in charge of the jade works (yü fu) who controlled the making of the mouth-jade, the garment used in the ceremony of recalling the soul of the dead, the angular pillow supporting the head of the corpse and a spatula for supporting the teeth (BIOT, Vol. I, p. 125).

Princes followed the observance of sending to their equals on their death, pieces of jade to be placed in the mouth of their deceased friend as the last honor to be rendered.   Special messengers were entrusted with this token who fulfilled their task as described in the Li ki (Tsa ki II, 31) as follows: "The messenger with the mouth-jade[3] holding a jade ring (pi) announced his message in these words, ' My humble prince has sent me [calling his name] with the mouth-jade.'   The assistant [to the son of the deceased] went into the house for report, and said in coming out, ' Our bereaved master [calling his name] is awaiting you.'   The bearer of the jade entered, ascended into the hall and gave his message; the son bowed to him [as sign of thanks] and touched the ground with his forehead [as sign of grief and mourning].   The bearer, kneeling, deposited the jade south-east of the coffin on a reed mat, or after interment, on a rush mat.   He then descended and returned to his place.   An adjutant in court-dress, but still wearing the shoes of

---

[1] Compare CHAVANNES, La sculpture sur pierre en Chine, p. 34.

[2] CHAVANNES, Se-ma Ts'ien, Vol. II, p. 481.

[3] A free and correct rendering of this term, as will be recognized from the specimens, would be "tongue-amulet."

mourning, ascended the hall by the steps on the western side, and kneeling, his face turned to the west, he took the jade ring. Then he descended the same western steps, going in an eastward direction." The mouth-jade was, accordingly, presented with rules of strict formality, and it is obvious from this passage that it could be presented even after the funeral had taken place without serving its purpose proper, and that also then the mourner was obliged to accept it; he doubtless kept it, but in what way, and to what end, is unknown. In a similar manner, also messengers with clothes to adorn the corpse, and others with the gift of a carriage and horses were despatched, communicating their messages in the same style as previously, the whole procedure being identical (*Tsa ki* II, 32, 33). But it is noteworthy that the bearer with the present of a chariot and a team of four yellow horses made his report by holding in his hands the jade tablet *kuei* which he afterwards deposited in the south-east corner of the coffin, whence the adjutant took it up to leave it for safe-keeping in a building situated in an easterly direction. It is further worth mentioning that it was the adjutant, a high official in immediate attendance of the prince, who cared for the jade ring *pi* and the jade tablet *kuei*, while his assistants took charge of the garments and the other gifts. A higher value was therefore attached to those. It is not expressly stated that the tablet *kuei* was intended for burial too, but since it was included among the objects given for this purpose, there is reason to believe that it was, and this conclusion would furnish a good explanation for the fact that such tablets have been found in tombs. On the other hand, it is clear from the accounts of the *Li ki* that the tablet is not a *han yü*, "a mouth-jade," but is distinct from it, moving on another line of thought. It had no reference to the body of the dead, but was a mark of honor bestowed on him.

Such were the customs of the feudal lords with one another, the offering of condolences, mouth-jade, grave-clothes, and chariots, and all this had to be accomplished on one and the same day, in the order prescribed (*Tsa ki* IV, 14).

A curious instance of an alleged or allegorical use of the mouth-jade in the case of live persons is narrated in the history of the kingdom of Wu, when King Fu Ch'ai (b. c. 494–472) joined the duke of Lu to attack the principality of Ts'i. At the point of giving battle, General Kung-sun Hia ordered his soldiers to chant funeral songs; another general requested his men to take into the mouth a piece of jade as used for a corpse, while still another bade his men carry a rope eight feet long to fetter the soldiers of Wu (A. Tschepe, Histoire du royaume de Ou, p. 121). It can hardly be surmised that the second clause is to be taken

Explanation of Pl. XXXVI.

Figs. 1–4. Plain Types of Tongue-Amulets.

Fig. 5. Tongue-Amulet carved in Shape of Realistic Cicada (*a* Upper, *b* Lower Face).

Figs. 6–9. Tongue-Amulets showing conventionalized Forms of Cicada.

in its real sense, for it would be difficult to see how a band of soldiers could be provided with these jade pieces at a moment's notice just before going to battle, unless we should suppose it a custom that every man should carry with him his mouth-jade, which is not very probable, and the general could hardly expect that a man while holding a piece of jade on his tongue could do efficient fighting. I therefore understand the sentence in a figurative sense meaning to say that the battle will be so fierce that every one should be prepared for death and burial.

The mortuary amulets in our collection described on the following pages were procured in Si-ngan fu from the private collection of a well known Chinese scholar and archæologist who has been engaged for many years in antiquarian researches with great success. For the definition of these objects, I entirely depend on his explanations which agree with the general opinions upheld in Si-ngan fu. It will be seen that there is not only the tongue-amulet mentioned in the *Chou li*, but a whole series of jade amulets serving also for the preservation of other parts of the body. The underlying idea evidently was to close up all apertures of the body by means of jade, the essence of the *yang* element which was to triumph over the destructive underground agencies of the *yin* element, and it is assumed that this full equipment of the body was developed in the Han period.[1] The characteristics of the pieces point to the same epoch. This is the most complete collection of this kind on record, and most of these types have not yet been described by Chinese archæologists.

Among the personal amulets worn by the corpse, those to be placed on the tongue are most important and frequently spoken of in the ancient texts. As all these amulets are imitative of bodily forms, those for the tongue are shaped in the outline of this organ. There are four types of them, the one plain, almost geometrically constructed, the other of a realistic design carved into the figure of a cicada, but simultaneously preserving the shape of a tongue. A series of nine pieces is illustrated on Plate XXXVI all in natural size, the four first being of the plain tongue-shaped type. The first three are made of the same

---

[1] The archæological evidence quite agrees with the literary researches of DE GROOT, The Religious System of China, Vol. I, pp. 271 *et seq*. The most important quotation for our purpose is that by KO HUNG: "If there is gold and jade in the nine apertures of the corpse, it will preserve the body from putrefaction." And T'AO HUNG-KING of the fifth century: "When on opening an ancient grave the corpse looks like alive, then there is inside and outside of the body a large quantity of gold and jade. According to the regulations of the Han dynasty, princes and lords were buried in clothes adorned with pearls, and with boxes of jade, for the purpose of preserving the body from decay." The stuffing of the corpse with jade took the place of embalming, except that it did not have the same effect. In the case of the Han Emperor Wu (B. C. 140–87), the jade boxes mentioned had their lids carved with figures of dragons, phenixes and tortoise-dragons (*l. c.*, Vol. II, p. 401).

ivory-colored [1] material, probably marble, which is decomposed and showing a rough surface in 1 and 3, while the original fine polish is preserved in 2. The substance of 3 has withered away so much that the ornamentation has disappeared and deep holes are eaten into the surfaces. The lines engraved on 1 and 2 explain themselves by serving the purpose of marking the parts of the tongue. In all these pieces, the medial portion is high and gradually sloping down towards the edges.

FIG. 196.
Lower Face of Tongue-Amulet shown in Plate XXXVI. Fig. 2.

In Fig. 1 the under surface is flat, and the tip is slightly turned upward. In Fig. 2, the lower side is shaped in the same manner as the upper one, but laid out with a different design of lines, as will be seen from Text-Figure 196.

The piece in Fig. 4, Plate XXXVI is of a uniformly pure milk-white jade, the two dark lines showing in the photograph being yellow in color. Rounded over the upper surface, it consists of two slanting portions on its lower side with a short incision cut horizontally into the medial line, in the same way as will be seen in Fig. 8 of this Plate.

Figures 5–9 of the same Plate show five variations of the cicada type, that in 5 being the most realistic, those in 8 and 9 being in an advanced stage of conventionalization. In Fig. 5 a, the two wings and the body are well designed; 5 b displays the lower face of the same specimen. All of this type have the two faces ornamented differently. The hardened earth incrustations which have penetrated into No. 6 will be recognized in the illustration. Both 5 and 6 are of grayish jade, and of excellent workmanship. No. 7 is remarkable for its size, its color, and its elegant technique. The color of jade is black in the two wings and the right upper portion, and dark-gray in the central and upper part. In this, as in so many other cases, we have occasion to admire the ingenuity and the color sense of the artist in carving the jade block in such a way that the colors were appropriately distributed, either to an artistic end, or as here, to lend an object its real colors, a realism of color and a color of realism. No. 8 is the smallest and plainest of this type which I know, and not ornamented on the obverse; it is of lustrous white jade with a slight greenish tinge. In the two slanting sides, it agrees with the plain tongue-shaped type, but the style of carving shows that here also the figure of the cicada is intended. No. 9 shows the specimen on its lower face which is of grayish jade, but with a very peculiar chocolate-brown

[1] Called by the Chinese "chickenbone-white" (chi ku pai).

JADE AMULETS FOR THE DEAD.

EXPLANATION OF PL. XXXVII.

Figs. 1-3.   Tooth-Shaped Tongue-Amulets.
Figs. 4-5.   Miniature Tongue-Amulets for Women and Children.
Figs. 6-8.   Umbilical Amulets.
Figs. 9-11.   Prism-Shaped Amulets.

portion in the upper end with a narrow bluish stripe below it. On the upper side, the two wings of the insect are brought out by lines engraved, as in the other specimens. Only two of them are provided with a contrivance by which they can be fastened. That in No. 5 has two small holes of about 2 mm in length drilled in the upper edge; they communicate in the interior and thus allow the passage of a wire or cord. The object in No. 6 is provided with a small perforated rounded handle.

Also the *Ku yü t'u* illustrates a *han yü* of the cicada shape, and Wu Ta-ch'êng has two of them in his collection. The opinion of archæologists is undivided in regard to this subject. Why the cicada was chosen for this amulet, seems not to be known. This idea may be connected with the *memento mori* brought out by the figures of a cicada and mantis on the Han jade buckles (see above p. 264).

But still more the peculiar manner of transformation of the insect from the larva to the pupa, well known to the ancient Chinese, may have a share in the shaping of this amulet. The young hatch out in a few weeks, drop to the ground, and may penetrate as deep as twenty feet below the surface. After a long subterranean existence, the pupa transformed from the larva crawls out of the ground, the skin splits, and the adult winged insect emerges. The observation of this wonderful process of nature seems to be the basic idea of this amulet. The dead will awaken to a new life from his grave, as the chirping cicada rises from the pupa buried in the ground.[1] This amulet, accordingly, was an emblem of resurrection.

The third type of tongue-amulets is represented on Plate XXXVII, Figs. 4 and 5. From their miniature size we may safely conclude that they were employed for children and women, judging from the fairly established rule that all objects relating to them are made on a smaller scale than those relating to man. They are flat, only 1–3 mm thick and not ornamented on the lower side; from a trapezoidal base in which two incisions are deeply cut merges the oval tongue-shaped part set off by two parallel engraved lines at the lower end to which another band of two lines corresponds in the upper portion. All these incisions are filled with a hardened reddish clay. It will be noticed that there are two perforations, one in the upper left part and another in the lower right part drilled through the incised line. Thus, these objects must have been fastened, and it seems plausible to infer that the silk thread

---

[1] The notion which the ancient Chinese affiliated with the cicada will be best gleaned from a passage in the philosopher Wang Ch'ung (FORKE, Lun-Hêng, Part I, p. 200): "Prior to its casting off the exuviæ, a cicada is a chrysalis. When it casts them off, it leaves the pupa state, and is transformed into a cicada. The vital spirit of a dead man leaving the body may be compared to the cicada emerging from the chrysalis."

passing through the hole on the one side was wound around a molar tooth on the left, and the other attached to a molar on the right side of the mouth.   The color of jade is dark-gray.

The pieces in Figs. 1, 2 and 3 of the same Plate XXXVI are explained also as tongue-amulets (*ya shê*), though they are rather modeled like teeth [1] and give the impression of being tooth-protectors.   The three objects show the same shape and the same design of meander slightly engraved into the surface; they are made of the same material which is a light grayish jade, and differ only in thickness which is 1, 4 and 3 mm respectively.

They are perforated near the base, and there is a notch cut into the lower edge just facing the perforation, another notch on the inner concave side, and two notches opposite on the convex side, so that the thread passing through the perforation must have been reeled over these notches.   This peculiar method leads me to think that pieces of this type could have been tied to a single tooth only, and that, taking their shape into consideration, they rather served for the preservation of the teeth.   The commentator KIA KUNG-YEN of the eighth century remarks that the mouth-jade supported the posterior molar teeth on both sides, and that, in the case of an emperor, it was in the shape of the circular disk *pi*, though on a smaller scale (BIOT, Vol. I, p. 125, Note 7; p. 492, Note 3).   Both these statements are improbable for technical reasons. The Chinese were and are practical people, and would not have committed themselves to the technical blunder of placing a circular object between the teeth; this opinion is, besides, such a late reflection that, also for this reason, it does not deserve much credence.[2]   There is a more trustworthy view on hand uttered by CHÊNG TUNG of the first century A. D.   The chief of the imperial jade factory (*yü fu*) of the Chou dynasty made, besides the tongue-amulet for the deceased sovereign, also an angular pillow for the support of the head of the corpse, and an angular spatula (*kio se*).   CHÊNG TUNG annotates that this spatula had seven corners, and that, according to the *I li*, it is used to support the teeth of the dead, whereupon the mouth-jade is placed on the tongue (BIOT, Vol. I, p. 125, Note 10).   Our three specimens in Figs. 1–3 of Plate XXXVII which are indeed spatulas come very near to this description and might be identified with these objects.

For the preservation of the eyes, a pair of oval pieces carved from

[1] Their similarity in shape with implements to loosen knots (Figs. 148, 149) will be noted.

[2] His statement that the mouth-jade of the Chou emperors was in the shape of the disk *pi* is doubtless suggested by the new regulation of the K'ai-yüan Code of the T'ang dynasty where it is stipulated for the first time that the disk should form the mouth-jade of the officials of the first, second and third ranks (see DE GROOT, The Religious System of China, Vol. I, p. 278).

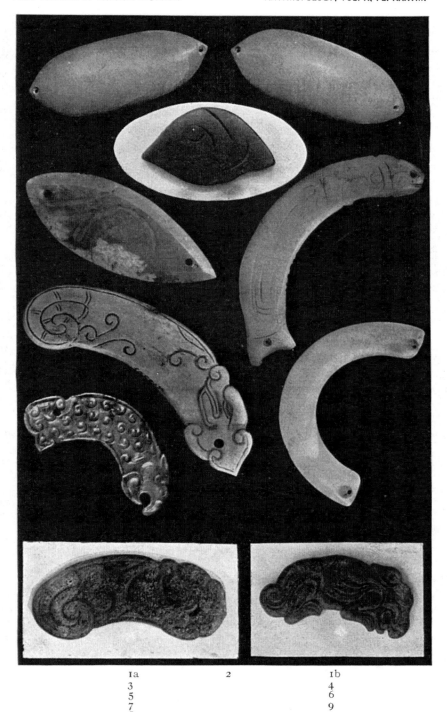

Ia   2   Ib
3   4
5   6
7   9
8

JADE AMULETS FOR THE DEAD.

EXPLANATION OF PL. XXXVIII.

Fig. 1a and b. Pair of Eye-Protecting Amulets.
Fig. 2. Presumably Eye-Amulet.
Fig. 3. Eye-Amulet with Design of Fish.
Figs. 4-7. Lip-Amulets, 4 and 7 in Shape of Fish.
Figs. 8-9. Amulets in the Shape of Monsters.

milk-white jade was used (Plate XXXVIII, Figs. 1 *a* and 1 *b*), pointed and perforated at both ends, the silk cords having presumably been tied all around the face and head. The two pieces constituting a pair — the only instance to my knowledge of a pair ever found — are plain, flat on the under side, and convex or arched over the surface. Their present designation is *ya yen-king* ("pressing the eyes"). The object in Fig. 3 of the same Plate carved from grayish jade with a chalk-white clayish mass spread in the lower section, which the Chinese wrongly attribute to the presence of mercury, served the same purpose and is evidently moulded in the shape of a fish, as indicated by the lines setting off the head; on the back, the same ornamentation is shown, except that, in place of one spiral, two are engraved side by side. This piece is flat on both sides with a uniform thickness of 2 mm.

The fish as an object covering the eye of the dead may be interpreted as the symbol of watchfulness.[1] Never closing its eyes, it is constantly wakeful. A padlock, protecting from thieves, is called a "fish-eye" (Pétillon, p. 497). The night rattle used in Buddhist monasteries is carved from wood in the shape of a carp (*mu li*, "wooden carp"). The "wooden fish" in the Buddhist temples is well known, and Pischel (Der Ursprung des christlichen Fischsymbols, p. 24) is quite right in stating that the Chinese explanation furnished to him by Dr. Franke from the *Chi yen*[2] does not agree with the Indian way of thinking; certainly because this idea is Chinese, and not Indian. Whether in Fig. 2 a fish is intended, I cannot assert positively; on the reverse, an eye is indicated by an engraved circle. Nor do I feel certain of identifying this object with an eye-amulet, though this is the Chinese explanation given me. It mainly deviates from the others of its kind in having a perforation only on one side in the extreme right tip drilled vertically through the lateral edges (not as in the others from top to bottom), which would warrant the conclusion that it was rather used as a girdle-pendant.

The amulets represented in Figs. 4–7 of the same Plate XXXVIII were placed on the upper lip to cover and preserve the mustache (*ya hu-tse*). They are curved in the shape of a half-circle, the two in 4 and 7 clearly imitating the form of a fish. The piece in 6 is plain, of a jade

---

[1] Compare the essay of Paul Carus, The Fish as a Mystic Symbol in China and Japan, in the July Number of *The Open Court*, 1911.

[2] The *San ts'ai t'u hui* and *T'u shu tsi ch'êng* have a different explanation as follows: "The 'wooden fish' is carved from wood. It has the shape of a fish, and is hollow. If one strikes its centre, it emits sounds. The Buddhist priests call it 'Jambuti.' It is a huge sea-fish which carries it. As it constantly feels an itching sensation, its fins are excited, and mountains and countries are shaken by it. For this reason they imitate its form and strike it, but this is merely idle talk. Nowadays it aids the Buddhist priests in chanting their Sanskrit prayers, and all make use of it."

milk-white in color, rounded on both sides, flattened at the ends which are perforated, and on the whole half-circular in shape. The jade in Fig. 4 is light-gray in color, that in 5 buff-colored. Some kind of monster is apparently intended; the tail is rolled up in a spiral, and the head is turned backwards; it looks, judging from the trunk, like that of an elephant or tapir.[1] The same carving is brought out with accurate agreement on both sides of this piece, and the work of engraving is very fine. This piece has only one perforation; it is 2 mm thick.

Figure 7 is a carving of gray jade with a few reddish specks; the fish intended is possibly a carp. The gills are indicated by cross-hatchings, and the scales by spirals, both sides displaying identical designs.

The two objects on Plate XXXVIII in Figs. 8 and 9 are added here only for the reason that they belong to this group from a typological viewpoint; it is doubtful whether they served the same purpose. It will be recognized that the design of the monster in Fig. 8 exactly tallies with the one in Fig. 5 of the same Plate, and that Fig. 9 exhibits a squatting monster of a similar type also with a curled-up spiral tail and elephant-head looking forward (not backward as in the two others); also two feet are delineated in this carving. The two pieces show identical designs on both faces, and are of a dark-green jade not identical with the modern Yünnan, Burmese and Turkistan green jade; they have been buried underground for a long period, being coated on their lower sides with a thick hardened layer of an earthy matter which through the action of some mineral has assumed a brownish-red tinge affecting partially also the upper face. They are 4–5 mm thick.

The three round objects represented in Figs. 6–8 on Plate XXXVII were used to be placed on the umbilic of the corpse (ya tu-ts'i). The symbolism with reference to this part of the body is self-evident. The piece in Fig. 6 of this Plate, of whitish jade with black veins, is a knob flat on the lower side, bearing a relief design in the centre which I take to be a symbolic representation of the navel itself, encircling the well known ornamental form of the character shou "longevity," which so frequently appears in connection with the dead, also nowadays painted on the coffin or embroidered on the draperies covering the catafalque. The piece in Fig. 7 is of gray jade interspersed with brownish specks (dark in the illustration). The central portion is slightly raised over the outer zone which is occupied by five engraved double spirals; four of these are grouped around a quadrangular figure; the same arrangement of ornamentation is applied to Fig. 8. Also in the piece in Fig. 1,

---

[1] Representations of both animals are frequent in the Chou and Han periods (see Chinese Pottery pp. 152, 170, 171, 207).

the spiral designs appear along the edge. The specimen in Fig. 2 has, corresponding to the raised circle on the upper surface, a circle incised on the reverse, in which three shallow oval cavities grouped around two apertures are cut out. In Fig. 3, only a circle is engraved on the back. This object is buff-colored, presumably due to underground decomposition. It will be noticed that these three umbilical amulets bear a certain resemblance to those explained by Wu as cap-buttons (Figs. 154, 155) which I think may rather belong to this class.

In Figs. 9–11 of Plate XXXVII, three tubes, octagonal in cut, are shown, said to have been inserted into the urethra of the corpse. Others explain that they were stuck into the nostrils, which is absurd as they are too long and too big for this purpose. These pieces should be distinguished from the jade signets called *kang-mao* described by Dr. Bushell in the work of Bishop which show a merely outward resemblance to this type, but are much shorter and perforated, while these objects are solid. That in Fig. 9 is of a brownish and bluish-black jade in an advanced state of decomposition, gradually tapering below, the diameter of the upper octagonal surface being 1.3 and that of the lower 1 cm. The pieces represented in Figs. 10 and 11 are carved from a yellowish-gray jade, uniformly thick, the sides of the former slightly curved in, those of the latter being straight.

IX. JADE OBJECTS USED IN DRESSING THE CORPSE

Before *post-mortem* rigidity (*rigor mortis*) [1] set in, it was necessary to place the corpse in the proper position for burial, to dress it with the grave-clothes, and to see that all parts of the clothing were in suitable shape and. remained unchangeable. A number of objects have therefore been devised to act as weights on the limbs, and as special care was taken in keeping the long sleeves in order, such pieces were especially made to be placed on the sleeves of the shroud, and are therefore designated by archæologists in Si-ngan fu *ya-siu, i. e.* objects pressing the sleeves. These objects were turned out of pottery, bronze, and jade. I had occasion to publish a piece of pottery of this kind,[2] without being aware, at that time, of its proper use. During my second expedition to China I obtained two other specimens of the same type, but of different ornamentation. They are here figured on Plate XXXIX. The one (Fig. 1) is a rather flat hollow disk of thin gray clay unglazed, with a diameter of 9.5 cm. The decoration is moulded in slight relief and arranged in concentric zones, on both sides identical, but with this difference that on the one side the star-figure in the inner zone is composed of eight triangles and that in the outer zone of nine, while on the opposite side the inner star-figure has only seven, but the outer twelve triangles. Also on the piece referred to there is a similar star having six points on the one and seven points on the other side, so that this difference must be intentional and have some meaning. The triangles of these stars are filled with dots arranged in the figure of a pyramid, while the triangles arising between them are each occupied with a design familiar to us under the name triskeles. The circle in the centre is filled with a three-leaved rosette. A large four-leaved rosette is the main ornament on both sides of the other *ya-siu* (Fig. 2) which is 12 cm in diameter, high in the centre (5.3 cm) and gradually sloping towards the edge. Both these pieces originate from graves of the Han period.

During the Chou dynasty, as the actual finds from the graves teach us, these weights for the corpse were of a much different shape and design. There are two beautiful specimens in our collection, both from

---

[1] An evanescent stiffening of all the muscles of the body occurring shortly after death and affecting the neck and lower jaw first, then the upper extremities, and finally reaching the lower limbs.

[2] Chinese Pottery of the Han Dynasty, Plate LIX, 2, where the note on p. 173 due to misinformation, has now to be cancelled.

306

CLAY DISKS OF HAN PERIOD, USED IN DRESSING THE CORPSE.

JADE AND BRONZE OBJECTS OF THE CHOU PERIOD, USED IN DRESSING THE CORPSE.

EXPLANATION OF PL. XL.

Fig. 1. Jade Knob with Relief of Frog.
Fig. 2. Cast-Bronze Knob with Relief of Frog.
Fig. 3. Bronze Cast of Realistic Tigerhead.
Fig. 4 *a* and *b*. Pair of Monsters in Hollow Bronze Cast.
Fig. 4 *c*. The Same, Side-View.
Fig. 5. Bronze Cast with same Design, on Smaller Scale, used for Women.

that period, the one being a *ya-siu* in jade, the other of bronze or copper, both of half-globular form and showing exactly the same design of a frog brought out in high relief. The one (Plate XL, Fig. 1) is finely carved from a light grayish-green and black-veined jade with a diameter of 6.5 cm over the circular basis and a height of 4 cm. The corresponding bronze piece is a solid cast with a diameter of 6.2 cm and a height of 3.5 cm. The only difference in the delineation of the two creatures is that the one in jade has four toes on each foot, the other three toes represented. It is important to note that each is provided with four feet which excludes the possibility of regarding it as the three-legged mythical frog *ch'an yü*, the emblem of the moon. Attention may be called right here to the small jade carving of a frog found in a grave of the Han period (Plate XLII, Fig. 2). This is also a four-footed frog with three toes, in a squatting position; the head is rather massive in proportion, the mouth being indicated by an incised half-circular line, the two eyes by two concentric circles. This piece is only 3.2 cm long and 2 cm high, carved from a pure white jade, but covered with an ivory-colored layer of hardened earth.

In the *Si king tsa ki*, it is on record: "The King of Kuang-ch'uan [1] opened the grave-mound of Duke Ling of Tsin [2] and found there a striped toad of jade (*yü ch'an-yü*) of the size of a fist and hollow inside, holding half a pint (5 *ko*),[3] and covered with a water-like gloss, as if it were new."

To return to our *ya-siu*, there remain three more interesting bronze specimens of the Chou period to be considered (Plate XL). All three are hollow casts moulded over a clay core which still sticks partially in the piece representing a tiger-head. I presume that the core was left inside intentionally to increase the weight of these objects. Of the first of these (Figs. 4 *a* and *b*) I obtained a pair and thus conclude that these pieces have all been made in pairs, naturally to cover the two sleeves. Another interesting fact may be gathered from a comparison of the two pieces in Figs. 4 and 5 which are identical in shape and design, but differ considerably in dimensions; the one is 6.5 cm long, 4.5 cm wide, and 5 cm high; while the other one is only 3 by 2.5 cm with a height of 2.4 cm. The bigger one was used in the burial of men, the smaller in the burial of women, according to Chinese information, which

---

[1] A kingdom of the Han period (CHAVANNES, Se-ma Ts'ien, Vol. II, p. 497, Vol. III, p. 99).

[2] B. C. 620–607 (CHAVANNES, *l. c.*, Vol. IV, pp. 311–316).

[3] Only added to impart an idea of the volume; it does not mean that the object in question served as a measure of capacity. — Regarding the folklore of the frog in China compare DE GROOT, Die antiken Bronzepauken (*Mitteilungen des Seminars für Or. Sprachen*, Vol. IV, 1, pp. 104–107); HIRTH, Chinesische Ansichten über Bronzetrommeln, pp. 27–32; CHAVANNES, Le T'ai Chan, p. 496.

I think is a plausible explanation.   The two pieces show traces of gilding
and resemble helmets in their shape and are moulded into the figure
of a curious monster which it is difficult to name.   It seems to me that
it is possibly some fabulous giant bird, for on the sides, two wings, each
marked with five pinions, are brought out; a long curved neck rises
from below and ends in a head on the top; it seems to terminate in a
beak, though the two triangular ears do not fit the conception of a bird.
A rectangular crest emerges above the head.   It will be noticed that
here the same principle of artistic arrangement in the parts of the
monster is followed as in the previous example of the frog.

In the smaller one of the two a flat bottom is inserted which is missing
in the other piece.   It had been also there, as may be seen from the
presence of four teeth inside near the base, for the purpose of holding
the bottom.

In the *ya-siu* of Fig. 3, Plate XL (4.5 cm long and wide, 3.5 cm high),
a tiger-head is produced with felicitous realism.   The open jaws exhibit
four fangs, a row of incisors and two pairs of the molars on either side.
The head rises from a rounded bronze plaque and is provided with a
large opening on either side near the base, apparently for the passage
of a cord.   It looks as if this piece had been fastened in this way around
the wrist of the corpse.   The religious significance of the tiger has
been discussed above p. 182.

I
2
3
JADE CARVINGS OF FISHES.

EXPLANATION OF PL. XLI.

Figs. 1-2. Mortuary Fishes, Han Period.
Fig. 3. For Comparison: Modern Carving of Fish with Lotus Flowers.

X. JADE CARVINGS OF FISHES, QUADRUPEDS AND HUMAN FIGURES IN THE GRAVE

We have seen that among the jade amulets placed on the corpse to prevent its decay the fish occurs on the eye and lip-amulets. But there are also instances of large separate carvings representing fishes which have no relation to the body, but have been placed in the coffin for other reasons.

On Plate XLI two mortuary jade fishes unearthed from graves of the Han period are figured. The one in Fig. 1 is a marvellous carving of exceedingly fine workmanship, all details having been brought out with patient care. It represents the full figure of a fish, both sides being carved alike, 20 cm long, 11 cm wide, and 2 cm thick, of a dark spinach-green jade. A small piece has been chipped off from the tail-fin. There is a small eye in the dorsal fin and a larger one below in the tail-fin. It is therefore likely that the object was suspended somewhere in the coffin; it is too large and too heavy (it weighs 1¼ pounds) to have served for a girdle-ornament. In this way, — with comparatively large bearded head and short body, — the Chinese represent a huge sea-fish called *ngao* (GILES No. 100). Such large and fine jade carvings are likely to have had a religious significance, and the following passage may throw some light on this subject.

"In the Han Palace Kun ming ch'ih a piece of jade was carved into the figure of a fish. Whenever a thunderstorm with rain took place, the fish constantly roared, its dorsal fin and its tail being in motion. At the time of the Han, they offered sacrifices to this fish in their prayers for rain which were always fulfilled." [1]

In Fig. 2 of the same Plate XLI, a fragment, perhaps only the half of the original figure, is represented carved in the shape of a fish of leaf-green jade clouded with white specks, on the lower face covered with a thick layer of hardened loess. It is 11.5 cm long, 4.2 cm wide, and 9 mm thick. [2]

While the religious symbolism formerly connected with the fish has almost disappeared it continues as a favorite ornament, and jade girdle pendants in the shape of fishes are still much in use. Fig. 3 of the same

[1] *Si king tsa ki*, quoted in *P'ei wên yün fu*, Ch. 100 A, p. 6 a.

[2] In the July number of the *Journal of the Anthropological Society of Tokyo* (Vol. XXVII, 1911), there is an article by Prof. S. TSUBOI describing some interesting figures of animals of chipped flint, one of them representing a well-formed fish (p. 132).

309

Plate XLI, represents such a modern carving of white jade showing a fish surrounded by lotus-flowers (9.8 cm long, 4 cm wide). The contrast between this modern and the two ancient pieces in design and technique is evident.

The butterfly carved from white and brownish-yellow jade (Plate XLII, Fig. 1) is a unique specimen among mortuary offerings. It is alleged by those who found it that it originates in the grave-mound of the famous Emperor Ts'in Shih (B. C. 246–211) near the town of Lin-tung which is 50 *li* to the east of Si-ngan fu. I am not fully convinced that this is really the case, though any positive evidence *pro* or *contra* this assertion is lacking; but there is no doubt that, judging from its appearance and technique, this is a burial object of considerable age and unusual workmanship, such as is likely to have been buried with a personage of high standing only. It is a flat carving (12.6 × 7.6 cm, 0.5 cm thick) both in open-work and engraved on both faces, the two designs, even in number of strokes, being perfectly identical. The work of engraving is executed with great care, the lines being equally deep and regular. We notice that a plum-blossom pattern is brought out between the antennæ of the butterfly; it is the diagram of a flower revealing a certain tendency to naturalism, which seems to bring out the idea that the butterfly is hovering over the flower. We further observe four designs of plum-blossoms, of the more conventional character, carved *à jour* in the wings. The case is therefore analogous to that illustrated on a Han bronze vase ("Chinese Pottery of the Han Dynasty," p. 283).

It is known that in modern times the combination of butterfly and plum-blossom is used to express a rebus (*mei tieh*) with the meaning "Always great age" (W. Grube, Zur Pekinger Volkskunde, p. 139).[1] It is difficult to say whether, in that period to which this specimen must be referred, this notion was already valid, though the possibility must be admitted in view of the early rebuses traced by A. Conrady (preface to Stentz, Beiträge zur Volkskunde Süd-Schantungs). It would, however, be erroneous to believe that the rebus in all cases presented the prius from which the ornament was deduced, for most of these ornamental components are much older and may even go beyond an age where the formation of rebuses was possible. The rebus was read into the ornaments, in well-nigh all cases; while other single ornaments were combined into complex compositions with the intention of bringing out a rebus. It is not the rebus which has created the ornaments, but it is the ornament which has elicited and developed the rebus; the rebus

[1] There is also the interpretation *hu-tieh nao mei* "the butterfly playfully fluttering around plum-blossoms" alluding to long life and beauty (*Ibid.*, p. 138, No. 15).

2   I   3

MORTUARY JADE CARVINGS.

JADE CARVING OF MONSTER *p'i-sieh*, HAN PERIOD.
PLATE XLIII.

has merely shaped, influenced and furthered the decorative compositions as e. g. occurring in the modern Peking embroideries figured by
Grube. In the present case, it is quite obvious that the association of
the butterfly with a floral design rests on natural grounds, and was, at
least not originally, provoked by a mere desire of punning, which is the
product of a subsequent development.

A very curious feature of this specimen is that the two upper large
plum-blossoms are carved out in loose movable rings turning in a deeply
hollowed groove but in such a way that they cannot be taken out, a
clever trick such as the later authors designate as "devil's work" (*kuei
kung*). This peculiarity certainly had also a significance with reference
to the mortuary character of the object. Such movable pieces are
designated by the Chinese as "living" (*huo*); so we have here two "living" plum-blossoms in distinction from the two "dead" plum-blossoms
below, and the two former might have possibly conveyed some allusion
to a future life.

The carving of the frog (Plate XLII, Fig. 2) has been mentioned
above p. 307.

In the ancient jades and bronzes the human figure is conspicuously
absent. And the jade carving representing a human figure (Plate XLII,
Fig. 3) and ascribed to the Han period is the only exception of this kind.
This figure, carved from a milk-white and black jade, is treated in an
almost geometric style. It is flat (9 mm thick, 6.4 × 2.3 cm). An
old man with long pointed beard is apparently intended; he is wearing
a round cap elevated over the hind-part of the head. The eyes are
marked by two incisions, and the brows by two slight depressions
above them. The nose is not represented nor are the ears. The mouth
is a line incised. The head is sitting right on the shoulders without a
neck. There is no intention to outline a body; even arms and hands
entirely disappear under the long gown, the folds and borders of which
are marked by engraved lines. The feet are not represented, but the
figure terminates in a trapezoidal base, on which two half-circles are
incised. These represent the uppers of the shoes as seen from the front.
Exactly the same trapezoidal base is found in a certain type of archaic
mortuary clay figures of men where the uppers of the shoes stick out
from the surface just above the lower edge of the base.

Nothing is known about the meaning of a figure like this one in the
grave, nor have I found as yet an allusion to this subject in a text of the
time of the Han dynasty.

The only specimen of the Han period that could be regarded as an
independent work of plastic art is the carving of a reclining monster
shown on Plate XLIII. It is cut out of a solid piece of onion-green jade

with layers of brown patches strewn in.   The head with long beard dropping on the breast is not unlike a dragon-head except the excessively large ears and the indentated crest.   A full set of teeth with four big fangs is represented in the open jaws.   The conventional character of the whole design is manifest.   A liberal use is made of the spiral — to express the nostrils, the ears, the cheek-bones, the joints of the legs [1] and the tail, and the curly bushes of hair covering the whole body.   The four feet stand out in high relief from the lower side and are each provided with four claws turned inward.   As alluded to above, this piece is clearly distinguished from all other jade carvings of the Han period in that it is not a flat plaque like those, but a full realistic sculptured figure, the other side not visible in the illustration being exactly the same as the one on view.

Such pieces are exceedingly rare and exceptional, and it was a lucky chance that I succeeded in securing this art-work from an old family of official standing in Si-ngan fu.   All Chinese there, competent to judge, place its date in the Han period; and, judging from the material of which it is made, a favorite jade of the Han time, and also from the style of execution, there is no doubt that this judgment is correct.

Unusual as is the workmanship of this carving, its dimensions and weight are also remarkable.   It measures 17 cm in length with a height of 12.2 cm and a width of 6.7 cm, and weighs 4¾ pounds.

In all likelihood, the figure of this monster is to be identified with the fabulous creature called p'i-sieh, the name of which means "something that wards off evil influences, a charm, an amulet." [2]   DE GROOT has shown that, because of the evil-dispelling attributes of the animal, stone images of it were placed upon the tombs from the Han down to the T'ang dynasty; stone p'i-sieh in connection with stone unicorns, elephants, horses and the like were erected in front of the mausolea of emperors and princes; stone tigers, sheep, men, pillars, and the like before those of officials.   It is therefore no matter of surprise that also miniature p'i-sieh of precious jade material were entombed in the graves of nobles during the Han dynasty; for the monster p'i-sieh was a favorite conception for jade carvings, and one of these (Fig. 197) derived from the Ku yü t'u would go as far back as beyond the Shang dynasty, if we could trust the statement there made that "according to a local tradition it was found in the grave of T'ai-k'ang (alleged B. C. 2188–2160), a farmer having struck against it with his hoe."   Judging from the design, as far as may be determined from these unsatisfactory

---

[1] Compare LAUFER, Felszeichnungen vom Ussuri (*Globus*, Vol. LXXIX, 1901, p. 70).

[2] Compare DE GROOT, The Religious System of China, Vol. III, pp. 1143 *et seq.*, and above p. 268.

Chinese sketches, I think it is not older than the Han time. It is further said in the text that in the period Yen-yu (1314-1321 A. D.) Chao Tse-ang [1] obtained this piece by purchase from Ch'êng Chih and used it as a paper-weight, — one of the many examples of how Chinese turn antiquities to a new practical mode of use.

The extravagance which the emperors of the Chou, Ts'in and Han dynasties exercised in the erection of their mausolea, and the wealth of treasures which they had interred in their vaults was stupendous. One

FIG. 197.
Ancient Jade Carving of the Monster *p'i-sieh*, used as Paper-Weight by the Painter Chao Mêng-fu
(from *Ku yü t'u*).

third of all the taxes of the empire is said to have been apportioned by the house of Han to being hoarded in the imperial graves. When the Emperor Wu died in B. C. 87, his mausoleum which was seventeen feet high and twenty feet square, with a mound of two hundred feet in diameter, was so filled up with treasures of all kinds that nothing more could be placed in it, and insurgents rifling the tomb were not able to carry off half of the valuables. Nine carriages were entombed with every emperor, and even live horses, leopards, and tigers, one hundred and ninety live animals being on record in one particular case. The reader may be referred to the description given by DE GROOT (The Religious System of China, Vol. II, Ch. IV). As all these graves were disturbed and pilfered at an early date, there is little hope that any remarkable spoils will ever come to light from them in the future, and we must be content with a few treasures which may impart some idea at least of the magnificence and glory of that Augustan age.

[1] Or Chao Mêng-fu, the famous painter, 1254-1322. An album containing six horse-paintings of his and dated 1305 is in our collection.

Nothing can be more erroneous than to picture the Chinese as grossly material or rationalistic people given only to worldly cares and affairs, as has been so often done. The grave and the life hereafter always stood in the centre of their thoughts and actions. The ancestors and the grave, both in mutual connection and causal dependence, are the groundwork and pillars of Chinese society and social development. This is also the only point in which a credit for idealism can be given to the eastern world. The rapid advance made by our civilization and the hopeful guaranty of its future are not due to our progress in technical matters and inventions, but to our idealism of thought, of work and activity, to the spiritual idealization of life. There is the one dominant ideal in the life of the Chinaman, — honorable or magnificent burial, a permanent coffin, and a well furnished grave. He plods and toils along his whole life with this great end in view, he saves his pennies up to enjoy a better existence in a better land, he lives indeed as much for the other as for this life.

XI. VASES OF JADE

Sacrificial vessels carved from jade were employed during the Chou period in the ancestral cult together with bronze vessels. Of great archæological importance in this respect is the brief paragraph relating to the offerings made in honor of Chou-kung in his ancestral temple in the kingdom of Lu (*Li ki, Ming t'ang wei*, 9) who was honored with the same ceremonies of the solemn sacrifice made by the emperor to his ancestors. The victim was a white bull. Of bronze vessels, they employed three, the bronze figure of a bull (*hi*), and the bronze figure of an elephant, both carrying the vase *tsun* on the back,[1] and the bronze vase *lei* with hill patterns. As vase (*tsun*) for the fragrant wine they employed the type called *huang mu* "yellow eye."[2] For the libations, they used the jade cup *tsan* provided with a handle in the shape of the great jade tablet *kuei*. To present the offerings, they used a jade tazza (*yü tou*)[3] and carved bamboo vases. As drinking-cup, they used the jade cup *chan* carved in the usual manner, and they added the cups called *san* and the cups called *kio*, both made of the jade called *pi*.[4] A cup of this type (*yü chan*) is illustrated in Fig. 198 after WU TA-CH'ÊNG who proposes this identification on the ground of the passage quoted.[5]

---

[1] It seems to me that this is the only possible definition doing justice to the archæological facts; there are no such things as "cups with the figure of a victim bull, of an elephant," as LEGGE translates, nor "vases on which an ox is represented," as proposed by COUVREUR who, however, adds also the translation "vase in the shape of an ox, of an elephant;" but it should be understood that the vase *tsun* is carried on the backs of the animals, the whole affair being made in one cast. These two vessels were doubtless used to receive the blood of the sacrificial bull, and are said to have originated at the court of the Chou (*Ming t'ang wei*, 18).

[2] The vases *tsun* are usually decorated with the conventionalized figures of the monster *t'ao-t'ieh*, the eyes of which are sometimes indicated by inlaid patches of gold. In *Ming t'ang wei*, 20, this vessel is ascribed to the Chou. I am inclined to think that *huang mu* in many cases designates the *t'ao-t'ieh* itself, merely being its epithet. Thus, in the *Ku yü t'u p'u* (Ch. 28, p. 2) the ornaments of a jade axe (reproduced in Fig. 2) are described as consisting of the *huang mu* and the cicada pattern, and as besides the latter only the *t'ao-t'ieh* is represented, it must be identical with the term "yellow eye."

[3] LEGGE translates: "The dishes with the offerings were on stands of wood, adorned with jade and carved." And COUVREUR: "des vases de bois ornés de jade." It is unnecessary to criticise these translations made without any regard to archæology. The *yü tou* are exactly what their name implies, *tou* or tazza made of jade. For illustrations see Han Pottery, pp. 188, 189. According to *Ming t'ang wei*, 28, they are connected with the house of Yin.

[4] LEGGE: "There were also the plain cups and those of horn, adorned with round pieces of jade." COUVREUR: "Les coupes additionelles étaient le *san* et la corne, dont le bord était orné de jade."

[5] It is worthy of note that, according to the *T'ao shuo* (BUSHELL, Description of Chinese Pottery, p. 96), those ancient jade cups were the prototypes of the porcelain cups made under the Sung, and that under the T'ang they were still turned out of white jade and designed for drinking wine.

315

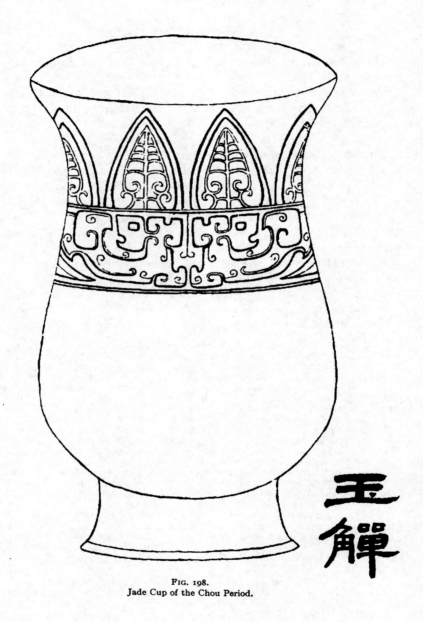

FIG. 198.
Jade Cup of the Chou Period.

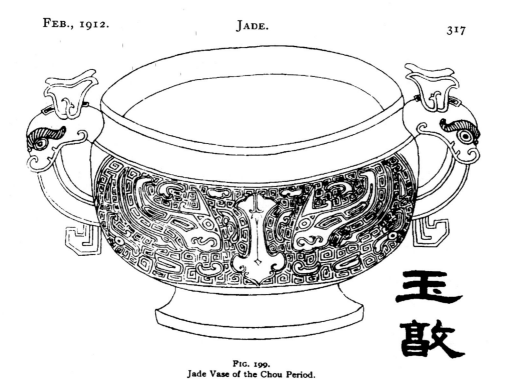

FIG. 199.
Jade Vase of the Chou Period.

FIG. 200.
Ancient Jade Bowl, dug up in Fêng-siang fu.

The *Chou li* mentions a plate of jade used in a ceremony when the feudal princes gave the oath of allegiance to the emperor. An ox was slaughtered on this occasion, and one of its ears cut off. The participants rubbed their lips with the blood as a sign of allegiance. A basin ornamented with pearls contained the ear, and the grain offered was placed on the jade plate; but according to another commentary no grain was offered in this ceremony, and the plate was used to collect the blood of the victim (BIOT, Vol. I, p. 126). According to Chêng Se-nung, this vessel was of the type called *tui*, and WU TA-CH'ÊNG has identified a jade *tui* in his collection with the Chou vessel for administering the oath (Fig. 199). "It is a red jade," he remarks; "the patterns are cleverly carved, and it is identical with the bronze vases *tsun* and *i* of the Shang and Chou periods; but while the bronze *tui* are sacrificial vessels, the jade *tui* are covenant vessels." Fig. 200 illustrates another bowl of the same type in his collection; it is carved from white jade of an ivory color and was excavated near the ancient city of Fêngsiang fu in Shensi Province.

If we can depend upon the *Li ki* (*Ming t'ang wei*, 27), the Hia dynasty would have availed itself of four vessels called *lien*, and the Yin dynasty of six vessels called *hu*, both words formed with the radical denoting "jade" and described by the commentaries as made of jade and holding millet and grain in the ancestral temple.

Ancient jade vases of the Chou and Han periods are now exceedingly rare, even in Chinese collections. I doubt if any exist outside of China. Our collection comprises a representative series of jade vases of later epochs which may convey an idea of the shapes and designs of the ancient pieces and may even surpass them in beauty of workmanship. As in the corresponding bronze vases of the same dates, all religious symbolism formerly connected with them is lost also in the vessels of jade, and the antique forms were simply imitated for artistic and decorative purposes.

The Mongol emperors kept jade jars of tremendous size to hold wine. One of these is described as being of black jade with white veins, and in accordance with these veins, fish and animals were carved on the jar. It was so big that it could hold more than thirty piculs of wine (BRETSCHNEIDER, Arch. and Hist. Researches on Peking, *Chinese Recorder*, Vol. VI, 1875, p. 319). Odoric de Pordenone mentions in the midst of the Great Khan's palace a jar of a certain precious stone called *merdacas*, all hooped round with gold, in every corner of which a dragon was represented as in the act of striking most fiercely; drink was conveyed into this vessel by conduits from the court of the palace. The word *merdacas* has not yet been explained. I believe it is a corruption

a   *b*
JADE VASE OF SUNG PERIOD, SIDE AND FRONT VIEWS.

JADE CUPS AND BOWL.

EXPLANATION OF PL. XLV.

Fig. 1. Jade Cup surrounded by Hydras and Plum-Tree in High Undercut Reliefs, Ming Period.

Fig. 2. Jade Bowl (View of Bottom) with Relief of Lotus, Ming Period.

Fig. 3. White Jade Loving-Cup, K'ien-lung Period.

from Mongol *erdeni kash*, which means "precious jade." We obtain a good idea of these wine-jars from a figure in the Sung Catalogue of Jades (Ch. 100, p. 3) which, according to the description in the text, was then known as the largest worked piece of jade. This vessel stood four feet four inches (Chinese) in height; it measured seven feet two inches in circumference, with a diameter of three feet six inches over the opening. It could hold eighty pints of wine, and is stated to have been a relic of the Tsin or T'ang period. A three-clawed dragon emerging from the sea and soaring into the clouds is carved on the surface of the jar which is of bright-white jade with moss-green marks and emerald-green speckles.

A vase (13 cm high and 5 cm wide), carved from gray jade sprinkled with russet spots, is shown in two views on Plate XLIV. It is carved with great ingenuity and full mastery of form, and it seems justifiable to date it, as Chinese judges propose, in the Sung period (960–1127 A. D.). The lower portion is occupied by the figure of a monster (face in *b*, profile in *a*) running around the four sides, its feet forming at the same time the feet of the vase. Being carved in high-relief, the impression is given that the monster carries or supports the vessel. A lizard-like hydra is climbing the edge and leaning on it with its front-feet. A band of triple scroll-work is laid around the body of the vase, and a band of square meanders borders the rim.

In the jade cup illustrated on Plate XLV, Fig. 1 (8 cm high), there are two such hydras ascending the wall of the cup leaning their chins on the brim. In the illustration, one is viewed from the back, and the head of the companion on the opposite side is visible. The two monsters are undercut and stand out as almost independent figures. In the same technique, the plum-tree branches and blossoms are treated, serving at the same time as handle. All details are worked out with minutest care. The head of a hydra is emerging, as if out of a mist, in the front. Two bats with hydra heads are in relief on the other side. Cloud patterns are spread along the lower edge and on the bottom. The artistic effect of this piece originating in the Ming period (1368–1644) is greatly heightened by the contrasts of color, the gray jade being interrupted in places by deep-brown and yellow tinges, the branches of the plum-tree being brown, the heads of the hydras purple-red, etc.; such features, however, cannot be adequately described, they must be seen and studied in the object itself.

Of the same period is the heavy jade bowl in Fig. 2 of the same Plate XLV, which is shown from its lower side. The carving is done in three layers. Below the rim an eight-petalled flower of conventional design is brought out, and over this one is raised in high relief a bunch of lotus-stems gracefully spread over the bottom, one stem with a large leaf whose rim is turned up, another with a seed-pod surrounded by two leaves, others surmounted by young buds. I regret that it is impossible to make an adequate reproduction which would do full justice to all the beauties of the design and its execution. The interior of the bowl is plain. It was excavated, and chemical effects underground seem to have brought about changes in the original coloring. It is 6.5 cm high with a diameter of 14.5 cm; the thickness of the three layers amounts to 6, 12, and 18 mm, respectively.

On Plate XLVI are illustrated a set of two vessels, a ewer and a covered vase, carved from a fine cloud-white translucid jade and attributed to the period of the Ming dynasty (1368–1644). The ewer is well known as an ancient type in bronze and pottery.[1] A dragon-head with open jaws springs forth from the handle which terminates in a fish-tail below. As this vessel served to pour out water over the hands in washing, the presence of the dragon doubtless had some realistic meaning, in that the dragon as the water-giving animal was supposed to spurt the water from its mouth. The three feet, triangular in section, are elegantly curved outward, and on the two outer sides decorated with engraved meander patterns surmounted by a monster's head (t'ao-t'ieh) between the two front-feet. The body of the vessel is laid out with a band on which two pairs of conventionalized reclining monsters in strongly geometric style are displayed in flat relief. The whole is a perfect piece of work. It is 14 cm high, and 16.7 cm wide from the tip of the snout to the handle.

The bottle-shaped vase (25 cm high, 10 cm wide) is decorated on both sides with blossoming peonies (mu-tan) and bamboo-leaves. On the neck are two Svastika enclosing the ornamental character shou "longevity." The Svastika stands here for the character wan "ten thousand," and we obtain a rebus reading chu ("bamboo") fu kuei[2] p'ing ("vase") liang wan shou, i. e. "With best wishes for happiness, honorable position, peace and numberless (twenty-thousand) years!"

The small wine-cup carved from white jade in Fig. 3 of Plate XLV and coming down from the K'ien-lung period (1736–1795) is an imitation in jade of the bronze sacrificial vessel i (GILES No. 5443) once used in the ancestral cult, but in modern times serving only as a loving-cup in the marital ceremony when bride and groom alternately drink wine from it. The two handles are formed into dragon-heads holding in their jaws the rim of the bowl and terminating in fish-tails. A threefold

[1] Compare Chinese Pottery of the Han Dynasty, pp. 131–132.
[2] Symbolized by the peony called fu kuei hua "flower of wealth and honorable position."

SET OF WHITE JADE EWER AND VASE, MING PERIOD.

INCENSE-BURNER CARVED FROM WHITE JADE IN OPEN-WORK, MING PERIOD.

YÜNNAN-MARBLE PLATE, MING PERIOD.

row of knobs, the ancient "grain" pattern, is laid around the body of the vessel which is 3:2 cm high with a diameter of 6.5 cm.

As regards technical skill and artistic taste, the incense-burner shown on Plate XLVII takes the first place in our collection. It is a superb work of the Ming period (1368–1644), and as all Chinese connoisseurs in Si-ngan fu agree, a unique production and certainly the finest of the kind which ever left that city. It was secured from the private collection of a high official in whose family it had been kept through many generations. It is composed of three pieces carved separately, the base, the bowl, and the cover, the three being joined in harmonious proportions. The handles with dragon-heads on the sides and a large peony blossom with leaves in high relief on the top, and the movable rings freely swinging in them are carved with the bowl out of the same living stone. The cover is laid down just over the dragon-heads and forms one piece with the flat dome by which it is surmounted. Base, bowl and lid, even the bottom, are carved throughout in open-work into a continuous leaf and floral design of peonies, apparently the imitation of a textile pattern.[1] The color of jade is light-gray, with a brownish tinge on the dome caused by the fumes from the incense burnt in the bowl, which is 15.4 cm high, with a diameter of 25 cm.

The large plate shown in Plate XLVIII is carved from Yünnan marble, known as stone of Ta-li fu (*Ta-li shih*) and is remarkable for its size (41 cm in diameter) as well as for its age. As far as I know, other ancient objects of this material have not been found. The piece in question was excavated in the environment of Si-ngan fu. Only the flaring rim has preserved its original jade-like polish, while it has disappeared altogether on the flat bottom where the originally white streaks have assumed underground a dirty-yellow color due to masses of loess falling and pressing on the surface. While it is impossible to fix a certain date for this object, which is void of any ornamentation, it may be generally assumed that it belongs to the Ming period or may even be older. The value of this kind of marble is attributed to its peculiar black stripes and clouds interrupting the white substance.[2] The Chinese are fond of cutting this stone out in slabs, round or square, to be used for screens, or to be inlaid in tables or chairs, and sawing it in such a way that, with some strain of imagination on the part of the intending purchaser, the black masses form veritable scenery with streams, hills and clouds. From this point of view, our plate is carved

[1] Compare BUSHELL, Chinese Art, Vol. II, p. 94. The subject of the derivation of patterns on pottery, metal and jade from textiles is deserving of a special monograph.

[2] Compare E. ROCHER, La province chinoise du Yün-nan, Vol. II, p. 259; F. DE MÉLY, Les lapidaires chinois, p. 6.

with admirable cleverness. The rim is girt with a wreath of black parallel strips; a black-veined band is thrown in a bold sweep across, suggesting a flock of startled birds hurriedly seeking shelter from the dark rain-clouds gathering above. On the lower side of the rim, the *lusus naturæ* has attained a still greater triumph, the black veins running parallel with the rim, being suggestive of a tiger-skin, or when viewed from a distance, of a snow-landscape filled with bare hill-ranges.

The flowervase shown on Plate XLIX, Fig. 1 (19 cm high), is carved from a Han jade, yellow-brown mottled with black streaks; it is unadorned, the artistic effect resting in the gracefulness of its shape and in the natural play of the various colors. It is a work of the K'ien-lung period (1736–1795). Figure 2 is also a flowervase of the same date and likewise produced from a red-mottled milk-white jade of the Han period (19 cm high). It is flat (3.5 cm wide) with rounded lateral sides. The floral design in flat relief is identical on both sides. A band of leaf ornaments is laid around the neck, and a fret of scrollwork around the rim. The two handles are shaped into dragon-heads.

The bowl (9.5 cm high, diameter of opening 26.5 cm), two views of which are shown on Plates L and LI, is carved from a jade-like serpentine with ornaments cut out as in cameo-work. It is posed on five low feet in the shape of cloud-patterns, the entire bowl with these feet being cut out of one piece of stone. The band of ornaments as seen in the illustration runs around the whole vessel. It is divided into two equal sections by two projecting knobs having the function of handles. The pattern consists of conventional forms of chrysanthemums and serrated leaves laid in elegant curves. On the bottom (Plate LI), two butterflies and a design of leaves are cut out in relief. The tips of the leaves on the left are rolled up into spirals and made into the appearance of fungi of immortality (*ling chih*), so that the rebus *ling tieh* is evidently brought out here, "may you reach high age!"

This piece was exposed to a fire which has caused a discoloration of the original leaf-green color into a pale yellow-white along the upper portion both outside and inside.[1] The date of this bowl is to be referred to the K'ien-lung period (1736–1795); it is a product of the great renaissance movement then going on in the perfection of design and technique.

---

[1] Mr. BISHOP (Vol. II, p. 240) describes a jadeite armlet (*cho-tse*) "showing the effects of fire action or heat to which it was exposed during a conflagration in 1878. Previous to the fire the coloration was a mottling of pea-green, gray, and brown. The purplish hue of the brown now seen on the armlet is due to stainings of the iron or bronze with which it was in contact during the fire. The heat has crackled the material throughout, and has completely cracked the ring at one side, the parting of the crack having a width of half a millimetre. The piece has suffered more on one side than the other, which still retains its polish. It is interesting as a well authenticated burnt piece of jadeite."

Explanation of Pl. XLIX.

Fig. 1. Plain Flowervase of Yellow Han Jade.

Fig. 2. Flowervase decorated with Floral Designs in Relief, carved from Han Jade.

I　　　　　　　　　　　　　　　　　　　　2

FLOWERVASES CARVED FROM HAN JADE, K'IEN-LUNG PERIOD.

FIELD MUSEUM OF NATURAL HISTORY.

GREEN BOWL WITH FLORAL RELIEFS IN CAMEO STYLE, K'IEN-LUNG PERIOD.

BOTTOM OF BOWL SHOWN IN PRECEDING PLATE.

I   3   2
FRUIT-DISHES OF WHITE JADE, K'IEN-LUNG PERIOD.

Three fruit-dishes of white jade are illustrated on Plate LII. Figures 1 and 2 are derived from a set of four identical in material, shape and design, carved in a very thin, transparent milk-white jade with clouded onion-green patches. Figure 1 shows the interior, Figure 2 the exterior of these dishes (4 cm high, diameter 13.3 cm), decorated with two zones of meanders encircling a floral wreath in flat relief. Figure 3 on Plate LII, one of a set of two (10 cm in diameter, 1.3 cm high), represents another type of dish, shallow, the edge being cut out in five petals as in certain metal mirrors of the T'ang period, with a raised circular portion in the centre which has a slight depression for the reception of small nuts or the like, while other kinds of fruit are grouped around in the outer compartment. Both sets are works of the K'ien-lung period (1736–1795).

While all of the ancient jade vases have their prototypes in a bronze vessel, and while it may be safely asserted that hardly any new forms of vessels were produced in jade, there are nevertheless two processes of technique applied to jade which we do not encounter in ancient bronze. The one is exemplified by the above jade censer entirely carved in open-work, the other by the bowls with high undercut reliefs, a style and process which seems to set in from the time of the Sung dynasty. We can but presume that the peculiar character of stone, and in particular of jade, easily lent and adapted itself to these two modes of work, and that they were first developed and cultivated in stone, not in metal. Under the Sung, also bronze vessels appear covered with undercut designs in relief, and there are fine examples of them in our collection, but I do not know of any early bronzes, Chou or Han, subjected to this kind of treatment. I am therefore inclined to assume that this was a later development (possibly going back to the days of the T'ang dynasty), and that this technique was transferred from jade to bronze. In both materials, it was extensively practised under the Ming and in the eras of K'ang-hi and K'ien-lung, and from the latter period, we possess also a great number of bronze vessels, especially censers and braziers, the bodies or covers of which are executed in open-work. In dealing with these productions, it will be necessary to refer to the corresponding jade pieces, and to make a careful comparison between the ornamentations of both.

XII. JADE IN THE EIGHTEENTH CENTURY

Marvellous works of jade were turned out during the K'ang-hi and K'ien-lung periods, many directly inspired by these emperors themselves and engraved with poems written by them, with their seals appended.

As a rule, the style of the ancients was followed. This is expressly testified by seals as, *e. g.*, *Ta Ts'ing K'ien-lung fang ku*, "Reign of K'ien-lung of the Great Ts'ing dynasty, imitating antiquity or the antique style" (BISHOP, Vol. II, p. 232). This does not always mean that the work in question is altogether a faithful reproduction of an antique, but only that the style and the spirit in the approved subject handed down from times of old have been preserved.

In the eighteenth century, jade bowlders were brought from Khotan to Peking, as we gather *e. g.* from an interesting poem composed by the Emperor K'ien-lung in 1774 and engraved on the bottom of a magnificent fish-bowl of nephrite; the poem opens with the words, "The colossal block was brought as a tributary offering from Khotan (*Ho-tien*), to be fashioned by skilful hands into a *wêng* (name of a type of vessel)-shaped bowl" (BISHOP, Vol. II, p. 232). From another imperial poem of the same date, we learn that bright pure jade was brought from the Yü-lung Valley, the Chinese name of the Yurungkash River in Turkistan which produces jade (*Ibid.*, p. 244).[1]

The Bishop collection, now in the Metropolitan Museum of New York, was very fortunate in securing a number of pieces originating from the Imperial Summer-Palace destroyed by Lord Elgin in 1860. Most of these were manufactured in the court-atelier for imperial use only and rank among works of the highest perfection which human skill may reach. Though we cannot boast of any such palace pieces, our collection is fairly representative of that memorable period in some fine and choice specimens all coming from the possession of families of high standing in Si-ngan fu and San-yüan of Shensi Province. It is thoroughly characteristic of those art-treasures amassed by the high official and of the taste displayed by him in the decoration of his private residence. These two sections in the Bishop collection and in our own felicitously supplement each other; the one is not rendered superfluous by the other, and serious students should apply themselves to the study of both.

---

[1] Occasionally, a nephrite bowl came also from India, as testified in a poem of the Emperor of the year 1770 engraved in a bowl of chrysanthemum shape (*Ibid.*, Vol. II, p. 250).

324

The conventional opinion on the decadence of Chinese art during the last centuries is not upheld even by a mere superficial examination of these two collections. The technical mastery in the carving of these jade pieces is as great as, perhaps even greater than, in any previous period of history, and they reveal a power of artistic composition and a harmony of form and taste unattained by European art-industry of the same age. True it is, they lack in originality and variability of ideas. The K'ien-lung epoch is weak in new inventions, and it is questionable whether any new creation arose during that period in the mind of any artist. The works of the past are copied, not slavishly and languidly, but with a zealous and fervent inspiration, with an honest desire to produce the best, with a truly artistic instinct. It was a gay and amiable art of a distinctly worldly imprint, certainly bare of that deep religious spirit which had instigated the great early masters to their transcendental and spiritually impressionistic motives. The emotional idealism and sentimentalism had vanished. The exigencies of the life here had come more and more to the front, and the old-time rigid sacredness was redeemed by a more human and social touch. No longer were the artists guided by ideas, but by considerations of taste, elegant forms, pleasing proportions and delicate treatment of ornamental details, though the subjects were still drawn from the ancient sources, but with a predilection for the genre.

While in its content Chinese art of the eighteenth century doubtless becomes somewhat wearisome and monotonous, it still maintains, on the other hand, such high standards and qualities and is so rich in great achievements as to render it sufficiently attractive to the foreign student. And it is perhaps the peculiar characteristics of the K'ien-lung era which have won for it so many foreign admirers. The collector of porcelains is aware of what is due to the accomplishments of the eighteenth century in this line; lacquer and ivory carving flourished at the same period and above all painting, the excellent qualities of which are justly insisted on by Hirth and made by him the starting-point for a study of Chinese painting in general. In no other age were literature and criticism cultivated to a higher degree, and under the patronage of these two big-hearted liberal emperors, K'ang-hi and K'ien-lung, the Chinese have produced masterpieces in printing, book-making and wood-engraving which may maintain an honorable place beside the best productions of the world at large. Where there are successful efforts in all fields of human activity, we are scarcely justified in speaking of a general decadence.

It would be preposterous to infer that the student of Chinese archæology could set his mind at rest over the development of the eighteenth century; not only, he cannot get along without it, but he should even make his start with a thorough knowledge of this period which is as necessary to him as his daily bread.   The great revival of antiquity led to a unique renaissance movement in literature and art; diligent searches for ancient books, manuscripts, and antiquities resulted in a widening of the horizon, in a deepening of thoughts and in a straining of intellectual forces unparalleled in China's long history.   The archæologist has every reason to look up to the deep endeavors of that epoch with a feeling of particular gratitude, as without them we should probably be forsaken or grope in the dark in more than one case.   To the vigilant wisdom of that generation we owe the preservation and tradition of numerous antiquities; many others, as bronzes, and the tiles and bricks of the Han period, were then brought to light and studied, and many ancient types which have long perished have come down to us solely in the reproductive and retrospective art of the eighteenth century. And that is exactly the point where the share of the archæologist in the harvest comes in.   He finds an unusually fecund field in the K'ien-lung epoch for the exercise of his wits in his particular domain.   The prototypes are lost, but the reproductions are there and must be utilized. On the following pages, jade sonorous stones and bells, *Ju-i* and other objects are discussed from an archæological point of view, while we are forced to refer to specimens of the K'ien-lung period, no older ones being in existence.   The conservative spirit of the Chinese thus becomes a substantial benefactor, and a good K'ien-lung reproduction is certainly better than a blank or a weak or poorly authenticated more ancient "original."   Where, and what is the original, after all?   Of these Chinese copies and copies of copies, the word of Holmes (The Autocrat of the Breakfast-Table) holds good: "A thought is often original, though you have uttered it a hundred times," and Emerson's saying: "When Shakespeare is charged with debts to his authors, Landor replies, 'Yet he was more original than his originals.   He breathed upon dead bodies and brought them into life.' "   Thus, it is no wonder that Carl Gussow of Munich could not believe Huang Hao's Red Carp of 1811 to be a copy, though expressly stated so by the artist on the painting; the entire conception, he thought, was so free and independent that it was bound to be an original (HIRTH, Scraps from a Collector's Note Book, p. 44).   As everything Chinese is pervaded by an atmosphere different from our own, so also a Chinese copyist is framed of a different mould; his work is creative reinvention, not purely receptive, but partaking of the spirit permeating the soul of the master.   Therefore, we may have confidence in studying archæology on the ground of the traditionary relics of the K'ien-lung epoch.

"The use of sonorous stone to make musical instruments," as Mr. J. A. VAN AALST (Chinese Music, p. 48) justly remarks, "may be said to be peculiar to China.[1] At all events, the Chinese were the first to give stone a place in music; their classics frequently mention the stone chime as being known by the ancient emperors and held in great esteem." They are mentioned on three occasions in the "Tribute of Yü" (in the *Shu king*) where also stones for polishing the musical stones occur.

There are two principal classes of sonorous stones, — the single stone and the compound stones. The former (*t'ê k'ing*) is a stone cut somewhat in the shape of a carpenter's square, but in the form of an obtuse angle with two limbs, the longer one called the "drum," the shorter, the "limb." It is suspended in a wooden frame by means of a silk cord passing through a hole bored at the apex. It is still employed during the ceremonies performed in the Confucian temples and struck with a hammer against the longer limb to give a single note at the end of each verse. In the compound stones (*pien king*), sixteen of the same type as the single stone, but on a smaller scale, are suspended in two equal rows on a wooden frame, all being of the same dimensions in length and breadth, differing only in thickness; the thicker the stone, the deeper the sound. Also this instrument serves in the Confucian temples, in connection with bell-chimes, the bell being struck at the beginning of each long note in the tune, and the stone at the end.[2]

All the jade gongs and bells in our collection, including one of rock-crystal, come down from the K'ien-lung epoch, the latter half of the eighteenth century, but they are of such eminent workmanship

---

[1] There are perhaps some exceptions in America (FISCHER, p. 28). Sonorous stones from Peru are referred to by C. ENGEL, A Descr. Catalogue of the Musical Instruments in the South Kens. Mus., p. 81. Prof. M. H. SAVILLE (The Antiquities of Manabi, Ecuador, p. 67, New York, 1907) relates after Suarez that "in Picoaza there was preserved, until a little while ago, a bell of the aborigines of that locality; it was a stone slab of black slate, a metre (little more or less) in height, and some centimetres wide; when this stone was suspended from one of its ends, the striking of it with another stone or with the hand produced a metallic and pleasant sound, which vibrated like that of a bell." Saville failed to find any traces of this stone, and thinks that it is probably being used as a metate in one of the houses in the village. A beautiful sonorous stone excavated by Dr. George A. Dorsey in Ecuador is in the collections of the Field Museum (see Publ. 56, *Anthr. Ser.*, Vol. II, No. 5, p. 259). There is, further, in this Museum (Cat. No. 70940) a trap signal from the Pomo Indians, California, consisting of two obsidian blocks for hanging in such a manner that the trapped deer strike them and announce their capture to the hunter. In the *Mémoires concernant les Chinois*, Vol. VI, p. 221, attention is called to black sonorous stones mentioned by Pliny.

[2] Compare VAN AALST, *l. c.*, pp. 48–49; DENNYS, Notes on Chinese Instruments of Music, p. 105 (*Journal China Branch R. As. Society*, Vol. VIII, 1874); A. C. MOULE, Chinese Musical Instruments, pp. 30–33 (*Ibid.*, Vol. XXXIX, 1908). AMIOT's Essai sur les pierres sonores still remains the most valuable contribution to this subject.

that they must be equal, if not superior, to any ancient specimens. It was then a renaissance period of art on all lines, greater than which no other previous age had seen, fostered by imperial patronage, enlisting the highest talent of the time. I doubt if at present any of these bells are in existence older than the eighteenth century. It is true the Bishop collection (Vol. II, p. 107) possesses a large flat gong of nephrite with a date-mark *Yüan-ting, i. e.* made in the period B. C. 116–111 of the Han Emperor Wu; but it is just this dated inscription which is apt to cause suspicion, as such inscriptions do not otherwise exist in ancient jade pieces and are not on record in any Chinese collection. I do not mean to throw any reflection on the great value and unusual beauty of this object, but it is a question open to discussion whether it really belongs to the period referred to.

During the T'ang dynasty the regulation obtained that in the worship of the Deities of Heaven and Earth stone was employed for the *k'ing*, while in the ancestral temple and in the palace *k'ing* of jade were utilized.

In 1764, the governor of Yarkand forwarded to the emperor K'ien-lung 39 large slabs weighing altogether 3975 catties, to make the peculiar musical stones called *k'ing*, besides a large supply of smaller slabs; and, the year after, sent a further large quantity for imperial use. The slabs were all quarried in the Mirtái Mountains, and sawn there by natives of Sungaria (BISHOP, Vol. I, p. 25).

The bell represented on Plate LIII though made as late as the K'ien-lung period (1736–1795) is a specimen of most brilliant workmanship. It was acquired from the private collection of a Chinese gentleman in the town of San-yüan north of Si-ngan fu. First of all, it is valuable for its material, being carved from a jade of the Han dynasty (*Han yü*). A brown-red tinge passing into light-yellow shades is strewn over a background of a glossy white which the Chinese designate as mutton-fat. Aside from the two pendants,[1] the entire bell is carved from one solid bowlder of jade, with walls 1 cm thick. It is hollowed out by means of the tubular drill,[2] as can be seen in the interior where, side by side, three cylindrical pieces unequal in length, have been taken out. It is 20 cm high, 14.6 cm wide at the base, and 3.8 cm thick, over the middle of the base.

In shape and design, this bell is an exact reproduction of one of the ancient bronze bells of the Chou period. The decoration on both faces is identical. The lower and upper edges are bordered by meander patterns. A band with a conventionalized monster's head (*t'ao-t'ieh*)

[1] They have been omitted in our Plate to ensure a larger reproduction of the bell.
[2] Compare BUSHELL, Chinese Art, Vol. I, p. 144.

BELL CARVED FROM HAN JADE, K'IEN-LUNG PERIOD.

RESONANT STONE, CARVED FROM JADE, IN SHAPE OF DRAGON, K'IEN-LUNG PERIOD.

BACK OF RESONANT STONE SHOWN IN PRECEDING PLATE.

RESONANT STONE, CARVED FROM JADE, K'IEN-LUNG PERIOD.

in flat relief is laid around the body, from which a girdle of leaf-ornaments is drooping. The handle is surrounded by two combatant dragons cut out in open-work.

The resonant stones are much favored as birthday presents or congratulatory gifts in general, as their designation *k'ing* (Giles No. 2208) is punned upon with another word *k'ing* of the same sound (No. 2211) meaning "good luck, happiness, blessings, to congratulate." The phrase *ki k'ing* "to strike the musical stone" is understood as "may blessings attend you" or "good luck and best wishes." This stone, therefore, forms part and parcel of the bride's dowry in Peking (Grube, Zur Pekinger Volkskunde, p. 32).

In the jade chimes of Plate LIV the upper smaller jade piece from which the resonant stone is suspended is composed entirely of designs intended as rebus. The peach as symbol of longevity stands for *shou* "long life;" the svastika on it means *wan* "ten thousand," the bat *fu* stands for *fu* "luck," the fungus of immortality for *ling* "high age," and the orchid *lan* for the verb *lan* "to come to an end." Thus, this ornament is read as the sentence: *wan shou fu lan ling*, "Numberless years and luck may come to an end only at old age!" (Compare Grube, *l. c.*, p. 138, No. 8).

The resonant stone (31 cm × 13.5 cm), of white and greenish jade, is carved in the shape of a scaly dragon surrounded by cloud-ornaments, *i. e.* the dragon soaring in the clouds as the beneficial sender of rain. Plate LIV shows the front-view, Plate LV the back where only the scaly body is visible, while the head is floating in the clouds and hidden under the cloud ornaments, a rather artistic conception.

In the specimen on Plate LVI (38 cm high), the original shape of the resonant stone is rather faithfully preserved; even the perforation in the apex has been retained. The additional ornaments are intended to be read as a rebus. The figure below the ring carved from a gray and brown jade is a Buddhistic emblem, one of the Eight Precious Objects, in Sanskrit çrīvatsa, in Chinese *p'an ch'ang*, and is used in the rebus for *ch'ang* "long" which is connected with the character *shou* "longevity" enclosed in a circle on the surface of the resonant stone; the bat *fu* represents again *fu* "luck" and the two fishes *yü* mean *yü* "abundance." The rebus therefore reads: *ch'ang shou fu, k'ing yu yü* "Long life and luck, blessings and abundance!" Despite this tendency, the single parts are harmoniously arranged and well proportioned. Both faces, also in the double-fish ornament, are carved alike.

The gong of sea-green jade on Plate LVII is composed of three parts connected by double chains of white jade, the links being cut out of one solid piece of stone. The lower plaque (19 cm × 8 cm) is the bell proper worked in relief and in open-work, both faces being decorated alike.   The two combatant dragons of geometric mould have their raised front-paws intertwined.   The figure of a bat resting on cloud-patterns is cut out in open-work below to furnish the character *fu* to be read together with the ornamental form of *shou* above, as "good luck and long life."   In the right and left upper corners two coiled hydras are joined in *à jour* carving.   The central plaque is framed by two conventionalized figures of elephants on the upper edge and two rampant hydras on the sides, interlacing their tails along the lower edge.   The picture of a pine-tree with a stag and the fungus of immortality is engraved on the front face, and that of a pine-tree with a garden-pavilion (*t'ing*) and sea-waves below it and a crane on the wing on the opposite face.

The upper brooches from which these gongs are suspended are of the same type as the girdle-ornaments, and the piece in the present specimen particularly recalls to mind the head-piece in the ancient girdle-pendants.

The gong on Plate LVIII consisting of two plaques is carved from rock-crystal; the two carvings are connected by silver chains with hooks formed into figures of bats and are suspended by means of a silver hook from a blackwood frame.   The shape and decorations of this gong are identical with the preceding one of jade; also here the same designs are cut out in relief on both faces, a task which requires more care and trouble in the translucent crystal than in jade.   The ornamental lines on the one face must exactly coincide with those on the opposite face, as otherwise the latter would be apt to shine through, and to disturb the harmony of the design.   The skill and accuracy of the Chinese worker in crystal is most admirable in this respect.

This one as well as the previous jade gongs are all works of the K'ien-lung period and fine specimens characteristic of the high accomplishments of the glyptic art of that period.

On Plate LIX, a jade resonant stone and a perforated disk (*kung pi*) of the same gray jade are represented, suspended from wooden stands (70.5 cm high) with a base shaped into a calabash (*hu-lu*), the symbol of fertility and numerous progeny (because of its many seeds). This is the way in which these objects are mounted and fixed up, when given as presents among officials, and in which they are set up as decorative objects in the mansion, adorned with vari-colored silken tassels. The sonorous jade (17.5 cm × 5.7 cm) has preserved its wedge-shape of old.   Four dragons in full figure are carved along the upper and lateral sides, and the figure of a fish below, *yü* "the fish" standing for *yü* "abundance," the whole ornament reading *ki k'ing yu yü*, "Luck,

RESONANT STONE, CARVED FROM JADE, K'IEN-LUNG PERIOD.

RESONANT STONE, CARVED FROM ROCK-CRYSTAL, K'IEN-LUNG PERIOD.

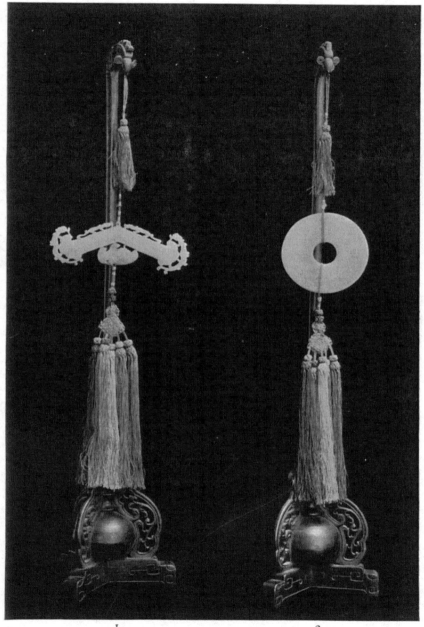

1   2

SET OF JADE RESONANT STONE AND DISKOID BADGE OF RANK, K'IEN-LUNG PERIOD

WHITE JADE CARVING OF MOUNTAIN SCENERY, K'ANG-HI PERIOD.

BACK OF JADE CARVING SHOWN IN PRECEDING PLATE.

WHITE JADE SCREEN, K'IEN-LUNG PERIOD.

BACK OF JADE SCREEN SHOWN IN PRECEDING PLATE.

blessing and abundance!" The disk (10.5 cm in diameter) is ornamented in conformity with the ancient approved style, the grain pattern (*ku*) on one side, and interlaced bands on the other. Also this set is credited to the K'ien-lung period.

Presents of jade objects were always fashionable in China and truly imperial gifts. Especially the arrival of a new offspring in the imperial family gave the occasion for bestowing upon him fine jade carvings implying wishes for long life. In the "Annals of the Kin Dynasty" (*Kin shih*), it is on record that in the 26th year of the period Ta-ting (1186 A. D.) a great-grandson was born to the emperor; in celebration of this event, a banquet was given in the K'ing-ho Palace, on which occasion the emperor presented the infant with a set of mountains carved in jade. Such sculptured jade landscapes are known under the name "longevity mountains" (*shou shan*).

An artificial hill of jade was erected in the palace of the Mongol emperors in Peking (BRETSCHNEIDER in *Chinese Recorder*, Vol. VI, 1875, p. 319).

A good example of such mountain-scenery is illustrated on Plates LX and LXI, the former showing the front-view, the latter the back of the carving. Rocks are piled up in graceful irregularity such as we are wont to see strewn around in Chinese gardens. Five clusters of fungi of immortality (*ling chih*) are growing out of the clefts of the rock, and two cranes, emblems of long life, add to the symbolic significance.[1] The bird standing below is turning its head back and looking up to its companion clinging with outspread wings to the edge of the rock. In Plate LXI the wings of this crane may be viewed; the rocks are covered with bamboo stems and leaves. This carving (19 cm × 15 cm; 3.5 cm thick at the base) is cut out of one solid piece of grayish jade and is a perfect masterpiece in every detail. Its date is in the K'ang-hi period (1662–1722).[2]

The screen figured on Plates LXII and LXIII in two views is carved from a white jade slab with light-green speckles and a few reddish veins, 0.6 cm thick, 30 cm long and 21.2 cm wide. It is enclosed by

[1] There is doubtless also a rebus intended in this subject to be read *chu ling hao shan*, i. e. "We pray for a life as long as that of the crane and the mountains."

[2] Dr. BUSHELL describes a similar type in the Bishop collection as follows (BISHOP, Vol. II, p. 216): "A little irregular piece carved in bold relief in the form of a mountain with trees and water in the usual style of a Chinese landscape, the pine occupying a conspicuous place. A stairway is cut in the hillside leading up to a pavilion with four pillars which is built upon a platform of rock above. In the foreground stand two figures in Taoist costume: an old man with a peach in his hand, representing Shou Lao, the god of longevity, accompanied by a youthful attendant carrying a branch upon his shoulder. At the back appears a similar scene with longevity emblems, including a pair of storks and a gigantic sacred fungus (*ling-chih*) growing from the rocks below."

a blackwood frame, the ornaments of which are inlaid with silver wire, and placed in a stand of carved blackwood, with a panel in open-work. On the one face of the slab (Plate LXII) two spotted deer, a stag reclining, and a doe walking are engraved with two fir-trees as background. Ripe cones are hanging from the branches. The fungus of immortality (*ling chih*) emerges also here from the ground and reminds us of the symbolical significance of the picture which implies a wish for old age and good income (*lu* "deer" by means of punning being the equivalent of *lu* "income"). The other face of the screen (Plate LXIII) is decorated with trunks of plum-trees, the branches being laden with blossoms. We observe that in both cases the artist has intentionally chosen to draw one tree, in horizontal position, in order to better fill the space.

The objects grouped on Plate LXIV are all works of the K'ien-lung period (1736–1795).

Figures 1 and 2 are ornaments fastened to the central part of a woman's girdle, both very finely carved from gray jade in open-work. The design in Fig. 1 (10 cm × 8 cm) is a goose with wings outspread covered and surrounded by lotus-flowers and leaves. The goose is a symbol of conjugal fidelity, and the lotus (*lien*), by way of punning with *lien* ("to join, to connect"), is suggestive of the notion of permanent ties. The object is therefore a love-token. The rectangular plaque in Fig. 2 (6.8 cm × 4.7 cm), cut out in two layers, displays on a diapered background, a pine-tree rising from the ground in the centre, bamboo-leaves on the left and plum-blossoms to the right; below, a stag and a doe facing each other in the shade of the pine-tree. The stag *lu* reads in the rebus *lu* "official salary, good income," the entire rebus being "we always pray for old age (symbolized by the pine-tree) with sufficient income!"

Figure 3 of Plate LXIV represents the cover to a round box for holding ink (7 cm in diameter), with a relief design of "the mother hydra watching two young ones"[1] embedded in ornamental clouds.

Figure 5 on Plate LXIV illustrates a paper-weight (*shu chên*) of white and red agate on which eight lizard-dragons mutually interlaced are carved out (7.3 cm × 6.5 cm, 4 cm high). The *Ku yü t'u p'u* contains a great many varieties of this type adorning the scholar's desk, carved into figures of animals.

Figure 4 on the same Plate is a purely decorative piece for the wall, carved in open-work from a gray-green jade with a layer of brown, representing a landscape of rocks with pine-tree, maple-tree, and fungus (13.2 cm × 8.7 cm).

A pair of jade flutes of the K'ien-lung period called "male" and

[1] Compare above p.276.

EXPLANATION OF PL. LXIV.

Figs. 1-2. Girdle Ornaments in Open-Work.
Fig. 3. Carved Ink-Box.
Fig. 4. Carving of Rock with Pine-Trees.
Fig. 5. Paper-Weight of Agate.

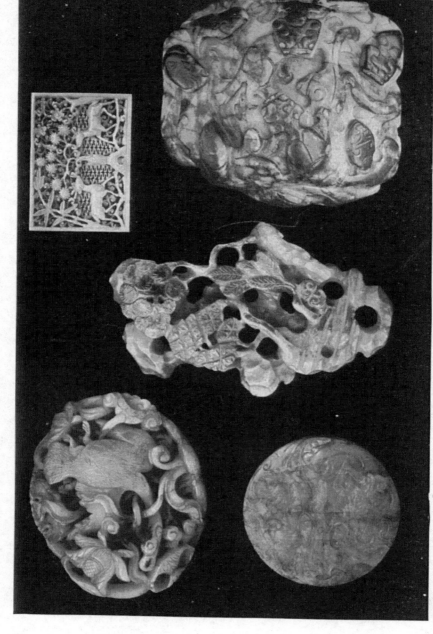

JADE WORKS OF THE K'IEN-LUNG PERIOD.

EXPLANATION OF PL. LXV.

Fig. 1. Pair of White Jade Flutes.
Figs. 2-3. Girdle Buckles.
Fig. 4. Ink-Slab with Design of Well.
Fig. 5. Relief of Flower-Vase.

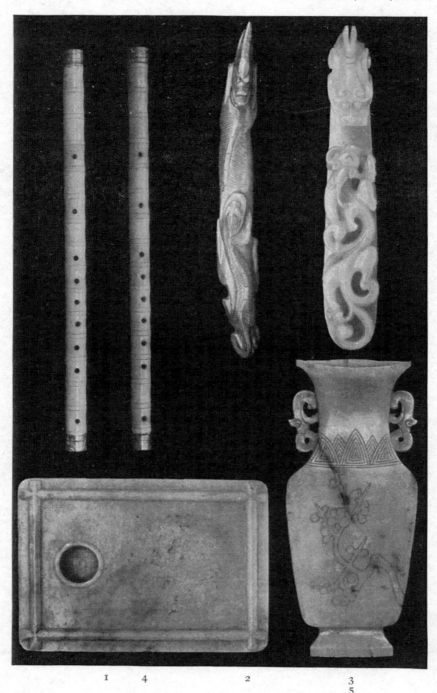

JADE WORKS OF THE K'IEN-LUNG PERIOD.

1　4　2　3
5

"female" flutes is illustrated on Plate LXV, Fig. 1. They are finely carved from a pure milk-white jade (49 cm long, diameter 2.5 cm) with separate mouthpieces of green jade. They are formed into the appearance of bamboo-stems, the joints being indicated by two parallel incised circles, i. e., they are direct imitations of bamboo flutes. Besides the nine holes visible in the illustration each piece has, further, two holes side by side on the lower face in the second joint from below.

Jade flutes are frequently alluded to in Chinese literature. The *Si king tsa ki* relates that at Hien-yang there was a jade flute with twenty-six holes. When the Emperor Kao-tsu first went to that place, he spied it in the treasury and played on it, whereupon mountains and groves with horses and chariots appeared in a mist, vanishing altogether when he ceased playing (BUSHELL in BISHOP, Vol. I, p. 49). The "Records of Liang-chou" say: "In the 2nd year of the period Hien ning (276 A. D.) brigands rifled the tomb of Chang Kiün and obtained a vessel of the type *tsun* carved from white jade, a musical pipe of jade (*yü siao*) and a jade flute (*yü ti*), each of these three being a particular object."

The Sung Catalogue of Jades figures a number of jade flutes and other musical instruments.

Figures 2 and 3 of Plate LXV represent two girdle-ornaments of the K'ien-lung period (1736–1795), both displaying the same motive "the mother dragon watching her young one" in different modes of technique, the one being a relief carving, the other being worked *à jour*. In Fig. 2 (14.5 cm long), the old dragon is cut out in full figure, and the cub is crawling up its body; the jade is gray with a layer of yellow on the upper face. The piece in Fig. 3 (14 cm long), of white jade, is a buckle of the same curved shape as those in Plate XXXIV, with a stud on the back which is decorated with a blossom diagram. The young dragon leaning on its beard is standing in full figure undercut on the surface of the buckle, a fungus of immortality at its side.

Figure 4 of the same Plate represents an inkslab carved from gray jade (12.5 cm × 9 cm, 2.3 cm high). We might call it an ink-well in a certain sense because the outline of a well-frame is brought out in the raised rectangle framing the surface, resembling the ancient form of the written character denoting a well. The round cavity, the well itself, serves as the receptacle for the liquid ink; the inkcake is rubbed on the flat surface of the stone.

Figure 5 of Plate LXV is not a real flower vase, as might appear from the illustration, but only the relief picture of one cut out of a solid flat slab of gray jade (16 cm × 6.6 cm, 1.5 cm thick). Pieces like this one are mounted on wooden panels and arranged on them together with other ornaments cut out of hard stone.   A branch with plum-blossoms is engraved on the surface.

The *Han Wu ku shih* [1] relates how the Emperor Wu built a temple and in the front-hall erected trees of jade, with branches made of red coral, leaves of green jade, flowers and seeds blue and red made of precious stones hollowed out in the middle like little bells, tinkling as they hung (Bushell in Bishop, Vol. I, p. 48).   A craftsman of the principality of Sung is credited with having cut jade or ivory into leaves of the paper-mulberry tree with such perfect resemblance that they might have been taken for natural products; but as each leaf required three years of labor, his compatriot, the philosopher Lieh-tse, ridiculed his vain ability (Pétillon, Allusions littéraires, p. 185).[2]   Under the present dynasty a pair of jade trees often forms part of the trousseau of a wealthy bride; the Chinese keep them covered with glass shades.

In the *Ku yü t'u p'u* no jade trees are figured or mentioned.   There is an interesting account in the "Annals of the T'ang Dynasty" (*T'ang shu*) in the notice regarding Tokharestan (T'u-ho-lo) to the effect that in 662 A. D. an "agate lamp-tree" (*ma-nao têng shu*), three feet high, was sent from there to the Chinese Court (Chavannes, Documents sur les Turcs occidentaux, p. 157, St. Pet., 1903).   The word *têng* written with the classifier for "metal," I take here, as is the case quite frequently, as a variant of *têng* "lamp," usually written with the classifier for "fire."   I understand the object in the sense that a regular tree was carved from agate, and that the ends of its branches were so made as to hold an oil-lamp or a candle.

A pair of such trees coming down from the K'ien-lung period (1736–1795) is in our collection (Plates LXVI and LXVII).   They are planted in pots worked in cloisonné enamel, the principal pattern consisting of five bats (*wu fu* = five blessings [3]), the larger central one being enclosed in a central medallion and connected with the figure of a resonant jade stone (*k'ing* = with congratulations or blessings).   The two outer and upper bats carry calabashes (*hu-lu*) suspended from bands, symbols of fertility and numerous progeny because of the numerous seeds which they contain.   Hence an appropriate design for a bridal couple.   For

---

[1] A record relating to the time of the Emperor Wu (b. c. 140–86) by Pan Ku, while others believe that it was compiled during the T'ang period (Bretschneider, Botanicon Sinicum, Part I, No. 138).

[2] According to the *Tarikh Djihan Kushai*, Buku Khan, the king of the Uighur, had a dream in which he saw a man dressed in white, who gave him a piece of jade in the form of a pine tree, and said to him; "As long as you are able to keep this piece of jade in your possession, you will rule over the four quarters of the globe."   His minister had the same dream.   (Bretschneider in *Journal China Branch R. Asiatic Society*, Vol. X, 1876, p. 202.)

[3] *Viz.* Old age, wealth, health, love of virtue, and a natural death.

JADE POMEGRANATE TREE IN JAR OF CLOISONNÉ ENAMEL, K'IEN-LUNG PERIOD.

JADE CHRYSANTHEMUMS IN JAR OF CLOISONNÉ ENAMEL, K'IEN-LUNG PERIOD.

I   2   3

SO-CALLED SCEPTRES OF GOOD LUCK, K'IEN-LUNG PERIOD.

EXPLANATION OF PL. LXVIII.

Fig. 1. Sceptre, Cast-Iron, Ornaments inlaid with Gold and Silverwire.
Fig. 2. Sceptre, carved from Wood inlaid with Three Plaques of White Jade.
Fig. 3. Sceptre, carved from White Jade with Three Reliefs.

the same reason the tree chosen (Plate LXVI) is the pomegranate (*shih-liu*, Giles No. 7258), the fruit of which is also an emblem of anti-race-suicide. Five large fruits of agate carved in forms surprisingly true to nature are hanging from the tree. The blossoming trees also express the sentence *tse sun ch'ang ts'ing*, "may your sons and grandsons always flourish!"

The leaves are cut out partially of jade, partially of turquoise, some are made of copper-foil. The petals composing the blossoms are all carved out of different kinds of jade, agate, and cornelian. To the right below there is a bush of red berries, and the fungus of immortality (*ling chih*). The other pot (Plate LXVII) is less uniform in the composition of the plant design, but richer in color. Eight bunches of various species of chrysanthemums are grouped here side by side, the petals of the blossoms being cut out of jade and coral. As the chrysanthemum is called *fu kuei hua* "flower of wealth and honorable position," the wish *fu kuei* is expressed by the gift of this flowerpot.

On Plate LXVIII, three objects are grouped, the one in Fig. 1 of iron, the two others of jade. These objects, peculiar to the Chinese, are known under the name *Ju-i*, a phrase meaning "as you desire, according to your wish." At the present time, on the occasion of a birthday or of New Year, this instrument is bestowed upon high dignitaries, or by the courtiers on the sovereign,[1] simply with the idea of implying good wishes. Formerly, it was made also of gold, silver, rhinoceros horn, bone, rock-crystal, bamboo, amber[2] and even iron, all of which have now grown out of fashion; it is now usually carved from wood[3] which is inlaid with jade plaques in the centre, and at the two ends, as shown in our Fig. 2, Plate LXVIII, while the specimen in Fig. 3 of the same Plate, entirely carved from jade, occupies an exceptional place.

---

[1] This appears *e.g.* from a *Ju-i* in the Bishop collection (Vol. II, p. 147) engraved with the inscription: "Made at the imperial manufactory. With reverential vows for a succession of fertile (*fêng*) years, and that throughout (*hien*) the world every wish be fulfilled! Respectfully presented (to the emperor) by his servant Wu King." The two words *hien* and *fêng* introduced into the stanza doubtless contain an allusion to the period *Hien-fêng* (1851–1861) which allows us to infer that this sceptre was presented to the emperor who reigned under this title. Dr. Bushell states that in the palace of the emperor a sceptre is placed in every reception-room upon a table before the throne, usually made of carved wood or red lacquer, with three plaques of jade inserted and often inlaid with jewels. It figures also among imperial presents to high dignitaries as a mark of special distinction.

[2] Sun K'üan, the later Wu Ta-ti (181–252 A. D., Giles, Biographical Dictionary, p. 613) is recorded to have owned a *Ju-i* of amber, while he still was prince of Wu. The story goes that he heard of P'an Fu-jên whose father he had condemned to death, being a great beauty. He ordered her portrait to be painted, and when the picture was brought before him, he was seized with such joyful surprise that he exclaimed: "This is a divine woman!" and struck the table with his amber *Ju-i* which thus broke to pieces.

[3] There are also specimens of porcelain, brass, gold-plated brass, and even glass.

A sceptre of this make is bestowed by the emperor on the empress on the occasion of the marriage ceremony (G. DEVÉRIA, Un mariage impérial chinois, p. 90. Paris, 1887). Much has been written about this curious type, but its real origin and history remains somewhat shrouded in mystery.

Prof. GILES (Introduction to the History of Chinese Pictorial Art, p. 159) expressed his judgment as follows: "Chao Si-ku, an archæologist of the thirteenth century, tells us that the *Ju-i* was originally made of iron, and was used ' for pointing the way ' and also ' for guarding against the unexpected,' *i. e.* for self-defence.[1] It was, in fact, a kind of blunt sword, and traces of basket-work are still to be found inside what must have been the sword-guard. Later on, when it had come to be merely a part of ceremonial regalia, other materials, such as amber, crystal, jade, lacquer, and bamboo were substituted for iron. This account is confirmed by more recent authorities, and comprises all we really know about the *Ju-i*, which to-day is frequently sent by friend to friend as a token of good wishes = May you have your heart's desire!" I think this is in general a fair statement of the case, and we may add that in regard to no other object of their culture are the accounts of the Chinese more unsatisfactory than anent the *Ju-i*.

The difficulty of archæological research lies in the fact that all the present specimens which can lay claim to a certain age do not go beyond the time of the reign of K'ien-lung (1736–1795), and that no ancient specimens of this type have been preserved; consequently, if we read in ancient texts of *Ju-i*, we have no guaranty that they are identical with those of the present day which, after all, in shape and design may be of recent date only; and as Prof. Giles justly points out, it was a kind of blunt sword, formerly. But how and when the transformation into the modern type was accomplished, escapes our knowledge. Furthermore, the expression *ju-i* may very well occur in an ancient text and simply mean "according to your wish," without having any reference to an instrument so named. It seems to me that this fact has sometimes been overlooked. Thus, *e. g.*, by J. EDKINS introducing his article ‚'The *Ju-i*, or Sceptre of Good Fortune" (*East of Asia Magazine*,

[1] In his recent article on Jade (*Adversaria Sinica*, No. 9, p. 321), Prof. GILES gives the following translation of this passage from the *Tung t'ien ts'ing lu:* "The men of old used the *Ju-i* for pointing or indicating the way, and also for guarding themselves against the unforeseen. It was made of wrought iron, and was over two feet in length, ornamented with patterns in silver either inlaid or overlaid. Of late years, branches of trees, which have grown into the shape required, and also pieces of bamboo, highly polished to resemble jade, and prepared without the aid of hatchet or awl, have been very much in vogue." "In support of the first clause of the above," Prof. GILES adds, "we find in history such passages as (fifth century): 'The Emperor pointed at him with his *Ju-i* and said;' 'The Emperor rapped on the table with his *Ju-i* in token of approbation,' etc."

Vol. III, 1904, p. 238) with the words: "The *Ju-i* is found in the biography of King Fang who died in the first century b. c. At an audience, he said to the Emperor Yüan-ti, 'I fear that though your Majesty acts in this way you will still not obtain what accords with your wish.' King Fang was a weather-prophet and a student of the Book of Divination, the *Yi king*. *Ju-i* means 'as you desire.' "

Dr. Bushell (Chinese Art, Vol. I, p. 148) stated that "the *Ju-i* sceptre derives its peculiar form from the sacred fungus called *ling-chih*,[1] the *Polyporus lucidus* of botanists, one of the many Taoist emblems of longevity." This is a reversion of the logical order of things. The *Ju-i* was in existence, and the fungus of immortality was one of the ornaments applied to it, but not instrumental in the production of the shape of the entire object. There are many *Ju-i* where this ornament is absent, as, e. g., the three specimens in our collection do not exhibit any trace of it. Nor am I inclined to accept as evidence the passage adduced by Bushell (Bishop, Vol. I, p. 49) from the "Biography of Hu Tsung" where we read that "during the Wu dynasty, when digging the ground, there was found a bronze casket in which was a *Ju-i* of white jade. The sovereign questioned Tsung who replied that Ts'in Shih-huang, on account of the existence of the spirit of the Son of Heaven in Nanking (?), had buried precious things there in several places to keep down the sovereign spirit."[2] There is no reason to adopt this personal interpretation of Tsung, and to credit the Ts'in period with the existence in it of the *Ju-i*.

The anecdote of Shih Ts'ung smashing a coral-tree with an iron *Ju-i* (Pétillon, Allusions littéraires, p. 240; Edkins, *l. c.*, p. 238) is of little, if any, historical value; but shows, according to Edkins, that the *Ju-i* is, in this case, an implement of destructive power influenced by Buddhist ideas. Edkins, referring to Eitel's "Handbook of Chinese Buddhism" (p. 130), makes it a point that the Chinese Buddhist term *ju-i shên* ("a body transmutable at will") relates to the magical power of assuming a body without dimensions and weight, and that the chief signification of *Ju-i* among the Buddhists is conquering power. If we stand on the solid basis of facts, we observe that the first representations of *Ju-i* of the type known to us make their appearance in the hands of Buddhist deities on paintings of the T'ang

---

[1] The same suggestion had already been proposed by Pétillon, Allusions littéraires, p. 241, Note.

[2] The passage, as quoted in Bushell's text, is much abbreviated, and in all probability, not adequately translated. The complete text will be found in the *T'u shu tsi ch'êng*, Section 32, Ch. 237, *Ju-i pu ki shih*. — Also Prof. Giles (*l. c.*, p. 320) refers to this passage as the earliest allusion in Chinese literature to the *Ju-i*; according to Giles, Hu Tsung died in 243 a. d.

period, thus, *e. g.* in the right hand of a Mañjuçrī by Wu Tao-tse.[1]   As
far as I know, this symbol never occurs in the Gandhāra sculptures nor
in any Chinese-Buddhist sculpture from the Wei down to the T'ang
period, but in painting exclusively.   In the case of Mañjuçrī, it might
be appropriate to assume that the *Ju-i* takes the place of the sword
which is his usual attribute; but a good many variations occur placing
certain reservations on too premature conclusions of this kind.   The
Japanese painter Sesshū (1420–1506) pictured a Mañjuçrī astride a
lion, holding a *Ju-i* in his left hand and nothing in his right, while he
is always holding the sword in his right.   In two other paintings, by
Minchō and Sanraku Kanō respectively (*Kokka*, Nos. 82 and 117), this
Bodhisatva is holding the *Ju-i* in his right and a book-roll in his left.
Again, in a Chinese painting ascribed to Chang Se-kung (*Kokka*, No. 149),
he is holding the *Ju-i* in his uplifted right and leaning its end on the
palm of his left.   Again turning to No. 168 of the same Journal, we
find a splendidly painted Mañjuçrī attributed to the Kose School of
the twelfth century, in which the attributes of the god are a sword in
his right and the sacred lotus-flower in his left.   It will be noticed here
that the stem of the lotus describes the same curve as the handle of
the *Ju-i*, and that the flower is shaped very much like the upper part
of our iron *Ju-i* (in Fig. 1, Plate LXVIII);[2] the lion, on whose back
the god is placed, has a lotus-blossom under each foot, the petals being
of a conventional geometric form, such as is found also in the upper
parts of the *Ju-i*.   The lotus-flower with long stem is frequent in the
hands of Bodhisatvas, conspicuous *e. g.* on the sculptures in the cave-
temples of Lung-mên in Honan (see *e. g.* CHAVANNES, Mission archéologique dans la Chine septentrionale, No. 397), and certainly in the
Indian sculptures, and in the Nepalese and Tibetan paintings (many
examples in A. FOUCHER, Étude sur l'iconographie bouddhique de
l'Inde, and L'art gréco-bouddhique; A. GRÜNWEDEL, Collection Uchtom-
ski, Part II, p. 13).   I do not mean to say that the Buddhist emblem
called *Ju-i* has developed from the lotus, though I think that the
alternation of both is suggestive.   But it is not necessary at all to
assume that the Chinese *Ju-i* in general is of Buddhist origin, as supposed *e. g.* by W. ANDERSON, Catalogue of Japanese and Chinese Paint-

---

[1] Celebrated Paintings of China, Vol. I, Plate II (Tokyo, Shimbi Shoin, 1907).
The upper part of this *Ju-i* shows a spiral-shaped cloud-pattern and a knob with
coral; the blade is adorned with four studs.   In the same volume is reproduced a
Samantabhadra by Ma Lin, holding in his right a *Ju-i* on which the figure of a
Buddha is represented.

[2] Compare A. FOUCHER, Étude sur l'iconographie bouddhique de l' Inde, p. 115,
Paris, 1900, and II. Part, p. 43, Paris, 1905.

ings, p. 32, Note).[1] It may very well be that the implement is Chinese in origin and even prebuddhistic, and that, as in so many other things, a kind of compromise took place, resulting in the assimilation and amalgamation of two ideas and two forms.

Also Prof. GILES (*l. c.*, p. 321) justly arrives at the conclusion that the prevalence of the lotus-flower as a decoration, though due of course to the influence of Buddhism, is scarcely sufficient evidence of "a religious origin" (*versus* Davis). There is no doubt that the original significance of this implement has been lost long ago. It seems to me that it may have grown out of one of the early jade emblems of the Chou period which, as we saw in Ch. II, were developed from ancient types of implements, and that in the beginning it was a symbol of light, generative power and fertility. The fact that on the occasion of his marriage the emperor writes the character for dragon (*lung*) on a slip of paper to be placed by four princesses in the palanquin of the empress, together with two jade *Ju-i* (DEVÉRIA, *l. c.*), is doubtless an outcome of that ancient idea. As said above, there are no ancient specimens left, and material of this kind must be awaited, before a satisfactory conclusion can be reached.

Figure 1 on Plate LXVIII represents a *Ju-i* of iron, which may be considered as one of the original forms of this instrument, the ornaments being incrustated with silver and gold wire (K'ien-lung period); presumably, the oldest type of this implement is preserved in this specimen. On the blade curved downward two dragons soaring in the clouds are playing around the sun-ball. On the handle, eight emblems are represented which are, — a fan consisting of a banana-leaf, a two-edged sword, the sacrificial vase *tsun*, a pair of castanets, a calabash, a flute, a blossoming flower, and a basket with handle.

The sceptre in Fig. 2 (47.5 cm long) is carved from blackwood on which three medallions carved from white jade are mounted with pictures of plum-blossoms, fungus and chrysanthemums in relief. That in Fig. 3 (40.7 cm long) is entirely carved from white jade with representations of the gods of luck rowing over a lake in boats with a basket supposed to contain their supernatural gifts. Rocks and pine-trees fill the scenery. These three varieties represent all the essential types of this implement.

---

[1] As is well known, the *Ju-i* occurs also in the hands of Taoist deities and priests, Buddhist monks and nuns, especially on commemorative portraits of the latter in temples. There is nothing of special interest in these that could throw light on the subject. The Japanese seem to have nothing to say regarding it; at least, in their great Buddhist Pantheon published under the title *Shoso Butsu zō dzu-i* (Vol. 5, p. 16), only an outline figure of the *Ju-i* (*nio-i*) is given, without any comment.

APPENDIX I

JADE IN BUDDHIST ART

In our collection there are several jade carvings of Buddhistic and Taoistic images which will be treated in a separate monograph dealing with Buddhist stone sculpture. Also the *Ku yü t'u p'u* (Ch. 98) figures six religious subjects executed in jade. One of these has a particular interest for us, as it is connected with the name of the great painter Yen Li-pên of the T'ang period, who worked in the latter half of the seventh century (GILES, Introduction to the History of Chinese Pictorial Art, p. 38; HIRTH, Scraps from a Collector's Note Book, p. 66). In the course of this study, we have had occasion to refer several times to the names of painters and to pictorial influence penetrating into the decorative motives of jades. To this tendency, which seems to set in from the beginning of the tenth century, we owe the preservation of some ancient pictures which would otherwise be lost. If not exact copies, they are authentic in so far that they preserve the style of the master and make us acquainted with the subject which he treated. As we were forced to expose the Sung Catalogue of Jades to severe criticism, it will be a matter of justice to credit it also with what is good in it, and to emphasize this merit of having transmitted to us a certain amount of valuable pictorial material.

The medallion-picture here reproduced in Fig. 201 is carved on a slab of jade light-green and white, two feet six inches (Chinese) long, two feet one inch wide, and 1.6 inches thick. The title given to the picture by the editors of the work is in the original on the upper right hand side outside of the frame, but here inserted inside in the upper left for technical reasons. It reads: "Ancient jade image of P'u-hien (Samantabhadra), the Great Master (Mahāsatva)." An inscription of seven lines is carved in the slab which may thus be translated: "In the period K'ai-p'ing (907–911 A. D.) of the Great Liang (*i. e.* Hou Liang) dynasty, from the imperial treasury, bestowed by imperial command as a dedication on the temple Hung ming (sze). Carved by the jade-cutter P'êng Tsu-shou. Picture of how the Great Master has his elephant washed. Style approaching that of Yen Li-pên."

In the palace of the Sung emperors, forty-five scrolls ascribed to this artist were still preserved (*Süan ho hua p'u*, Ch. 1, p. 8); and, though the subject of the present picture is not mentioned there in the list of his

341

works, we may presume that a copy or copies of that motive had survived up to that date, or at least to the period alluded to in the inscrip-

FIG. 201.
"Brushing the Elephant of Samantabhadra,"
in the style of the Painter Yen Li-pên.
Jade Carving of the Period 907–911 A. D.
(from *Ku yü t'u p'u*).

tion which is that usurped by the rebel Chu Wên (854–914 A. D., GILES,
Biographical Dictionary, p. 188), who assassinated the last emperor

of the T'ang dynasty and mounted the throne in 907 as first emperor of the Later Liang dynasty. The Sung editors pass their verdict on the value of this reproduction as follows: "In this carving of the Great Master Samantabhadra, the subject of the painting 'Washing the Elephant' by Yen Li-pên of the T'ang period is imitated. There, the image of the Buddha, the gods (*Deva*) with their attendants, the servants of the elephant, the elephant itself and another quadruped are all represented. But the most clever representation, though excellent in its merits as a copy, cannot reach the original. In the method of carving, however, it is of perfect workmanship."

This judgment is worthy of note, for it shows in what high estimation the work of Yen was held in the Sung period, and that, as I understand, the reproduction in question repeats the composition and style, but not the true spirit or individual touch of the original. It should be added that, according to the Sung Catalogue of Painters (*Süan ho hua p'u*, Ch. 1, p. 5 b), the same subject had been painted by Chang Sêng-yu of the sixth century.

Prof. GILES (*l. c.*) has given a brief description of this subject after a woodcut inserted in the *Fang-shih mo p'u* of 1588 where some changes are introduced, and has remarked that it is not easy to say to what this picture refers. The explanation afforded by the *Ku yü t'u p'u* renders it sufficiently clear. We now know that it is the question of Samantabhadra's elephant, which symbolizes care, caution, gentleness, and a weighty dignity (EDKINS, Chinese Buddhism, p. 385). This Bodhisatva is usually represented as mounted on an elephant and grouped into a triad with the image of Buddha in the centre and that of Mañjuçrī on a lion's back. A Nepalese miniature depicting him astride an elephant is reproduced in A. FOUCHER's "Étude sur l'iconographie bouddhique de l'Inde" (Plate VI, No. 2; Paris, 1900). The cult of this god who symbolizes goodness and happiness is localized on the famous mountain Ngo-mei in Sze-ch'uan Province, where is erected in the temple Wan-nien sze a colossal statue of the elephant, cast of white copper and twelve feet high, surmounted by the image of the Bodhisatva enthroned on a bronze lotus-flower.[1] Each of his feet rests on a lotus of bronze, in the

---

[1] Described by E. C. BABER, Travels and Researches in Western China, pp. 32–33, and A. J. LITTLE, Mount Omi and Beyond, p. 63, London, 1901. The six tusks of the elephant mentioned by him and noticed by W. ANDERSON (Catalogue of Japanese and Chinese Paintings, p. 81) on a Japanese scroll of the eighteenth century seem to be a later addition, suggested by the legend of the six-tusked elephant in whose shape Buddha entered the womb of his mother (compare SPEYER, *Zeitschrift d. Deutschen Morgenl. Ges.*, Vol. LVII, 1903, p. 305). Yen's painting could certainly not be cited as evidence for the fact that Samantabhadra's elephant originally had only two tusks, for his aim was to delineate a lifelike scene. But from the beautiful religious painting by Wu Tao-tse preserved in the temple Tōfuku-ji near Kyōto (Celebrated Painters of China, Vol. I, Plate III, Tōkyo, 1907) we glean the fact

same way as lotus-flowers are carved under his feet in our picture. This statue is celebrated all over Tibet and alluded to in Tibetan history (G. HUTH, Geschichte des Buddhismus in der Mongolei, Vol. II, p. 414, Strassburg, 1896). There, it is referred to a prophecy related in the Sūtra of the Predictions of the Elephant-Mountain. This notice is important, as it will possibly lead along the right track in discovering the legend which forms the basis of the picture under consideration.

The Bodhisatva has alighted from the animal and is standing on the left of it with folded hands; his smiling face is bearded, and his head is tonsured. A flower seems to merge from behind his left shoulder.[1] He wears the long flowing monk's garb. There is a monk in front of him, gazing at the elephant, turning his shaven head to the spectator. A boy is carrying on his right shoulder a package of sacred books surmounted by a flaming jewel, and is leaning his left on his bent knee in order to relieve the weight of his burden. In the background on the left, we notice three worshippers praying with their hands folded, a monk and two laymen, or gods, as supposed in the *Ku yü t'u p'u*. The animal turning its nose with a certain admiration toward the elephant has one horn and a scaly body; it is doubtless introduced as an inferior creature to illustrate the superiority of the sacred elephant in all his glory. His head is bridled, but he seems to feel quite cheerful over the situation. A queer-looking attendant pours streams of water out of a jar over his back, and a youthful boy in a kneeling posture, clad only with an apron, is engaged in sweeping his back with a broom.

The composition of this little picture is admirable. The elephant's brushing is placed in the centre of the scene, and everything radiates from this action, all participants fixing their attention on this point, either adoring or admiring the jolly monster. Simplicity predominates, and superfluous additions liable to detract the attention of the looker-on are wisely discarded. There is no scenery for background, except the ornamental clouds hanging above and stretching below. The unity of the composition is strictly adhered to in the accentuation of the one sacred act, a touch of serene humor being spread over the whole. Anoth-

---

that this was really the case under the T'ang. Notable here is the bright intelligent smile in the elephant's face; he is squatting on the ground and wears gold earrings. The Bodhisatva, of white skin, adorned with all royal ornaments, double earrings, a feminine hairdressing with gold pin, coral brooch and lotus, is seated on the elephant's back, the left foot hanging down, the right one drawn up, reading in a book of brown leaves inscribed with characters in gold. — Of Chinese literature on the Ngo-mei shan, the *Ngo shan t'u shuo* (2 Vols., 1889) deserves special mention; it contains a series of good wood-engravings depicting all the scenery and temples of this place of pilgrimage.

[1] Judging from the painting of Wu Tao-tse, it is a lotus stuck into his hair. Note the difference between the two pictures: there, he is the god in full apparel and of feminine appearance; here, he is the bearded monk with tonsure and without jewels.

er nicety of the composition is the posture of the elephant taken from the front which allowed the artist to centre it correctly, and to elevate its back so high that the sweeper rises into prominent view. Yen Li-pên surely was an artist who knew what he wanted, and who could carry his intentions into effect. Painted in colors, his work must have created a lasting impression. The elephant is certainly animated by "life's motion;" his head is finely modeled, his drooping ears, trunk and tusks are true to nature, and he seems to enjoy the ticklish sensations from his shampoo. The attempt to mark the folds in the skin of the pachyderm is no less remarkable, and the painter seems to have made earnest studies of the animal from life. Altogether, this picture presents an intimate genre-scene of Buddhist art, an offshoot of the epoch of the T'ang, such as no other of this class has survived, and the Sung Catalogue of Jades deserves our thanks for its preservation.

In modern wood-engraving, this motive has been frequently copied.[1] We alluded to the cut in the *Fang-shih mo p'u* (Ch. 5, p. 21 b) which is a poor makeshift, and Prof. Giles (*l. c.*) was quite right in the remark that it is not easy to gather from this woodcut that the painter was a great artist in our sense of the term. It is here reproduced in Fig. 202 for no other reason than to afford an instructive comparison between a good and a dead copy of an ancient painting. It is not stated from which source this copy is taken; the legend in the upper right corner refers it to T'ang Yen Li-pên. The elephant has turned here into an automatic machine, and all figures bear a stiff wooden character; all spirituality is lost. Note the emaciated arm and leg and the horrible hand of the boy shouldering a box (supposed to contain sacred books), the exaggerated flames of the jewel, the insipid change in the costumes, head-dresses and faces of the two laymen, the wrong attitude of the man pouring the water, and the caricature of the sweeper who is standing on the animal's back, instead of kneeling, and touching with the end of his broom a cloud on which a book-case is hovering, — an additional flatness. Luckily, the bad quality of this picture is exceptional in

---

[1] In a modern wood-engraving printed in Nanking and representing a sermon of Buddha before the assembly of monks, Mañjuçrī is riding on the lion's back in the foreground, and to his left an elephant is being vehemently scrubbed. Hokusai has in his *Mangwa* (Vol. 13, p. 20) the sketch of an elephant washed by six men. The believers in the superiority of Japanese over Chinese art should not fail to look up this freakish caricature, and to compare it with Yen's natural creation. Hokusai's elephant is provided with bear-claws! The objection that he had had no chance to see a live specimen is not valid; he had occasion enough to observe good models in Buddhist art. And see at the same time on p. 23 the combination of a camel with a cocoanut-tree! Nobody would think that this creature with sharp eagle-claws should represent a camel, if the name were not printed beside it. And then there was a time when a shallow mind like Hokusai could be considered in Europe as a revelation of East-Asiatic art.

the *Fang-shih mo p'u*, which is a work of great merit and value for the study of decorative and pictorial motives.

It is not yet ascertained how old this form of Samantabhadra mounted on the elephant is, and there is reason to believe that it originated in China, not in India. In the Lamaist iconography of Tibet this form is strikingly absent, and the god is usually represented as sitting on the

FIG. 202.
"Brushing the Elephant"
(Woodcut from *Fang-shih mo p'u*).

seed-pod of a lotus-flower (see A. GRÜNWEDEL, Mythologie des Buddhismus, p. 140). Also in China, a sitting image of his has been preserved, likewise a carving in jade illustrated in the *Ku yü t'u p'u* (Fig. 203) and preceding the one just described. It is entitled: "Miraculous Buddhist image, of ancient jade, representing the Great Master P'u-mên (Samantabhadra)." It is carved out of a slab of pure-white flawless jade, measuring two feet four inches in height, two feet in width, two inches and two-tenths in thickness. On this picture, the Bodhisatva is conceived of as a hermit or recluse seated under a rock-shelter, or in a cave on a heap of rushes, or on a rush-mat, the alms-bowl at his right and a flowervase with a bare twig in it behind, his person being

古玉普門大士天然佛像

FIG. 203.
Ancient Jade Carving representing the Bodhisatva Samantabhadra
(from *Ku yü t'u p'u*).

enveloped by clouds; regarding the style of the latter even the editors remark that it is pictorial. It is evident that this representation is copied from a painting. The snail-like curls of the head and beard are noticed in the text. We further remark the smile of the face, — for he is, as his name implies, the Good One, — the earring, the long drawn-out ear, the eye of wisdom on the forehead, and the slight bending forward of the body. It is curious that the hands are not outlined, but disappear under the robe, as if to avoid the chilly mist of the clouds.

The story of this image is reported as follows: "In the period Hi-ning (1068–1078 A. D.) of the Sung dynasty, the Empress-Dowager Hüan-jên was an adherent of Buddha and commanded Kao K'an with the office of *nei-shih tu-chih* to take along and offer imperial incense, and to represent the Court on the island of P'u-t'o [1] in the worship of the Great Master Samantabhadra and the Bodhisatva Avalokiteçvara. In the cave Ch'ao-yin ("The Sound of the Tide") he proclaimed the imperial will, when suddenly the voice of thunder sounded in this cave, accompanied by a torrent of water which brought this image to light. Kao K'an hurriedly received the image and reported to the throne. The Empress Dowager received it respectfully in the palace in order to establish a regular cult for this image, for it was a heavenly most precious gift."

The Ch'ao-yin cave mentioned in this text is illustrated and briefly described in the Chinese Chronicle of the Island (*P'u-t'o shan chi*, edition of 1739, Ch. 1, p. 9). The jade carving of the image must have existed before the year 1068. If I am not mistaken, the style of drawing here displayed is that of the Buddhist painters of the T'ang period, and the artistic inspiration underlying the composition seems rather to testify in favor of than to militate against such a supposition. Creative power in the production of Buddhist subjects seems to have prevailed much stronger under the T'ang than under the Sung. Also the Sung tradition that this image represents Samantabhadra need not be questioned. We have thus, to recapitulate, three well authenticated types of this Bodhisatva coming down from the T'ang epoch, — the purely religious form of the cult represented by the painting of Wu Tao-tse depicting him as *the* Bodhisatva, the genial human monk by Yen Li-pên, and the happy meditating recluse by an unknown artist; the two latter indubitably personal inventions of individual masters.

I may be allowed to add in Fig. 204 a fourth variety which is reproduced from a Japanese wood-engraving made after a painting of Sesshū (1420–1506). This Samantabhadra forms a triad with Çākyamuni and Mañjuçrī to whom we alluded above in the notes on the *Ju-i*. Though revealing many points of resemblance with the picture of Wu

---

[1] The famous island in the Chusan Archipelago, east of Ningpo, devoted to the cult of Avalokiteçvara (*Kuan-yin*); described by G. SMITH (Narrative of an Exploratory Visit to Each of the Consular Cities of China, pp. 264–278, New York, 1847), J. EDKINS (Chinese Buddhism, pp. 259 *et seq.*) and many others. I spent a week there in July 1901. When passing through Calcutta in March 1908, I happened to meet a Buddhist monk from that island who had traveled the whole distance, speaking no other language than Chinese, for the purpose of collecting among his countrymen funds for rebuilding the temples of P'u-t'o.

FIG. 204.
Picture of Samantabhadra by Sesshū
(from Japanese Wood-Engraving).

Tao-tse, there are many traits stamping the work of Sesshū as an independent production. The position in which the elephant is drawn, and the bold dashes of the brush marking its massive head and trunk betray a self-conscious genius. The new feature, from an iconographic point of view, is the conception of the Bodhisatva who is plainly clad, without any jewels and hair-ornaments, and wears his hair flowing down his shoulders. This cannot be interpreted as a feminine feature, as the Mañjuçrī in this group is represented in the same style of hair-dressing. It is difficult to guess what the artist's intention really was. Samantabhadra here appears neither as the Bodhisatva in the traditional form nor as the monk, but simply as the reader of a Buddhist text on a scroll. In our collection, there is an ancient wood-carving, probably of the Ming period, from a temple in Si-ngan fu, representing Samantabhadra sitting on a recumbent elephant, also with long flowing hair, but not reading nor holding any attribute.[1]

When I visited the sacred isle of P'u-t'o ten years ago, I was shown in one of the temples (P'u-tsi sze) a jade-carved statue of the goddess Kuan-yin. Almost life-size, she is represented gracefully reclining, resting her chin on the right palm, sleeping, in the posture of Buddha's Nirvāṇa. The body is dressed in gorgeous silk attire, and the head is painted in colors. The image is kept under a glass case, and I saw it shortly before sunset when the last sunrays produced a marvelous effect on the snow-white transparent jade. I was informed that this work had but recently been executed in Canton at a cost of 10,000 Mexican dollars through subscriptions raised by a pious community. It is not only one of the most magnificent works of sculpture ever executed in China, but also the most lifelike piece of statuary I have ever seen. It inspires an impression which cannot be forgotten, and is a living proof that art is still alive in China, if opportunities are offered.

[1] A Japanese wood-carving of Samantabhadra sitting on a lotus and posed on a standing elephant see in *Annales du Musée Guimet, Bibl. d'études*, Vol. VIII, Pl. XI. A painting of the same type where the elephant seems to be wading through water in SEI-ICHI TAKI, Three Essays on Oriental Painting, Plate I, London, 1910.

APPENDIX II

THE NEPHRITE QUESTION OF JAPAN

It is well known that among the antiquities of early Japan which may be dated roughly from a few centuries B. C. down to the sixth and seventh centuries A. D., two kinds of ornamental stones are prominent,— the *kudatama*, oblong perforated cylinders, and the *magatama*, curved or comma-shaped beads, both referred to in the Kōjiki and Nihongi and found in large numbers in the ancient graves. They were presumably strung and worn as necklaces. The *magatama* were made of various stone material of which N. G. MUNRO (Prehistoric Japan, p. 456) enumerates the following list:— agate, jasper, chalcedony, serpentine, steatite, quartz, crystal, glass, jade, chrysoprase and nephrite; with the remark that the three latter are not found in Japan. Also W. G. ASTON (Nihongi, Vol. I, p. 49) gives nephrite as one of the materials for *magatama*, and adds that some of these materials do not occur in Japan. H. v. SIEBOLD, I believe, was the first to suspect the Chinese origin of these nephrite *magatama*, and the geologist E. MILNE, first brought to light the fact that this mineral is never met with in Japan. But all authors express themselves in a general way, and none has thought it worth taking up the investigation of the problem involved. Even W. GOWLAND, one of the best connoisseurs of Japanese antiquity, is content to say: "The stones of which magatama are made are rock-crystal, steatite, jasper, agate, and chalcedony, and more rarely chrysoprase and nephrite. The last two minerals are not found in Japan" (GOWLAND, The Dolmens and Burial Mounds in Japan, *Archæologia*, 1897, p. 478). O. NACHOD (Geschichte von Japan, Vol. I, p. 144, Gotha, 1906) repeats the fact, without formulating the problem.

But those engaged in the archæology of Japan do not explain to us either another fact no less extraordinary, that is the large number of glass beads in the dolmens. Of 1108 beads discovered by Gowland in one of them, there were 791 of glass, all dark-blue, with the exception of a few green or amber colored, seventeen of silver, a hundred and twenty-three of baked clay, a hundred and thirty-three of steatite, and forty-one of jasper. To any one acquainted with the history of glass, it must be clear at the outset that the ancient Japanese cannot have manufactured these colored glass beads, but must have received them from an outside continental source.

351

The first tangible historical facts relating to actual glass manufacture in Japan refer us to the year 1570, when a foreign artisan settled in Nagasaki, and taught the natives there how to blow glass; and in the period Kwanyei (1624–1643), the arrival of Chinese artisans at Nagasaki gave the industry a great stimulus. They taught the Chinese methods of blowing glass, and the art, spreading throughout the country, was practised at Kyōto, Osaka, and Yedo (J. L. BOWES, Notes on Shippo, p. 12, London, 1895). Certainly, the mirror of cloisonné enamel (figured by BOWES on p. 14) in the Imperial Treasure-House (*Shosoin*) of Nara, is, despite the claims to native workmanship by Japanese connoisseurs, a work of Persian origin, as plainly shown by the style of the floral ornamentation. We know that the Chinese made their first acquaintance with glass during their intercourse with the Roman Orient about the time of our era, but that they did not learn how to make it before the fifth century. There was, no doubt, a lively trade in colored glass beads going on between the anterior Orient and the Far East during the first centuries A. D., and it is possible that they reached Japan in this manner. But it can hardly be presumed that the Japanese became acquainted with the process of making glass earlier than the Chinese, and it is even open to doubt whether glass beads were made, as stated by Bowes, in the time of the Emperor Shomu (724 A. D.). Either the Japanese glass beads of early historic times have been imported from the Roman Orient, and most probably by way of China, or if the claim of indigenous manufacture can be sustained with any plausible evidence, these antiquities and the graves from which they come, cannot be as old as they are supposed to be by Japanese archæologists and their foreign followers. There would be many other reasons to believe that the remains of the so-called protohistoric age of Japan cannot go back to any great antiquity, but seem to go down as far as a period between the second and eighth centuries; but this is not the place to discuss this problem. We were merely obliged to raise this question, in order to obtain a correct point of view in estimating the possible age of the nephrite magatama.

The nephrite magatama, if they exist, occur only in small numbers. It is too well known how difficult it is, even for a specialist, to recognize nephrite at a glance, without experimental investigations, for unconditional credence to be applied to the definitions of laymen. He who has studied Fischer's careful book on the subject is aware of the numerous disappointments to which the premature labeling of specimens as jades in museums and private collections has been subjected. And what, in the course of a century, has not been taken for nephrite![1] For this reason, we must look upon the Japanese nephrite magatama with a certain feeling of diffidence. The monumental work of Mr. Bishop has paved the way also in this question and supplies us with accurate evidence for the occurrence of jadeite in a Japanese magatama.

The Bishop collection (Vol. I, p. 231), fortunately, contains one ancient specimen from Japan. "It is a curved bead, or *magatama*, of light emerald-green jadeite with dead-oak-leaf stainings, pierced for suspension as a pendant or as part of a necklace, where it is supposed to have been strung with a number of the tubular beads called *kudatama*. No jade has hitherto been found *in situ* in any part of Japan, so that all beads dug up there are presumed to be of exotic origin." The specimen in question is figured and described in the work of BISHOP, Vol. II, p. 113.

Dr. BUSHELL (in BISHOP, Vol. I, p. 47) translates the following passage: "The *Tu yang tsa pien*, a work of the end of the ninth century (WYLIE, Notes, p. 194) records that during the T'ang dynasty the kingdom of Japan presented to the emperor an engraved gobang board of warm jade, on which the game could be played in winter without getting cold, and that it was most highly prized. Thirty thousand *li* east of Japan is the island of Tsi-mo, and upon this island the Ninghia Terrace, on which terrace is the Gobang Player's Lake. This lake produces the chess-men which need no carving, and are naturally divided into black and white. They are warm in winter, cool in summer, and known as cool and warm jade. It also produces the catalpa-jade, in structure like the wood of the catalpa-tree, which is carved into chess-boards shining and brilliant as mirrors." This account is evidently fabulous, and as stated by Wylie, the work in question, chiefly occupied with an account of rare and curious objects brought to China from foreign countries between 763 and 872 A. D., contains many statements having the appearance of being apocryphal. It may well be doubted that real jade is meant there, and even if this were the case, no evidence for the natural occurrence of jade in Japan or the adjacent islands would accrue from this text.

However easy the nephrite problem of Japan may look on the surface, it is difficult to decide from what source, how, and when the

---

[1] In a small pamphlet, "Uebersicht und Bemerkungen zu von Siebold's Japanischem Museum" (without year and place), p. 4, also *jade* is listed among the mineral products of Japan. Either this is a case of mistaken identity or a case of recent importation from China. Mr. BISHOP (Vol. I, p. 184) remarks: "A small number of worked jade objects have also come from Japan, but probably in the course of commerce from China, as we have the explicit statement of Mr. Wada that jade is not found in that country."

nephrite or jadeite material was transmitted to Japan. These jewels may go back, after all, to an early period when historical intercourse between Japan and China was not yet established; they represent two clearly distinct and characteristic types, such as are not found in the jewelry of ancient China. If the Japanese magatama and kudatama would correspond to any known Chinese forms, it would be possible to give a plausible reason for the presence of jade in the ancient Japanese tombs; but such a coincidence of types cannot be brought forward. Nor is it likely that similar pieces will be discovered in China, as necklaces were never used there either anciently or in modern times. We must therefore argue that the two Japanese forms of ornamental stones were either indigenous invention or borrowed from some other non-Chinese culture sphere in southeastern Asia the antiquities of which are unknown to us. It seems plausible to presume that these jewels were first cut in materials found on the soil of Japan, and later on also from nephrite brought over from the mainland. But so far, all indications are lacking as to the channels through which, and as to the time when, such a trade might have been carried on.

BIBLIOGRAPHY

The papers relating to stone implements and quoted in the first chapter, as well as publications which have but occasionally been cited, are not here listed again. For the sake of completeness, also books and papers not quoted on the preceding pages, but containing references to this subject, have been here included. The student of Jade will find everything that is necessary in the works of FISCHER and BISHOP. The reader seeking general information may turn to the papers of EASTER and GILES, or look up the subject in the books on Precious Stones by FARRINGTON, KUNZ, and BAUER.

M. AMIOT, Essai sur les pierres sonores de Chine. *Mémoires concernant les Chinois*, Vol. VI, pp. 255–274 (2 plates). Paris, 1780.
>This essay contains also valuable notes on Chinese jade in general.

ARZRUNI, Nephrit von Schahidulla-Chodja im Küen-Lün Gebirge. *Zeitschrift für Ethnologie*, Vol. XXIV, 1892, pp. 19–33.

A. W. BAHR, Old Chinese Porcelain and Works of Art in China. London, 1911.
>Figures a few articles of Jade on Plates CXIII — CXV.

J. DYER BALL, Things Chinese, or Notes connected with China. Fourth edition. Hongkong, 1903.
>Article "Jade," pp. 356–358.

ÉDOUARD BIOT, Le Tcheou-li, ou Rites des Tcheou. 2 vols., and 1 volume Table analytique. Paris, 1851.

G. C. M. BIRDWOOD, The Industrial Arts of India. London, 1880.
>On jeweled Jades of India: Vol. II, p. 32, with two plates.

HEBER R. BISHOP, Investigations and Studies in Jade. New York, 1906. Privately printed. 2 vols. (Copy in the Library of the Field Museum.)
>Compare review by CHAVANNES, *T'oung Pao*, 1906, pp. 396–400.

S. BLONDEL, Le Jade. Étude historique, archéologique et littéraire sur la pierre appelée *yü* par les Chinois. Paris, 1875. English translation in *Annual Report of Smithsonian Institution*, 1876, pp. 402–418.

S. BLONDEL, L'art capillaire dans l'Inde, à la Chine et au Japon. *Revue d'Ethnographie*, 1889, pp. 440–441.
>Some notes on Jade hair-spangles.

S. W. BUSHELL, Chinese Art. Vol. I. London, 1904.
>Ch. VII. Carving in Jade and Other Hard Stones, pp. 134–151.

S. W. BUSHELL, Description of Chinese Pottery and Porcelain, being a translation of the T'ao Shuo. Oxford, 1910.
>Contains also useful references to Jade vessels.

ÉDOUARD CHAVANNES, Les mémoires historiques de Se-ma Ts'ien. 5 vols. Paris, 1895–1905.

355

ÉDOUARD CHAVANNES, Le T'ai Chan. Essai de monographie d'un culte chinois. Appendice: Le dieu du sol dans la Chine antique. Paris, 1910.

G. W. CLARK, Kwiechow and Yün-nan Provinces. Shanghai, 1894.
　　Remarks on the Jade trade from Burma to Yün-nan, p. 52.

S. COUVREUR, S. J., Chou King. Texte chinois avec une double traduction en français et en latin. Ho kien fou [China], 1897.

S. COUVREUR, S. J., Dictionnaire chinois-français. Ho kien fou, 1890.

S. COUVREUR, S. J., Li Ki ou Mémoires sur les bienséances et les cérémonies. Texte chinois avec une double traduction en français et en latin. 2 vols. Ho kien fou, 1899.

G. DUMOUTIER, Les symboles, les emblèmes et les accessoires du culte chez les Annamites. Paris, 1891.

S. E. EASTER, Jade. The *National Geographic Magazine.* Vol. XIX, Washington, 1903, pp. 9–17.

O. C. FARRINGTON, Gems and Gem Minerals. Chicago, 1903.
　　Jade, pp. 165-167.

HEINRICH FISCHER, Nephrit und Jadeit nach ihren mineralogischen Eigenschaften sowie nach ihrer urgeschichtlichen und ethnographischen Bedeutung. Second edition, Stuttgart, 1880 (first edition, 1875). Two colored plates.

A. J. C. GEERTS, Les produits de la nature japonaise et chinoise. 2 vols., Yokohama, 1878.

H. A. GILES, Jade. *Nineteenth Century,* 1904, pp. 138–145.

H. A. GILES, Jade. *Adversaria Sinica,* No. 9. Shanghai, 1911, pp. 312-322.

H. A. GILES, A Chinese Biographical Dictionary. London, 1898.

H. A. GILES, A Chinese-English Dictionary. Second edition, revised and enlarged. Fascicules I-IV. Shanghai, 1909–1911.
　　For the portion not yet out, the first edition is quoted. Fascicule V arrived after the completion of the manuscript.

H. A. GILES, An Introduction to the History of Chinese Pictorial Art. Shanghai, 1905.

W. R. GINGELL, The Ceremonial Usages of the Chinese, B. C. 1121, as prescribed in the "Institutes of the Chow Dynasty Strung as Pearls;" or, Chow Le Kwan Choo. Translated from the Original Chinese, with Notes. London, 1852.
　　Quoted for reference to Chinese illustrations.

E. GORER and J. F. BLACKER, Old Chinese Porcelain and Hard Stones. London, 1911.
　　Not yet seen.

F. GRENARD, Mission scientifique dans la Haute Asie, Vol. II. Paris, 1898.
　　Jade of Khotan, pp. 187-188. Plate III contains four modern Jade objects of Khotan.

J. J. M. DE GROOT, Les fêtes annuellement célébrées à Émoui. 2 Vols. Paris, 1886. *Annales du Musée Guimet*, Vols. XI and XII.

J. J. M. DE GROOT, The Religious System of China. Vols. I–VI. Leiden, 1892–1910.

B. DU HALDE, A Description of the Empire of China. 2 vols. London, 1738.
Note on Jade, Vol. I, p. 16 (quoted on p. 26).

CH. DE HARLEZ, I-Li, Cérémonial de la Chine antique. Paris, 1890.

CH. DE HARLEZ, La religion et les cérémonies impériales de la Chine moderne, d'après le cérémonial et les décrets officiels. Bruxelles, 1893–94.

SVEN HEDIN, Durch Asiens Wüsten. Leipzig, 1899.
Antiquities of Nephrite from Khotan. Vol. II, p. 28.

A. HEDINGER, Das wirkliche Ende der Nephritfrage. *Globus*, Vol. LXXXIX, 1906, pp. 357–358.

W. L. HILDBURGH, Chinese Methods of Cutting Hard Stones. *Journal of the R. Anthropological Institute*, Vol. XXXVII, 1907, pp. 189–195.

W. L. HILDBURGH, Chinese Imitations of Hard Stones. Reprinted from *Journal of Society of Arts*, London, 1906 (2 pages).

F. HIRTH, The Ancient History of China to the End of the Chou Dynasty. New York, 1908.

F. HIRTH, Chinesische Ansichten über Bronzetrommeln. Leipzig, 1904.

F. HIRTH, Chinesische Studien. München, 1890.

R. LOGAN JACK, The Back Blocks of China. London, 1904.
Jade speculation in Yün-nan, p. 216.

STANISLAS JULIEN, Hoei-Lan-Ki, ou l'histoire du cercle de craie, drame en prose et en vers, traduit du chinois et accompagné de notes. London, 1832.
A list of metaphorical phrases composed with the word "Jade" is given on pp. XIII–XV.

ATHANASE KIRCHER, La Chine illustrée. Amsterdam, 1670.
Jade of Khotan, p. 87.

B. KOBERT, Ein Edelstein der Vorzeit und seine kulturhistorische Bedeutung. Stuttgart, 1910. Ten plates (compare review by MAX BAUER in *Centralblatt für Mineralogie, Geologie und Paläontologie*, 1911, p. 431).

T. DE LACOUPERIE, The Jade Eastern Traffic. *Babylonian and Oriental Record*, Vol. III, 1889, pp. 99–104.

JAMES LEGGE, The Chinese Classics. Second edition. Seven volumes. Oxford, 1893.

JAMES LEGGE, The Sacred Books of China. The Texts of Confucianism. Parts III and IV. The Li Ki, Oxford, 1885. *Sacred Books of the East*, Vols. XXVII, XXVIII.

J. H. STEWART LOCKHART, A Manual of Chinese Quotations. Hong-kong, 1903.
Quotations relating to Jade, pp. 394-404.

W. F. MAYERS, The Chinese Reader's Manual. Shanghai, 1874.
Articles on Jade, pp. 99-100, 283-284.

F. DE MÉLY, Les lapidaires de l'antiquité et du moyen âge. Tome I. Les lapidaires chinois. Paris, 1896.

A. B. MEYER, Jadeit- und Nephrit-Objekte. A. Amerika und Europa. B. Asien, Oceanien und Afrika. *Publikationen des Ethnographischen Museums Dresden.* Leipzig, 1882 and 1883.

A. B. MEYER, Die Nephritfrage kein ethnologisches Problem. Berlin, 1883. English translation: The Nephrite Question in *American Anthropologist*, Vol. I, 1888, pp. 231-242.

A. B. MEYER, Neue Mitteilungen über Nephrit. *Globus*, Vol. LXXXVI, 1904, pp. 53-55.

M. PALÉOLOGUE, L'art chinois. Paris, 1887.
Les pierres dures, pp. 155-177.

P. CORENTIN PÉTILLON, Allusions littéraires. 2 vols., Shanghai, 1895, 1898. *Variétés sinologiques*, Nos. 8, 13.
On Jade see in particular pp. 234-250.

M. ABEL-RÉMUSAT, Histoire de la ville de Khotan, suivie de recherches sur la substance minérale appelée par les Chinois pierre de Iu, et sur le Jaspe des anciens. Paris, 1820.

BARON RICHTHOFEN'S Letters, 1870-1872. Shanghai (no date).
A few remarks on the Jadeite of Yün-nan, p. 138.

ÉMILE ROCHER, La province chinoise du Yün-nan. Paris, 1880.
Brief note on the Jadeite of Yün-nan. Vol. II, p. 260.

F. W. RUDLER, On the Source of the Jade used for Ancient Implements in Europe and America. *Journal of the R. Anthropological Institute*, Vol. XX, 1890, pp. 332-342.

L. SCHERMAN, Berichte des Kgl. Ethnographischen Museums in München I (1908). München, 1909.
P. 81. ·Collection of 171 Chinese stone carvings acquired by King Ludwig I of Bavaria, sixty-five years ago. Figures of a jade ring engraved with two dragons and four bats, and of a beautiful carving of malachite representing Bodhidharma.

G. SCHLEGEL, Uranographie chinoise. 2 vols. and atlas. Leiden, 1875.

J. D. E. SCHMELTZ, Ethnographische Musea in Midden-Europa. Leiden, 1896.
Figures of two *Ju-i* and some remarks on these implements, p. 30.

H. GRAF VON SCHWEINITZ, Orientalische Wanderungen in Turkestan und im nordöstlichen Persien. Berlin, 1910.
Illustration of the famous nephrite sarcophagus of Timur in the Gur-Emir of Samarkand, p. 99.

J. G. SCOTT and HARDIMAN, Gazetteer of Upper Burma and the Shan States. Part I, Vol. II, Rangoon, 1900.
On the Jade Mines of Burma, pp. 277-289.

F. PORTER SMITH, Contributions towards the Materia Medica and Natural History of China. Shanghai, 1871.
> Brief note on Jade, p. 124.

J. L. SOUBEIRAN et DABRY DE THIERSANT, La matière médicale chez les Chinois. Paris, 1874.
> Brief note on Jade, p. 36.

YAGI SŌZABURŌ, Kōko Benran (in Japanese). Tokyo, 1902.
> On Chinese Jade, pp. 149-158.

M. AUREL STEIN, Sand-Buried Ruins of Khotan. London, 1904.
> Jade-mining of Khotan, pp. 233-236, with illustration of Jade pit and diggers.

OSMOND TIFFANY, The Canton Chinese, or the American's Sojourn in the Celestial Empire. Boston, 1849.
> A book teeming with useful information. On Canton Jade works, pp. 101-102.

A. TSCHEPE, Histoire du royaume de Ou. Shanghai, 1876. *Variétés sinologiques*, No. 10.

A. TSCHEPE, Histoire du royaume de Tch'ou. Shanghai, 1903. *Variétés sinologiques*, No. 22.

A. TSCHEPE, Histoire du royaume de Tsin. Shanghai, 1910. *Variétés sinologiques*, No. 30.

A. TSCHEPE, Histoire du royaume de Han. Shanghai, 1910. *Variétés sinologiques*, No. 31.

CH. E. DE UJFALVY, Expédition scientifique française en Russie, Vol. III, Paris, 1880.
> On Jade implements found in Siberia near Tobolsk, Barnaul and Yakutsk, p. 146.

WADA, Die Schmuck- und Edelsteine bei den Chinesen. *Mitteilungen der Deutschen Gesellschaft für Natur- und Völkerkunde Ostasiens*, Vol. X, No. 1, Tokyo, 1904.

S. WELLS WILLIAMS, The Middle Kingdom. New York, 1901.
> Few notes on Jade, Vol. I, pp. 309-310, 312.

A. WOLLEMANN, Das Ende der Nephritfrage. *Globus*, Vol. LXXXIII, 1903, pp. 144-145.

A. WYLIE, Notes on Chinese Literature. New edition. Shanghai, 1901.

H. YULE, The Book of Ser Marco Polo. Third edition, by H. Cordier. London, 1903.
> Jade of Khotan, Vol. I, pp. 191, 193.

CATALOGUES

No completeness has been attempted here. It is impossible to secure all auction catalogues.

THE ART INSTITUTE OF CHICAGO. General Catalogue of Sculpture, Paintings and Other Objects. Chicago, 1910.
> Note on Jade, pp. 141-142.

360   BIBLIOGRAPHY.

The Heber R. Bishop Collection of Jade and Other Hard Stones.
    The Metropolitan Museum of Art, Handbook No. 10.   (No date.)
T. Brinckmann, Führer durch das Hamburgische Museum für Kunst
    und Gewerbe.   Hamburg, 1894.
        Brief description of two small Chinese cups of green nephrite, p. 589.
Catalogue De Luxe of the Art Treasures collected by Thomas E.
    Waggaman.   New York, 1905.
        Carved Jade, Nos. 489–550 (the pages are not numbered).   One of the
    curiosities of this catalogue is No. 490 "*Maori* Jade Ring, dated K'ien-lung
    period."
Catalogue of Collection of A. D. Vorce.   New York, 1905.
        Jades and Agates, Nos. 48–99.
Catalogue of the Collection of Chinese Exhibits at the Louisiana Pur-
    chase Exposition, St. Louis, 1904, pp. 129, 130.
B. Hirschsprungs Samling.   Copenhague, 1911.
        Jade, p. 61.
Port Catalogues of the Chinese Customs' Collection at the Austro-
    Hungarian Universal Exhibition, Vienna, 1873.   Shanghai, 1873,
    pp. 18, 426.
The Auguste F. Chamot Collection.   New York, 1907.
        Jades Nos. 1–70 (Interesting).
E. Deshayes, Petit guide illustré au Musée d'Ennery, Paris, 1908.
        Jade, pp. 46–47.   Figure of winged lion of jade, p. 28.
Führer durch das Museum für Völkerkunde, 14th ed.   Berlin, 1908.
        Chinese Jades, pp. 249–250.   Interesting because of dated pieces of the
    K'ien-lung period.
Yamanaka and Co., Illustrated Catalogue of Antique and Modern
    Chinese and Japanese Objects of Art.   New York, 1905.
        Jades, pp. 29–38, 1 plate.

INDEX

Aalst, J. A. v., 327.
Acupuncture, stone needles for, 66, 67.
Agaric, 209.
Agate, ring of, 167; paper-weight of, 332; tree of, from Tokharestan, 334; fruits and petals of flower, carved from, 335.
Agriculture, differentiated from hoe-culture, 48–49.
Ainu, grooved stone hammer of, 51, 52.
Alchemy, notions regarding jade connected with, 296.
Amber, *Ju-i* of, 335, 336.
America and Asia, historical relations between, 52.
Amiot, on jade of Shensi and Shansi, 26; on resonant stones, 327.
Amulets, jade celts worn as, 45; of jade, for the protection of the dead, 294–305.
Anderson, J., on stone implements of Yün-nan, 30–32.
Anderson, W., 338, 343.
Andree, R., 65, 69.
Annam, jade of, 24.
Antelope, design of, on jade buckle, 268.
Anthropomorphic conceptions in ancient Chinese religion, 121, 174–185.
Aquatic plants, design of, on girdle-pendant, 238, 249.
Archæology, Chinese, methods of, 15, 17, 22.
Archer's thumb-ring, 283–284.
Arrow-heads, of stone, in Hui-wu, 25; of flint, found in Mongolia, 34; of jade, 34; in Liao-tung, 35, 58; of stone, mentioned in *Shu king*, 55; of the Su-shên, 57–59, 68; in Korea, 59; in Kiang-si, 59; in Sze-ch'uan, 60; figured and described in *Kin-shih so*, 60–62.
Aston, W., 59, 351.
Astronomical instruments, of jade, 104–107, 112.
Autumn, symbolized by the tiger, 175.
Axe, of stone, found in mound near Kal-gan, 32–34; emblem of sovereign, 45; stone axes mentioned in Chinese records, 60, 63; ceremonial jade axes, 41–43.

Baber, E. C., on stone implements of Sze-ch'uan, 32; on phallicism, 99; on Ngo-mei shan, 343.

Back-scratcher, 254–255.
Badges of rank, 84–88, 164, 330.
Balas ruby, 109.
Ball, V., 76.
Bamboo pattern, 251, 255, 333.
Bamboo tablets, discovered in grave, 21; as writing-material, 114, 115, 117.
Banana leaves, design of, on jade ornament, 206.
Band-ornaments, on jade disk, 163; realistic meaning of, in connection with dragon, 164.
Bats, symbol of happiness, existence of design of in Han period doubtful, 168; on court-girdle of the T'ang dynasty, 287, 289; on modern specimens, 329, 330, 334.
Bauer, M., 111, 155.
Biot, E., translator of *Chou li*, 15; on south-pointing chariot, 113.
Bird, head on jade disk, 160; in connection with dragon, symbol of clouds, 162, 164, 168, 226, 236–239.
Bird-cage, of jade, from Turkistan, 292.
Bishop, H. R., collection of and work on jade, 6–8; on localities of jade in China, 23, 24; on discoloration and decomposition of jade, 27; celts of, 34; jade book of, 118; alleged jade wheel-nave of, 123; *kuei pi* of, 168; correct explanation for jade horse of, 247; on the action of fire on jade, 322; Palace jade pieces of, 324; Han neph-rite gong of, 328; jade mountain of, 331; *Ju-i* of, 335; on jadeite magatama of Japan, 353.
Black jade, 25, 37, 60, 96, 99, 120, 130, 131, 137, 245.
Blood-stone, 39.
Blue jade, 120, 229, 292.
Boar's tooth, 203, 254.
Bodding, P. O., 77.
Bodmer-Beder, A., investigations of jades of Switzerland of, 3.
Bogoras, W., 52.
Books, composed of jade slabs, 118.
Boston Museum of Fine Arts, 202.
Bowes, J. L., 352.
Bowlders, of jade, 26.
Boxes of jade, buried with corpse to preserve its flesh, 21, 299.
Brass, girdles of, 286.
Brockhaus, A., 222, 243.

361

THE

JOURNAL

OF THE

ROYAL ASIATIC SOCIETY

OF

GREAT BRITAIN AND IRELAND

FOR

1913

QUOT RAMI TOT
ARBORES

A. D. MDCCCXXIII INST.

PUBLISHED BY THE SOCIETY

22 ALBEMARLE STREET, LONDON, W.

M DCCCC XIII

JADE. A study in Chinese Archæology and Religion. By BERTHOLD LAUFER. 68 plates, 6 of which are coloured, and 204 text-figures. Chicago, U.S.A., February, 1912.

In 1907 the authorities of the Field Museum of Natural History, of Chicago, commissioned Dr. Laufer to carry on research work and make collections in Tibet and China, under an endowment provided by Mrs. T. B. Blackstone of that city. Dr. Laufer went, saw, and collected. On his return it was decided to work up the Chinese material in a series of monographs. This handsome volume of 370 pages is the first of them, and, even were no other to follow, both the Field Museum and the author would well deserve congratulations, the former on the selection of so keen and competent an agent, and the latter on the success with which he has carried out his quest and the subsequent researches demanded by the specimens acquired.

Singanfu, *alias* Hsianfu, the capital city of Shensi Province, appears to have proved a rich mine of antiquarian treasures for Dr. Laufer, who was well advised to explore, and well financed to exploit, this ancient home of wealth.

The plan of the book is, after preliminary matter, to divide and classify the specimens of jade secured in China for the Museum into the various categories of use and application to which Chinese culture from the earliest period has put objects of this fascinating stone, or rather stones ; and while doing so, to discuss the various points of custom and belief which they illustrate and help to explain. The book is thus partly a *catalogue raisonnée* of the jade exhibits in the Field Museum, and partly a series of studies of Chinese antiquity as it discloses itself in these characteristic relics.

The way of the reader is greatly eased and lightened by the very numerous illustrations. Among these

Dr. Laufer has most appropriately included a number of the drawings in the late Wu Ta-ch'êng's *Ku Yü T'u K'ao*, "Investigations into Ancient Jades with Illustrations." Happy that land whose ancient jades can so well stand investigations. As an admirer of that great scholar in another branch of learning, I cannot refrain from quoting the words, both generous and just, in which Dr. Laufer speaks of him (Introduction, p. 13): "Wu Ta-ch'êng is not bound by the fetters of the past and not hampered by the accepted school traditions. With fair and open mind he criticizes the errors of the commentators to the *Chou li*, the *Ku Yü T'u P'u*, and many others, and his common sense leads him to new and remarkable results not anticipated by any of his predecessors. Because my own collection is a counterpart of his, being made from an archæological, not an artistic point of view, I could choose no better guide for the interpretation of this collection than him; I have followed him with keen admiration and stand to him in the relation of a disciple to his master."

Another excellently true appreciation of the absurd figures of ceremonial and other antiquarian objects evolved (like a certain camel elsewhere) from the inner consciousness of the Sung dynasty scholars, will be found on p. 16 of the Introduction. I, too, have often wondered "that such figures could find their way into foreign books (Biot, Pauthier, Zottoli, Legge, Couvreur) . . . without a word of comment or criticism".

The whole Introduction is a valuable and interesting essay, but I must pass on to give some sketch of the scope of the chapters that follow, twelve in number. The first is devoted to Jade (whether Jadeite or Nephrite) and other stone implements, and figures numerous chisels, hammers, knives, axes, and hatchets of jade, attributed to the Chou dynasty, and mostly discovered in Shensi province. Among them is one, illustrated on p. 43, of

which, but for the perforation, the miniature in my
collection, figured on Plate V, B, of my paper on Chinese
Writing in the Chou Dynasty, in the Journal for October,
1911, might almost be a model. Dr. Laufer treats all this
part of the subject in a most interesting way. Then
come other chapters treating of Jade symbols of Sovereign
Power; of Astronomical Instruments; of the stone, used
as writing material; of its use in religious worship for
images of the cosmic deities, Earth, Heaven, North,
East, South, and West, and of the Dragon, a long and
valuable contribution to a difficult and obscure subject.
Chapters vi to xii deal respectively with Jade Coins and
Seals; Personal Ornaments; Amulets of the Dead;
Objects used in dressing the corpse; Carvings of animal
and human figures in the grave; Jade Vases; and, lastly,
of Jade in the eighteenth century. These headings will
give an idea of the scope of the work. For the manner
of it Dr. Laufer brings a trained intelligence and great
keenness to his task, but above all a certain refreshing
and vivid sense of reality, so that in his hands the things
of the past lose that ancient and fish-like savour that is
apt to hang about them, and are made to appeal to us
as guests of a rational curiosity, not as dim ghosts of
a distant and distasteful antiquity.

In the course of these pages there naturally occur
a number of passages translated from native authors.
Dr. Laufer's renderings of these are not in all cases
satisfactory. The Chinese written language is a hard
taskmaster, and demands before all a long experience
which probably the author's other studies and occupations
have prevented him from devoting to it. I shall only
therefore mention one instance, and that simply because
Dr. Laufer has been led to infer a phallic symbolism
through misunderstanding of the text. On p. 44 he
illustrates from the *Chin Shih So*, of the brothers Fêng,
two ancient bronze hatchets, and writes: "The latter

(Fig. 6) is interesting with reference to the jade dance-axes in exhibiting a more primitive form of the triangular pattern, and it is very interesting to take note of the interpretation of the brothers Fêng that this ornament is a *yang wên*, 'a pattern of the male principle.'" To this he appends the note: "They expressly deny that it has the function of a written character. The Chinese wording certainly means in our language a phallic emblem." This statement is gravely erroneous. The passage from the *Chin Shih So* is reproduced with the figure, and it really runs thus: "Probably used in ancient times as a ceremonial weapon. The face has the figure ꓘ in relief [*yang-wên*], which is probably the character *yüch*, 'battle-axe.' [Note by the brother Fêng Yün-]P'êng. With regard to the figure ꓘ it is an ornament, and not necessarily a character."

But this is a mere speck in an admirable contribution to knowledge, which I greatly hope will, in due course, be followed by the others projected by the author.

L. C. HOPKINS.

A CHINESE–ENGLISH DICTIONARY IN THE CANTONESE DIALECT. By Dr. E. J. EITEL. Revised and enlarged by I. G. GENÄHR. Hong-Kong: Kelly & Walsh, 1912.

The Cantonese-English Dictionary, the first half of which was reviewed in a recent number of the JRAS., is now completed.

It contains 8,349 Chinese characters, as against 8,092 in Williams's Tonic Dictionary and 10,644 in Eitel's. The first edition of Professor Giles's Mandarin Dictionary has 13,848 characters. In Dr. Wells Williams's Tonic Dictionary there are 707 different syllables given; in Eitel's 731; and the number has not been increased in this latest issue. There are as many as 780 syllables